DIFFERENTIAL TRANSFORMATION METHOD FOR MECHANICAL ENGINEERING PROBLEMS

DIFFERENTIAL TRANSFORMATION METHOD FOR MECHANICAL ENGINEERING PROBLEMS

MOHAMMAD HATAMI

DAVOOD DOMAIRRY GANJI

MOHSEN SHEIKHOLESLAMI

ELSEVIER

AMSTERDAM • BOSTON • HEIDELBERG • LONDON
NEW YORK • OXFORD • PARIS • SAN DIEGO
SAN FRANCISCO • SINGAPORE • SYDNEY • TOKYO
Academic Press is an imprint of Elsevier

Academic Press is an imprint of Elsevier
125 London Wall, London EC2Y 5AS, United Kingdom
525 B Street, Suite 1800, San Diego, CA 92101-4495, United States
50 Hampshire Street, 5th Floor, Cambridge, MA 02139, United States
The Boulevard, Langford Lane, Kidlington, Oxford OX5 1GB, United Kingdom

Notices
Knowledge and best practice in this field are constantly changing. As new research and experience broaden our understanding, changes in research methods, professional practices, or medical treatment may become necessary.

Practitioners and researchers must always rely on their own experience and knowledge in evaluating and using any information, methods, compounds, or experiments described herein. In using such information or methods they should be mindful of their own safety and the safety of others, including parties for whom they have a professional responsibility.

To the fullest extent of the law, neither the Publisher nor the authors, contributors, or editors, assume any liability for any injury and/or damage to persons or property as a matter of products liability, negligence or otherwise, or from any use or operation of any methods, products, instructions, or ideas contained in the material herein.

Library of Congress Cataloging-in-Publication Data
A catalog record for this book is available from the Library of Congress

British Library Cataloguing-in-Publication Data
A catalogue record for this book is available from the British Library

ISBN: 978-0-12-805190-0

For information on all Academic Press publications visit our website at https://www.elsevier.com/

Publisher: Joe Hayton
Acquisition Editor: Brian Guerin
Editorial Project Manager: Carrie Bolger
Production Project Manager: Mohana Natarajan
Cover Designer: Matthew Limbert

Typeset by TNQ Books and Journals

DEDICATED TO:

All the Iranian fathers and mothers who resist and defended with whole existence against the assault to our homeland during the 1980–88 years that brought safety and comfort for us these days to live in love and peace.

CONTENTS

PREFACE

Great Lord of Life and Wisdom! In thy Name
Which to transcend no flight of thought may claim!

Shahnameh, Abul-Qasim Ferdowsi (940–1020 C.E.)

Many phenomena in the nature need to be modeled mathematically or numerically for better perception of its physic and enhance the researchers to solve the possible problems. Mechanical engineering, chemical engineering, petroleum, energy crisis, industrial heat exchangers, boilers, engines, etc., all contain some studies which need mathematical modeling. During most of these modelings, ordinary differential equations or partially differential equations will appear and need powerful solution methods, numerically or analytically. Current book introduces the differential transformation method (DTM) as one of the most powerful mathematic/analytic methods for solving these differential equations. The contents of the current book are able to benefit engineers, researchers, and graduate students who want to develop their knowledge in basic phenomena of all sciences, especially mechanical engineering. In the introductory chapters (Chapters 1 and 2), DTM is introduced in simple and complicated versions including all improvements and developments. In other chapters (Chapters 3–8), application of DTM on various examples in solid and fluid mechanics is demonstrated, and several examples of recently published papers from high-quality journals are included to illuminate the subject. The authors are very much pleased to receive the readers' comments and amendments on the materials of the book. Finally, we would like to express our sincere thanks to the staff of books publishing at Elsevier for their helpful support. We hope this book will be of great benefit to you.

Mohammad Hatami
Assistant Professor of Mechanical Engineering Department, Esfarayen University of Technology, Esfarayen, North Khorasan, Iran

CHAPTER 1

Introduction to Differential Transformation Method

1.1 INTRODUCTION

Most of the problems in mathematics and nature are inherently nonlinear. For solving and analyzing these problems, analytical and numerical methods must be applied. homotopy analysis method (HAM), homotopy perturbation method (HPM), adomian decomposition method (ADM), weighted residual method (WRM), etc., are some common and classical analytical methods have been presented in the literature for solving nonlinear problems [1–76].

The differential transformation method (DTM) is an alternative procedure for obtaining an analytic Taylor series solution of differential equations. The main advantage of this method is that it can be applied directly to nonlinear differential equations without requiring linearization and discretization, and therefore, it is not affected by errors associated with discretization. The concept of DTM was first introduced by Zhou [77], who solved linear and nonlinear problems in electrical circuits. DTM due to the following advantages has been used by many researchers and they tried to improve and increase its accuracy, which is discussed in this book.

1. Unlike perturbation techniques, DTM is independent of any small or large quantities. So, DTM can be applied no matter if governing equations and boundary/initial conditions of a given nonlinear problem contain small or large quantities, or not.
2. Unlike HAM, DTM does not need to calculate auxiliary parameter $\hbar_1$, through h-curves.
3. Unlike HAM, DTM does not need initial guesses and auxiliary linear operator, and it solves equations directly.
4. DTM provides us with great freedom to express solutions of a given nonlinear problem by means of Padé approximant and Ms-DTM or other modifications.

This chapter introduces DTM generally and contains the following:

Differential Transformation Method for Mechanical Engineering Problems
ISBN 978-0-12-805190-0
http://dx.doi.org/10.1016/B978-0-12-805190-0.00001-2

1.3 Multistep Differential Transformation Method
1.4 Hybrid Differential Transformation Method and Finite Difference Method
1.5 Differential Transformation Method Applying on Initial-Value Problems and Ordinary Differential Equations
1.6 Two-Dimensional Differential Transformation Method for Partial Differential Equations
1.7 Differential Transformation Method—Padé Approximation
1.8 Differential Transformation Method on Singular Two-Point Boundary Value Problem

1.2 PRINCIPLE OF DIFFERENTIAL TRANSFORMATION METHOD

For understanding the method's concept, suppose that $x(t)$ is an analytic function in domain D, and t_i represents any point in the domain. The function $x(t)$ is then represented by one power series whose center is located at t_i. The Taylor series expansion function of $x(t)$ is:

$$x(t) = \sum_{k=0}^{\infty} \frac{(t - t_i)^k}{k!} \left[\frac{d^k x(t)}{dt^k} \right]_{t=t_i} \quad \forall t \in D \tag{1.1}$$

The Maclaurin series of $x(t)$ can be obtained by taking $t_i = 0$ in Eq. (1.1). expressed as [77]:

$$x(t) = \sum_{k=0}^{\infty} \frac{t^k}{k!} \left[\frac{d^k x(t)}{dt^k} \right]_{t=0} \quad \forall t \in D \tag{1.2}$$

As explained in Ref. [77], the differential transformation of the function $x(t)$ is defined as follows:

$$X(k) = \sum_{k=0}^{\infty} \frac{H^k}{k!} \left[\frac{d^k x(t)}{dt^k} \right]_{t=0} \tag{1.3}$$

Where $X(k)$ represents the transformed function, and $x(t)$ is the original function. The differential spectrum of $X(k)$ is confined within the interval $t \in [0, H]$, where H is a constant value. The differential inverse transform of $X(k)$ is defined as follows:

$$x(t) = \sum_{k=0}^{\infty} \left(\frac{t}{H} \right)^k X(k) \tag{1.4}$$

It is clear that the concept of differential transformation is based upon the Taylor series expansion. The values of function $X(k)$ at values of argument k are referred to as discrete, i.e., $X(0)$ is known as the zero discrete, $X(1)$ as the first discrete, etc. The more discrete available, the more precise it is possible to restore the unknown function. The function $x(t)$ consists of the T-function $X(k)$, and its value is given by the sum of the T-function with $(t/H)k$ as its coefficient. In real applications, at the right choice of constant H, the larger values of argument k the discrete of spectrum reduce rapidly. The function $x(t)$ is expressed by a finite series and Eq. (1.4) can be written as:

$$x(t) = \sum_{k=0}^{n} \left(\frac{t}{H}\right)^{k} X(k) \tag{1.5}$$

Some important mathematical operations performed by DTM are listed in Table 1.1.

Example A: As shown in Fig. 1.1, a rectangular porous fin profile is considered. The dimensions of this fin are length L, width w, and thickness t. The cross section area of the fin is constant, and the fin has temperature-dependent internal heat generation. Also, the heat loss from the tip of the fin compared with the top and bottom surfaces of the fin is assumed to be negligible. Since the transverse Biot number should be small for the fin to be effective, the temperature variations in the transverse direction are

Table 1.1 Some Fundamental Operations of the Differential Transform Method [77]

Origin Function	Transformed Function
$x(t) = \alpha f(x) \pm \beta g(t)$	$X(k) = \alpha F(k) \pm \beta G(k)$
$x(t) = \dfrac{d^m f(t)}{dt^m}$	$X(k) = \dfrac{(k+m)!F(k+m)}{k!}$
$x(t) = f(t)g(t)$	$X(k) = \sum_{l=0}^{k} F(l)G(k-l)$
$x(t) = t^m$	$X(k) = \delta(k-m) = \begin{cases} 1, & \text{if } k = m, \\ 0, & \text{if } k \neq m. \end{cases}$
$x(t) = \exp(t)$	$X(k) = \dfrac{1}{k!}$
$x(t) = \sin(\omega t + \alpha)$	$X(k) = \dfrac{\omega^k}{k!} \sin\left(\dfrac{k\pi}{2} + \alpha\right)$
$x(t) = \cos(\omega t + \alpha)$	$X(k) = \dfrac{\omega^k}{k!} \cos\left(\dfrac{k\pi}{2} + \alpha\right)$

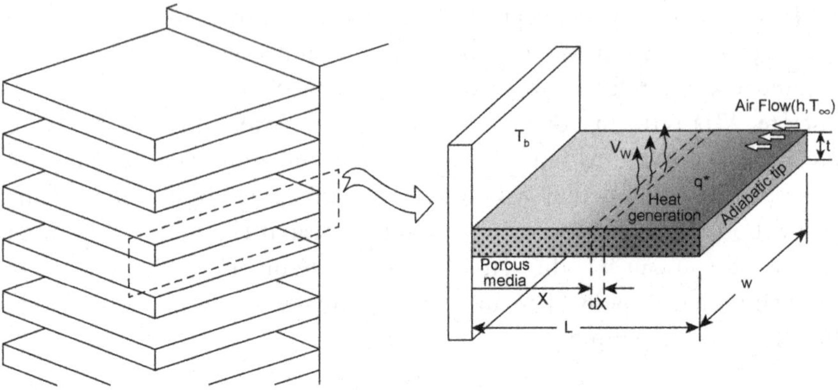

Figure 1.1 Schematic of convective porous fin with temperature-dependent heat generation.

neglected. Thus heat conduction is assumed to occur solely in the longitudinal direction [12].

Energy balance can be written as:

$$q(x) - q(x + \Delta x) + q^{*} \cdot A \cdot \Delta x = \dot{m} c_p [T(x) - T_\infty] + h(p \cdot \Delta x)[T(x) - T_\infty] \tag{1.6}$$

The mass flow rate of the fluid passing through the porous material is:

$$\dot{m} = \rho \cdot V_w \cdot \Delta x \cdot w \tag{1.7}$$

The passage velocity from the Darcy's model is:

$$V_w = \frac{gK\beta}{\nu(T - T_\infty)} \tag{1.8}$$

Substitutions of Eqs. (1.7) and (1.8) into Eq. (1.6) yield:

$$\frac{q(x) - q(x + \Delta x)}{\Delta x} + q^{*} \cdot A = \frac{\rho \cdot c_p \cdot g \cdot K \cdot \beta \cdot w}{\upsilon} [T(x) - T_\infty]^2 + hp[T(x) - T_\infty] \tag{1.9}$$

As, $\Delta x \to 0$ Eq. (1.9) becomes:

$$\frac{dq}{dx} + q^{*} \cdot A = \frac{\rho \cdot c_p \cdot g \cdot K \cdot \beta \cdot w}{\upsilon} [T(x) - T_\infty]^2 + hp[T(x) - T_\infty] \tag{1.10}$$

Also from Fourier's Law of conduction:

$$q = -k_{eff} A \frac{dT}{dx} \tag{1.11}$$

Where A is the cross-sectional area of the fin $A = (w \cdot t)$ and k_{eff} is the effective thermal conductivity of the porous fin that can be obtained from following equation:

$$k_{eff} = \varphi \cdot k_f + (1 - \varphi)k_s \tag{1.12}$$

where φ is the porosity of the porous fin. Substitution Eq. (1.11) into Eq. (1.10) leads to:

$$\frac{d^2T}{dx^2} - \frac{\rho \cdot c_p \cdot g \cdot K \cdot \beta}{t \cdot k_{eff} \cdot \nu}[T(x) - T_\infty]^2 - \frac{h \cdot p}{k_{eff} \cdot A}[T(x) - T_\infty] + \frac{\overset{*}{q}}{k_{eff}} = 0 \tag{1.13}$$

It is assumed that heat generation in the fin varies with temperature as Eq. (1.14):

$$\overset{*}{q} = \overset{*}{q}_\infty(1 + \varepsilon(T - T_\infty)) \tag{1.14}$$

Where $\overset{*}{q}_\infty$ is the internal heat generation at temperature T_∞.

For simplifying the above equations some dimensionless parameters are introduced as follows:

$$\theta = \frac{(T - T_\infty)}{(T_b - T_\infty)}, \quad X = \frac{x}{L}, \quad M^2 = \frac{hPL^2}{k_0 A}, \quad S_h = \frac{Da \cdot x \cdot Ra}{k_r}\left(\frac{L}{t}\right)^2$$

$$G = \frac{\overset{*}{q}_\infty}{hP(T_b - T_\infty)}, \quad \varepsilon_G = \varepsilon(T_b - T_\infty) \tag{1.15}$$

where S_h is a porous parameter that indicates the effect of the permeability of the porous medium as well as buoyancy effect, so higher value of S_h indicates higher permeability of the porous medium or higher buoyancy forces. M is a convection parameter that indicates the effect of surface convecting of the fin. Finally, Eq. (1.13) can be rewritten as:

$$\frac{d^2\theta}{dX^2} - M^2\theta + M^2G(1 + \varepsilon_G\theta) - S_h\theta^2 = 0 \tag{1.16}$$

In this research we study finite length fin with insulated tip. For this case, the fin tip is insulated so that there will not be any heat transfer at the insulated tip and boundary condition will be,

$$\theta(0) = 1$$
$$\left.\frac{d\theta}{dX}\right|_{x=1} = 0 \tag{1.17}$$

Properties of Porous Materials are presented in Table 1.2.

Table 1.2 Properties of Porous Materials

Porous Material	K_s (W/K–m)	C_p (kJ/(kg–K))	ρ (kg/m^3)
Aluminum	218	0.91	2700
Si_3N_4	25	0.5	2300

Now we apply DTM from Table 1.1 into Eq. (1.16) to find $u(t)$:

$$(k+1)(k+2)\Theta(k+2) - S_h\left(\sum_{l=0}^{k}\Theta(l)\Theta(k-l)\right) - (M^2 - \varepsilon_G GM^2)\Theta(k) + GM^2\delta(k) = 0 \tag{1.18}$$

Rearranging Eq. (1.18), we have

$$\Theta(k+2) = \frac{S_h\left(\sum_{l=0}^{k}\Theta(l)\Theta(k-l)\right) + (M^2 - \varepsilon_G GM^2)\Theta(k) - GM^2\delta(k)}{(k+1)(k+2)} \tag{1.19}$$

and boundary condition transformed form is,

$$\Theta(0) = 1, \quad \Theta(1) = a \tag{1.20}$$

where a is an unknown coefficient that must be determined. By solving Eq. (1.19) and using boundary conditions, the DTM terms are obtained as

$$\begin{aligned}
\Theta(2) &= -\frac{1}{2}GM^2 + \frac{1}{2}S_h + \frac{1}{2}M^2 - \frac{1}{2}\varepsilon_G GM^2 \\
\Theta(3) &= \frac{1}{3}S_h a + \frac{1}{6}M^2 a - \frac{1}{6}M^2 a\varepsilon_G G \\
\Theta(4) &= -\frac{1}{8}S_h\varepsilon_G GM^2 - \frac{1}{12}S_h GM^2 + \frac{1}{12}S_h^2 + \frac{1}{8}S_h M^2 - \frac{1}{12}\varepsilon_G GM^4 \\
&\quad + \frac{1}{12}S_h a^2 - \frac{1}{24}GM^4 + \frac{1}{24}M^4 + \frac{1}{24}M^4\varepsilon_G G^2 + \frac{1}{24}\varepsilon_G^2 G^2 M^4 \\
\Theta(5) &= -\frac{1}{20}S_h aGM^2 + \frac{1}{12}S_h^2 a + \frac{1}{12}S_h aM^2 - \frac{1}{12}S_h a\varepsilon_G GM^2 \\
&\quad + \frac{1}{120}M^4 a - \frac{1}{60}a\varepsilon_G GM^4 + \frac{1}{120}a\varepsilon_G^2 G^2 M^4
\end{aligned} \tag{1.21}$$

and etc.

Now by substituting Eq. (1.21) into Eq. (1.17) and using boundary condition the "a" coefficient and then $\theta(X)$ function will be obtained. For Si_3N_4 with following coefficient,

$$G = 0.4;\ M = 1;\ Da = 0.0001;\ \frac{L}{t} = 10;\ Ra = 10000;\ \varepsilon_G = 0.6;$$
$$L = 1,\ k_s = 954; \tag{1.22}$$

S_h can be calculated. After five iterations in DTM series, temperature distribution will be obtained as,

$$\begin{aligned}\theta(x) = {} & 1 - 0.2430983438x + 0.1524109014x^2 - 0.03280383661x^3 \\ & + 0.01079942879x^4 - 0.001716344389x^5 + 0.0004283472660x^6 \\ & - 0.00007114658871x^7\end{aligned} \tag{1.23}$$

Fig. 1.2 shows the validation of this solution.

1.3 MULTISTEP DIFFERENTIAL TRANSFORMATION METHOD

Multistep differential transformation method (Ms-DTM) due to some advantages is applied in physical application. For example, Ms-DTM due to small time steps has a powerful accuracy especially for initial-value problems. Also, because it's based on DTM, does not need to small parameter, auxiliary function and parameter, discretization, etc., versus other analytical methods. For perception of Ms-DTM basic idea, consider a general equation of n-th order ordinary differential equation [29],

$$f\left(t, y, y', \ldots, y^{(n)}\right) = 0 \tag{1.24}$$

subject to the initial conditions

$$y^{(k)}(0) = d_k, \quad k = 0, \ldots, n-1. \tag{1.25}$$

To illustrate the DTM for solving differential equations, the basic definitions of differential transformation are introduced as follows. Let $y(t)$ be analytic in a domain D, and let $t = t_0$ represent any point in D. The function $y(t)$ is then represented by one power series, whose center is located at t_0. The differential transformation of the k-th derivative of a function $y(t)$ is defined as follows:

$$Y(k) = \frac{1}{k!}\left[\frac{d^k y(t)}{dt^k}\right]_{t=t_0}, \forall t \in D. \tag{1.26}$$

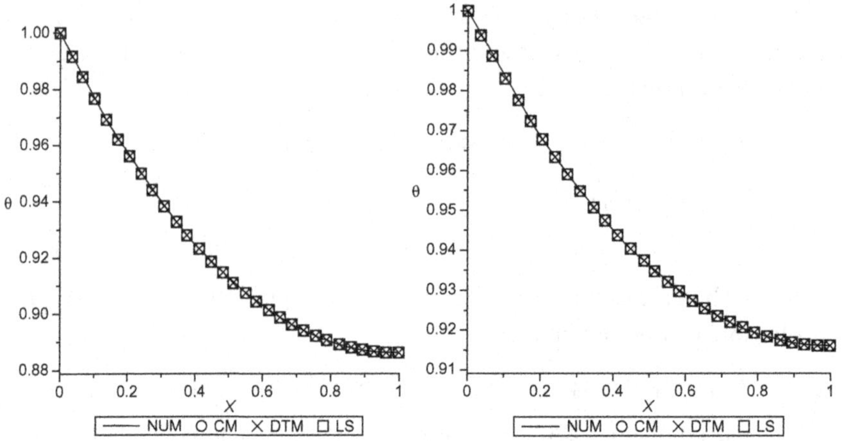

Figure 1.2 Temperature validation among Collocation, DTM, LS, and numerical methods for (left) Si_3N_4 (right) AL.

In Eq. (1.26), $y(t)$ is the original function and $Y(k)$ is the transformed function. As in [29] the differential inverse transformation of $Y(k)$ is defined as follows:

$$y(t) = \sum_{k=0}^{\infty} Y(k)(t - t_0)^k, \quad \forall t \in D. \tag{1.27}$$

In fact, from Eq. (1.26) and Eq. (1.27), we obtain

$$y(t) = \sum_{k=0}^{\infty} \frac{(t - t_0)^k}{k!} \left[\frac{d^k y(t)}{dt^k}\right]_{t=t_0}, \quad \forall t \in D. \tag{1.28}$$

Eq. (1.28) implies that the concept of differential transformation is derived from the Taylor series expansion. From the definitions of Eq. (1.26) and Eq. (1.27), it is easy to prove that the transformed functions comply with the following basic mathematics operations (see Table 1.1).

In real applications, the function $y(t)$ is expressed by a finite series and Eq. (1.28) can be written as

$$y(t) = \sum_{k=0}^{N} Y(k)(t - t_0)^k, \quad \forall t \in D. \tag{1.29}$$

Eq. (1.29) implies that $\sum_{k=N+1}^{\infty} Y(k)(t - t_0)^k$ is negligibly small.

Let $[0, T]$ be the interval over which we want to find the solution of the initial-value problem. In actual applications of the DTM, the approximate solution of the initial-value problem can be expressed by the finite series,

$$y(t) = \sum_{k=0}^{N} b_k t^k, \quad t \in [0, T]. \tag{1.30}$$

Assume that the interval [0, T] is divided into N subintervals $[t_{i-1}, t_i]$, $i = 1, 2\ldots,M$ of equal step size $h = T/M$ by using the nodes $t_i = ih$. The main ideas of the multistep DTM are as follows. First, we apply the DTM to Eq. (1.24) over the interval $[0, t_1]$, we will obtain the following approximate solution,

$$y_1(t) = \sum_{k=0}^{N} b_{1k} t^k, \quad t \in [0, t_1] \tag{1.31}$$

using the initial conditions $y_1^{(k)}(0) = d_k$. For $i \geq 2$, at each subinterval $[t_{i-1}, t_i]$ we will use the initial conditions $y_i^{(k)}(t_{i-1}) = y_{i-1}^{(k)}(t_{i-1})$ and apply the DTM to Eq. (1.24) over the interval $[t_{i-1}, t_i]$, where t_0 in Eq. (1.26) is replaced by t_{i-1}. The process is repeated and generates a sequence of approximate solutions $y_i(t)$, $i = 1, 2,\ldots,M$ for the solution $y(t)$,

$$y_i(t) = \sum_{k=0}^{N} b_{ik}(t - t_{i-1})^k, \quad t \in [t_{i-1}, t_i] \tag{1.32}$$

In fact, the multistep DTM assumes the following solution,

$$y(t) = \begin{cases} y_0(t), & t \in [0, t_1] \\ y_1(t), & t \in [t_1, t_2] \\ \cdots & \\ y_M(t), & t \in [t_M, t_{M+1}] \end{cases} \tag{1.33}$$

Example A: Consider a particle that slides along a surface that has the shape of a parabola $z = cr^2$ (see Fig. 1.3). Following assumptions are considered for particles motion modeling:

- Particle is at equilibrium.
- The particle rotates in a circle of radius R.
- The surface is rotating about its vertical symmetry axis with angular velocity ω.

By choosing the cylindrical coordinates r, θ, and z as generalized coordinates. The kinetic and potential energies are [29],

$$\begin{aligned} T &= \frac{1}{2} m\left(\dot{r}^2 + r^2\dot{\theta}^2 + \dot{z}^2\right) \\ U &= mgz \end{aligned} \tag{1.34}$$

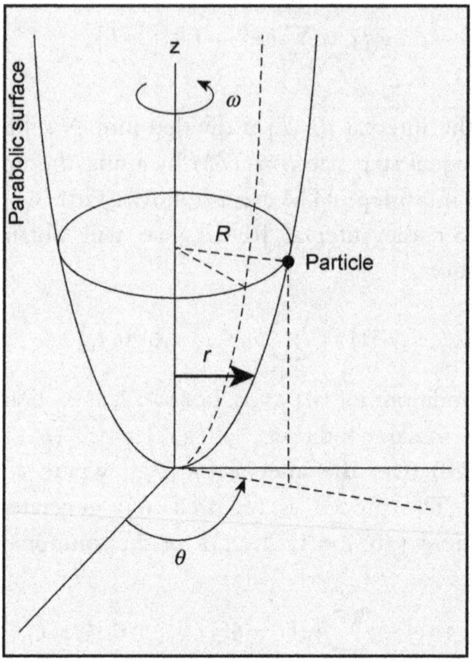

Figure 1.3 Schematic view of a spherical particle on a rotating parabolic surface.

We have in this case some equations of constraints that we must take into account, namely

$$z = cr^2 \quad \dot{z} = 2c\dot{r}r \tag{1.35}$$

and

$$\theta = \omega t \quad \dot{\theta} = \omega \tag{1.36}$$

Inserting Eq. (1.36) and Eq. (1.35) in Eq. (1.34), we can calculate the Lagrangian for the problem

$$L = T - U = \frac{1}{2}m(\dot{r}^2 + 4c^2r^2\dot{r}^2 + r^2\omega^2) - mgcr^2 \tag{1.37}$$

It is important to note that the inclusion of the equations of constraints in the Lagrangian has reduced the number of degrees of freedom to only one, i.e., r. We now calculate the equation of motion using Lagrange's equation

$$\begin{aligned}\frac{\partial L}{\partial r} &= m(4c^2 r\dot{r}^2 + r\omega^2 - 2gcr)\\ \frac{d}{dt}\frac{\partial L}{\partial \dot{r}} &= m(\ddot{r} + 4c^2 r^2 \ddot{r} + 8c^2 r\dot{r}^2)\end{aligned} \tag{1.38}$$

and

$$\ddot{r}(1 + 4c^2 r^2) + \dot{r}^2(4c^2 r) + r(2gc - \omega^2) = 0 \tag{1.39}$$

It is considered that $2gc - \omega^2 = \varepsilon^2$ and so

$$\ddot{r} + 4c^2 \ddot{r} r^2 + 4c^2 r\dot{r}^2 + \varepsilon^2 r = 0 \tag{1.40}$$

It's considered that initial particle position is in radius A, and its initial velocity is zero. So, its initial conditions are:

$$r(0) = A, \quad \dot{r}(0) = 0 \tag{1.41}$$

For solving the particle motion on a rotating parabola by an efficient, fast, and high accurate method, Eq. (1.40) is solved by Ms-DTM,

$$\begin{aligned}&(k+2)(k+1)R_j(k+2)\\ &- 4c^2 \sum_{k1}^{k} \sum_{l=0}^{k1} R_j(l)R_j(k1-l)(k-k1+1)(k-k1+2)R_j(k-k1+2)\\ &+ \varepsilon^2 R_j(k) + 4c^2 \sum_{k1=0}^{k} \sum_{l=0}^{k1} R_j(l)(k1-l+1)R_j(k1-l+1)R_j(k-k1+1)\\ &(k-k1+1) = 0\end{aligned} \tag{1.42}$$

With initial condition as,

$$\begin{aligned}&R_0(0) = A, && R_0(1) = 0\\ &R_i(0) = r_{i-1}(t_i), \quad R_i(1) = r'_{i-1}(t_i), && i = 1, 2, \ldots, K \leq M\end{aligned} \tag{1.43}$$

Since, the procedure of solving Eq. (1.34) is autonomous of constants A, ε, and c; so, for generalization and simplification of problem for future cases with different physical conditions the constants, which represent physical properties are assumed to be as the following,

$$A = c = \varepsilon = 1. \tag{1.44}$$

After solving Eq. (1.34), and using initial conditions Eq. (1.25) and Eq. (1.36), position of the particle, $r(t)$, will be appeared as following equation for each 0.25 s time step,

$$
r(t)=\begin{cases}
r_0(t)=1-\dfrac{1}{10}t^2-\dfrac{11}{3000}t^4-\dfrac{553}{2250000}t^6-\dfrac{59363}{3150000000}t^8-\dfrac{10256209}{7087500000000}t^{10} & t\in[0,0.25)\\
r_{0.25}(t)=1.006293-0.050230616t-0.1013895(t-0.25)^2-0.003744513(t-0.25)^3-0.00390231(t-0.25)^4-0.38551e-3(t-0.25)^5 & \\
\quad -0.279966e-3(t-0.25)^6-0.40481e-4(t-0.25)^7-0.23109e-4(t-0.25)^8-0.396e-5(t-0.25)^9-0.18719e-5(t-0.25)^{10} & t\in[0.25,0.5)\\
r_{0.50}(t)=1.025707-0.10188t-0.105738(t-0.5)^2-0.0079821(t-0.5)^3-0.0046757(t-0.5)^4-0.88127e-3(t-0.5)^5 & \\
\quad -0.398166e-3(t-0.5)^6-0.99646e-4(t-0.5)^7-0.383726e-4(t-0.5)^8-0.10192e-4(t-0.5)^9-0.32931e-5(t-0.5)^{10} & t\in[0.5,0.75)\\
r_{0.75}(t)=1.05996-0.1565587t-0.113642(t-0.75)^2-0.013349(t-0.75)^3-0.00621702(t-0.75)^4-0.0016488(t-0.75)^5 & \\
\quad -0.65612e-3(t-0.75)^6-0.206483e-3(t-0.75)^7-0.7254e-4(t-0.75)^8-0.2103e-4(t-0.75)^9-0.5254e-5(t-0.75)^{10} & t\in[0.75,1.0)\\
\cdots & \\
r_{9.0}(t)=-2.63269+0.374569t-0.1812(t-9)^2+0.053389(t-9)^3-0.024782(t-9)^4+0.0101267(t-9)^5 & \\
\quad -0.00325035(t-9)^6-0.1527698e-3(t-9)^7+0.001640656(t-9)^8-0.0020994(t-9)^9+0.0020165285(t-9)^{10} & t\in[9.0,9.25)\\
r_{9.25}(t)=-1.88514+.29261t-0.1490558(t-9.25)^2+0.033957(t-9.25)^3-0.014987(t-9.25)^4+0.00580704(t-9.25)^5 & \\
\quad -0.00229197(t-9.25)^6+0.72248e-3(t-9.25)^7-0.8865e-4(t-9.25)^8-0.14027e-3(t-9.25)^9+0.19074e-3(t-9.25)^{10} & t\in[9.25,9.5)\\
r_{9.5}(t)=-1.2385+0.223613t-0.1284215(t-9.5)^2+0.0219744(t-9.5)^3-0.009517(t-9.5)^4+0.0032032(t-9.5)^5 & \\
\quad -0.001265(t-9.5)^6+0.43899e-3(t-9.5)^7-0.1384e-3(t-9.5)^8+0.2617e-4(t-9.5)^9+0.85178e-5(t-9.5)^{10} & t\in[9.5,9.75)\\
r_{9.75}(t)=-0.6551+0.162983t-0.11507(t-9.75)^2+0.0141168(t-9.75)^3-0.0064938(t-9.75)^4+0.001773667(t-9.75)^5 & \\
\quad -0.70295e-3(t-9.75)^6+0.22445e-3(t-9.75)^7-0.78165e-4(t-9.75)^8+0.22326e-4(t-9.75)^9-0.5106e-5(t-9.75)^{10} & t\in[9.75,10]
\end{cases} \tag{1.45}
$$

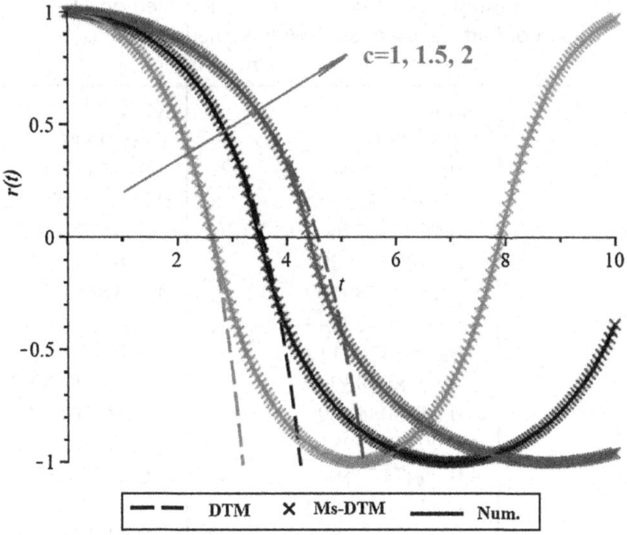

Figure 1.4 Multistep Differential Transformation Method efficiency in particle motion analysis, compared with Differential Transformation Method and numerical solution.

Eq. (1.45) is depicted in Fig. 1.4 and is compared with obtained result from DTM. The values are presented in Table 1.3. Fig. 1.4 shows the particle's position for three different c constants when $A = \varepsilon = 1$.

1.4 HYBRID DIFFERENTIAL TRANSFORMATION METHOD AND FINITE DIFFERENCE METHOD

The differential transform is defined as follows [78]:

$$X(k) = \frac{1}{k!}\left[\frac{d^k x(t)}{dt^k}\right]_{t=t_0} \tag{1.46}$$

where, $x(t)$ is an arbitrary function, and $X(k)$ is the transformed function. The inverse transformation is as follows

$$x(t) = \sum_{k=0}^{\infty} X(k)(t - t_0)^k \tag{1.47}$$

Table 1.3 Multistep Differential Transformation Method's Values for Position of Particle, Compared With Numerical Results

t	*r*(*t*)	
	Num.	Ms-DTM
0	1.0000000	1.0000000
0.5	0.974766908	0.9747669183
1.0	0.896067130	0.8960671597
1.5	0.753057776	0.7530578586
2.0	0.5189251820	0.518925278
2.5	0.13321789149	0.133218454
3.0	−0.33404132749	−0.333782926
3.5	−0.64064202117	−0.639815924
4.0	−0.82788893947	−0.8266833025
4.5	−0.9393086789	−0.9377677280
5.0	−0.9926709981	−0.9907876234
5.5	−0.9947172580	−0.9924488554
6.0	−0.94568648462	−0.9429458988
6.5	−0.83945793817	−0.836078069
7.0	−0.65951111277	−0.655144210
7.5	0.36531058744	0.359190250
8.0	0.09376192157	0.100601846
8.5	0.49425716809	0.497276130
9.0	0.73793715747	0.738436224
9.5	0.88704545678	0.885813784
10	0.97044161417	0.9677503517

Substituting Eq. (1.46) into Eq. (1.47), we have

$$x(t) = \sum_{k=0}^{\infty} \frac{(t-t_0)^k}{k!} \left[\frac{d^k x(t)}{dt^k} \right]_{t=t_0}$$

The function $x(t)$ is usually considered as a series with limited terms, and Eq. (1.47) can be rewritten as:

$$x(t) \approx \sum_{k=0}^{m} X(k)(t-t_0)^k \tag{1.48}$$

where, m represents the number of Taylor series' components. Usually, through elevating this value, we can increase the accuracy of the solution.

Although the DTM series solution has a good approximate of the exact solution, but this series is diverged for greater areas. For this reason, the multistep DTM is used. Based on this technique, the solution domain is divided to some subdomains.

To solve a differential equation in the domain [0, T] using multistep DTM, this domain is divided to N sections. We suppose the subdomains are equal, and length of each subdomain will be $H = T/N$. So, a separate function is considered for every subdomain as follows:

$$x(t) = \begin{cases} x_1(t) & , \quad t \in [t_1, t_2] \\ \vdots & \\ x_i(t) & , \quad t \in [t_i, t_{i+1}] \\ \vdots & \\ x_N(t) & , \quad t \in [t_N, t_{N+1}] \end{cases} \tag{1.49}$$

where $t_i = (i-1)H$. Multistep DTM for every sub domain defined as:

$$X_i(k) = \frac{H^k}{k!}\left[\frac{d^k x_i(t)}{dt^k}\right]_{t=t_i} \tag{1.50}$$

The inverse multistep DTM is

$$x_i(t) = \sum_{k=0}^{\infty} X_i(k)\left(\frac{t-t_i}{H}\right)^k$$

To solve the partial differential equation $u(y, t)$ in the domain $t \in [0, T]$ and $y \in [y_{firs}, y_{end}]$ using hybrid multi-step differential transformation method (MDTM) and FDM, we apply finite difference approximate on y-direction and take MDTM on t. The time domain is divided to N_t sections. We suppose the time subdomains are equal, and length of each subdomain is $H = T/N_t$. So there is a separate function for every subdomain as follows:

$$u_i(y,t) = \begin{cases} u_1(j,t), & t \in [t_1, t_2], \quad 1 \le j \le N_y + 1 \\ \vdots & \\ u_i(j,t), & t \in [t_i, t_{i+1}], \quad 1 \le j \le N_y + 1 \\ \vdots & \\ u_{N_t}(j,t), & t \in [t_N, t_{N_t+1}], \quad 1 \le j \le N_y + 1 \end{cases} \tag{1.51}$$

where $t_i = (i-1)H$ and N_y is the number of cells in y-direction.

Example A: The fluid is assumed to be flowing between two infinite horizontal plates located at the $y = \pm h$ planes. The upper plate moves with a uniform velocity U_0 while the lower plate is kept stationary. The two plates are assumed to be electrically insulating and kept at two constant temperatures, T_1 for the lower plate and T_2 for the upper plate with $T_2 > T_1$. A constant pressure gradient is applied in the x-direction. A uniform magnetic field B_0 is applied in the positive y-direction while the induced magnetic field is neglected by assuming a very small magnetic Reynolds number (See Fig. 1.5). The Hall effect is taken into consideration and consequently a z-component for the velocity is expected to arise. The fluid motion starts from rest at $t = 0$, and the no-slip condition at the plates implies that the fluid velocity has neither a z- nor an x-component at $y = -h$ and $y = h$. The initial temperature of the fluid is assumed to be equal to T_1. Since the plates are infinite in the x- and z-directions, the physical quantities do not change in these directions and the problem is essentially one-dimensional. The flow of the fluid is governed by the Navier—Stokes equation, which has the two components [78],

$$\rho\frac{\partial u}{\partial t} = -\frac{dP}{dx} + \mu\frac{\partial^2 u}{\partial y^2} + \frac{\partial \mu}{\partial y}\frac{\partial u}{\partial y} - \frac{\sigma B_0^2}{1+m^2}(u + mw) \tag{1.52}$$

$$\rho\frac{\partial w}{\partial t} = \mu\frac{\partial^2 w}{\partial y^2} + \frac{\partial \mu}{\partial y}\frac{\partial w}{\partial y} - \frac{\sigma B_0^2}{1+m^2}(w - mu) \tag{1.53}$$

where ρ is the density of the fluid, μ is the viscosity of the fluid, v is the velocity vector of the fluid $= u(y, t)i + w(y, t)j$, σ is the electric conductivity of the fluid, m is the Hall parameter given by $m = \sigma\beta B_0$, and β is the Hall

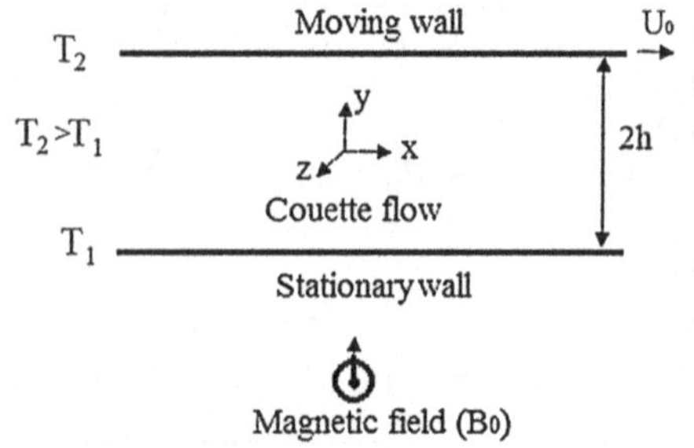

Figure 1.5 Schematic of the problem.

factor [78]. The energy equation describing the temperature distribution for the fluid is given by:

$$\rho c_p \frac{\partial T}{\partial t} = \frac{\partial}{\partial y}\left(k\frac{\partial T}{\partial y}\right) + \mu\left[\left(\frac{\partial u}{\partial y}\right)^2 + \left(\frac{\partial w}{\partial y}\right)^2\right] + \frac{\sigma B_0^2}{(1+m^2)}(u^2+w^2) \tag{1.54}$$

where T is the temperature of the fluid, c_p is the specific heat at constant pressure of the fluid, and k is thermal conductivity of the fluid. The viscosity of the fluid is assumed to vary exponentially with temperature and is defined as $\mu = \mu_0 f(T) = \mu_0 \exp(-a(T - T_1))$. Also the thermal conductivity of the fluid is varying linearly with temperature as$k = k_0 g(T) = k_0[1 + b(T - T_1)]$. The problem is simplified by writing the equations in the nondimensional form. To achieve this define the following nondimensional quantities [78],

$$\widehat{y} = \frac{y}{h}, \ \widehat{t} = \frac{tU_0}{h}, \ \widehat{P} = \frac{P}{\rho U_0^2}, \ (\widehat{u}, \widehat{w}) = \frac{(u, w)}{U_0}, \ \theta = \frac{T - T_1}{T_2 - T_1}, \ \alpha = -\frac{d\widehat{P}}{d\widehat{x}} \tag{1.55}$$

$f(\theta) = e^{-a\theta}$, a is the viscosity parameter. $g(\theta) = 1 + b\theta$ b is the thermal conductivity parameter. $\mathrm{Re} = \dfrac{\rho U_0 h}{\mu_0}$ is the Reynolds number. $Ha^2 = \dfrac{\sigma B_0^2 h^2}{\mu_0}$, Ha is Hartmann number. $\Pr = \frac{\mu_0 c_p}{k_0}$ is Prandtl number and $Ec = \dfrac{U_0^2}{c_p(T_2 - T_1)}$ is Eckert number.

In terms of the above nondimensional quantities, the velocity and energy Eqs. (1.52)–(1.54) read;

$$\frac{\partial u}{\partial t} = \alpha + \frac{1}{\mathrm{Re}} f(\theta)\frac{\partial^2 u}{\partial y^2} + \frac{1}{\mathrm{Re}}\frac{\partial f(\theta)}{\partial y}\frac{\partial u}{\partial y} - \frac{1}{\mathrm{Re}}\frac{Ha^2}{1+m^2}(u + mw) \tag{1.56}$$

$$\frac{\partial w}{\partial t} = \frac{1}{\mathrm{Re}} f(\theta)\frac{\partial^2 w}{\partial y^2} + \frac{1}{\mathrm{Re}}\frac{\partial f(\theta)}{\partial y}\frac{\partial w}{\partial y} - \frac{1}{\mathrm{Re}}\frac{Ha^2}{1+m^2}(w - mu) \tag{1.57}$$

$$\begin{aligned}\frac{\partial \theta}{\partial t} &= \frac{1}{\mathrm{RePr}} g(\theta)\frac{\partial^2 \theta}{\partial y^2} + \frac{1}{\mathrm{RePr}}\left(\frac{\partial g(\theta)}{\partial y}\right)\left(\frac{\partial \theta}{\partial y}\right) \\ &+ \frac{Ec}{\mathrm{Re}} f(\theta)\left[\left(\frac{\partial u}{\partial y}\right)^2 + \left(\frac{\partial w}{\partial y}\right)^2\right] + \frac{EcHa^2}{\mathrm{Re}(1+m^2)}(u^2+w^2)\end{aligned} \tag{1.58}$$

The boundary and initial conditions for components of velocity and temperature are:

$$
\begin{aligned}
&IC's\rightarrow\begin{cases} u(y,0)=0 \\ w(y,0)=0 \\ \theta(y,0)=0 \end{cases} \\
&BC's\rightarrow\begin{cases} u(-1,t)=0,\ u(1,t)=1 \\ w(-1,t)=0,\ w(1,t)=0 \\ \theta(-1,t)=0,\ \theta(1,t)=1 \end{cases}
\end{aligned}
\tag{1.59}
$$

Once the values of the velocities and temperature are obtained, the friction coefficient and Nusselt number will be determined. The local skin friction coefficient at the lower wall is,

$$
C_f=\frac{2}{\mathrm{Re}}\frac{\partial U}{\partial y}\bigg|_{y=-1} \tag{1.60}
$$

And the local Nusselt number for the lower wall is defined as,

$$
Nu_x=\frac{\partial \theta}{\partial y}\bigg|_{y=-1} \tag{1.61}
$$

The solution of the system of Eqs. (1.56)–(1.58) can be assumed as the following form:

$$
\begin{aligned}
&\text{for}\quad 1\le i\le N_t,\ \ 1\le j\le N_y+1 \\
&u_i(j,t)=\sum_{k=0}^{m}U_i(j,k)\left(\frac{t-t_i}{H}\right)^k \quad t\in[t_i,t_{i+1}] \\
&w_i(j,t)=\sum_{k=0}^{m}W_i(j,k)\left(\frac{t-t_i}{H}\right)^k \quad t\in[t_i,t_{i+1}] \\
&\theta_i(j,t)=\sum_{k=0}^{m}\Theta_i(j,k)\left(\frac{t-t_i}{H}\right)^k \quad t\in[t_i,t_{i+1}]
\end{aligned}
\tag{1.62}
$$

After taking the second order accurate central finite difference approximation with respect to y and applying MDTM on Eqs. (1.56)–(1.58), the following recurrence relations can be obtained:

for $1 \le i \le N_t,\ 1 \le j \le N_y - 1$

$$\begin{aligned}
U_i(j,k+1) &= \\
&\frac{H}{k+1}\left\{\alpha\delta(k) + \frac{1}{\mathrm{Re}\Delta y^2}\sum_{r=0}^{k} F_i(j,k-r)(U_i(j+1,r) - 2U_i(j,r) + U_i(j-1,r))\right. \\
&+\frac{1}{4\mathrm{Re}\Delta y^2}\sum_{r=0}^{k}(F_i(j+1,k-r) - F_i(j-1,k-r))(U_i(j+1,r) - U_i(j-1,r)) \\
&\left.-\frac{Ha^2}{\mathrm{Re}(1+m^2)}(U_i(j,k) + mW_i(j,k))\right\}
\end{aligned} \tag{1.63}$$

$$\begin{aligned}
W_i(j,k+1) &= \frac{H}{k+1}\left\{\frac{1}{\mathrm{Re}\Delta y^2}\sum_{r=0}^{k} F_i(j,k-r)(W_i(j+1,r) - 2W_i(j,r)\right. \\
&+W_i(j-1,r)) \\
&+\frac{1}{4\mathrm{Re}\Delta y^2}\sum_{r=0}^{k}(F_i(j+1,k-r) - F_i(j-1,k-r))(W_i(j+1,r) - W_i(j-1,r)) \\
&\left.-\frac{Ha^2}{\mathrm{Re}(1+m^2)}(W_i(j,k) - mU_i(j,k))\right\}
\end{aligned} \tag{1.64}$$

$$\begin{aligned}
\Theta_i(j,k+1) &= \frac{H}{k+1}\left\{\frac{1}{\mathrm{Re\,Pr}\Delta y^2}(\Theta_i(j+1,k) - 2\Theta_i(j,k) + \Theta_i(j-1,k))\right. \\
&+\frac{b}{\mathrm{Re\,Pr}\Delta y^2}\sum_{r=0}^{k}\Theta_i(j,k-r)(\Theta_i(j+1,r) - 2\Theta_i(j,r) + \Theta_i(j-1,r)) \\
&+\frac{1}{4\mathrm{Re\,Pr}\Delta y^2}\sum_{r=0}^{k}(\Theta_i(j+1,k-r) - \Theta_i(j-1,k-r))(\Theta_i(j+1,r) - \Theta_i(j-1,r)) \\
&+\frac{Ec}{4\mathrm{Re}\Delta y^2}\sum_{r=0}^{k}\sum_{s=0}^{r} F_i(j,s)[(U_i(j+1,r-s) - U_i(j-1,r-s))(U_i(j+1,k-r) \\
&-U_i(j-1,k-r)) \\
&(W_i(j+1,r-s) - W_i(j-1,r-s))(W_i(j+1,k-r) - W_i(j-1,k-r))] \\
&\left.+\frac{EcHa^2}{\mathrm{Re}(1+m^2)}\sum_{r=0}^{k}[U_i(j,r)U_i(j,k-r) + W_i(j,r)W_i(j,k-r)]\right\}
\end{aligned} \tag{1.65}$$

where $F_i(j, k)$ and $G_i(j, k)$ are the differential transform of the functions $f(\theta)$ and $g(\theta)$, respectively. Applying MDTM on initial conditions in Eq. (1.59), we have:

$$\begin{aligned} &\text{for} \quad 1 \le j \le N_y + 1 \\ &U_1(j,0) = 0, \quad W_1(j,0) = 0, \quad T_1(j,0) = 0 \end{aligned} \tag{1.66}$$

The boundary conditions in Eq. (1.59) can be transformed as follow:

$$\text{for} \quad 1 \le j \le N_t$$

$$BC\text{'s} \quad \text{for } u(y,t) \rightarrow \begin{cases} U_i(1,k) = 0, \ k \ge 0 \\ U_i(N_y+1,0) = 1, \ U_i(N_y+1,k) = 0, \ k \ge 1 \end{cases} \tag{1.67}$$

$$BC\text{'s} \quad \text{for } w(y,t) \rightarrow \begin{cases} W_i(1,k) = 0, \ k \ge 0 \\ W_i(N_y+1,k) = 0, \ k \ge 0 \end{cases} \tag{1.68}$$

$$BC\text{'s} \quad \text{for } \theta(y,t) \rightarrow \begin{cases} \Theta_i(1,k) = 0, \ k \ge 0 \\ \Theta_i(N_y+1,0) = 1, \ \Theta_i(N_y+1,k) = 0, \ k \ge 1 \end{cases} \tag{1.69}$$

For solving the problem in whole of the time subdomains, we must use the continuity condition in each time subdomain. These conditions can be expressed as:

$$\begin{aligned} &\text{for} \quad 2 \le j \le N_y, \ 2 \le i \le N_t \\ &U_i(j,0) = \sum_{k=0}^{m} U_{i-1}(j,k), \quad W_i(j,0) = \sum_{k=0}^{m} W_{i-1}(j,k) \\ &\Theta_i(j,0) = \sum_{k=0}^{m} \Theta_{i-1}(j,k) \end{aligned} \tag{1.70}$$

The accuracy of this method is shown in Fig. 1.6.

1.5 DIFFERENTIAL TRANSFORMATION METHOD APPLYING ON INITIAL-VALUE PROBLEMS AND ORDINARY DIFFERENTIAL EQUATIONS

In this section, applying the DTM on initial-value problems (IVPs) and ordinary differential equations (ODE) are discussed [79]. The differential equation for the initial-value problem can be described as

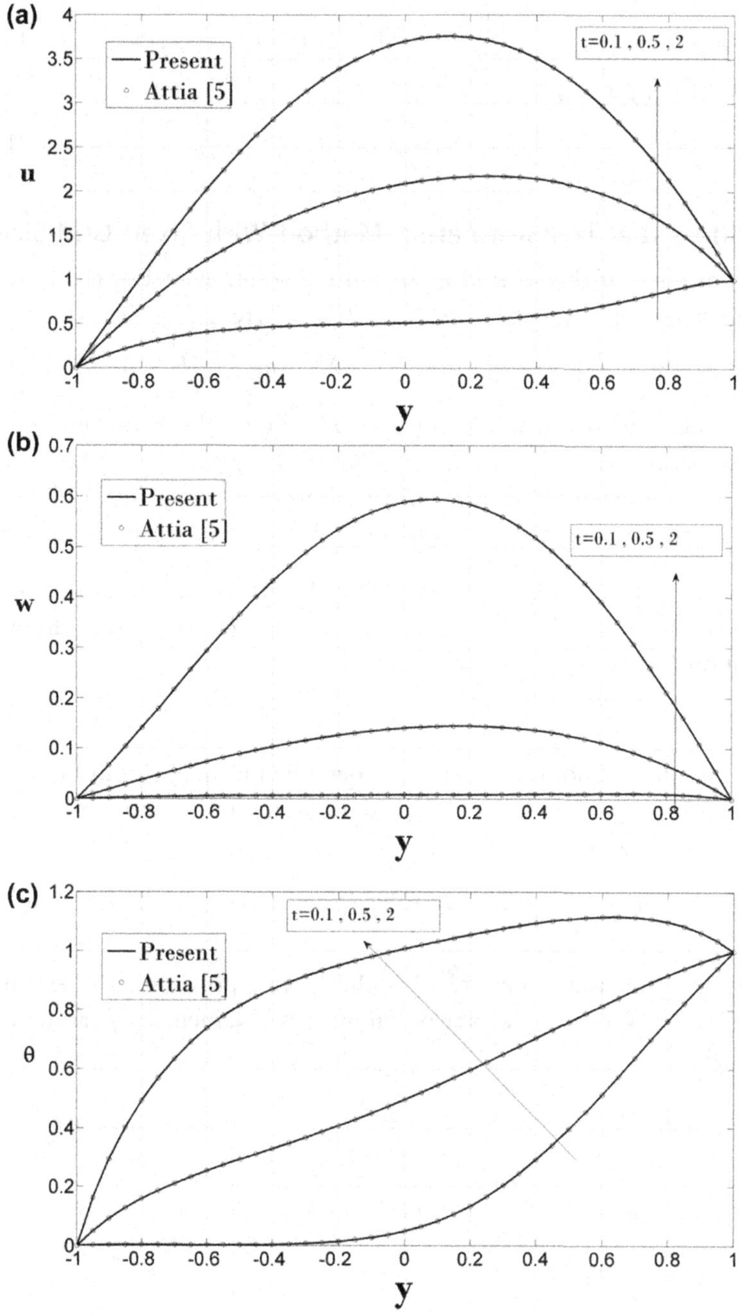

Figure 1.6 Comparison between Hybrid Differential Transformation Method with Ref. [5] when $\alpha = 5$, $\mathrm{Pr} = 1$, $\mathrm{Re} = 1$, $Ec = 0.2$, $Ha = 1$, $m = 3$, $a = 0.5$, $b = 0.5$.

$$\frac{dy}{dt} = f(t, y), \; a \leq t \leq b \tag{1.71}$$

with initial condition

$$y(a) = \alpha. \tag{1.72}$$

A. Differential Transformation Method With Fixed Grid Size

The objective of this section is to find the solution of Eq. (1.71) at the equally spaced grid points $\{t_0, t_1, t_2, \ldots, t_N\}$, where

$$t_i = a + ih, \quad \text{for each } i = 0, 1, \ldots, N, \text{ and } h = (b - a)/N \tag{1.73}$$

The domain of interest $[a, b]$ is divided into N subdomains and the approximation functions in each subdomain are $y_i(t)$; $i = 0, 1, 2, \ldots, N-1$, respectively. Taking the differential transformation of Eq. (1.71), the transformed equation describes the relationship between the spectrum of $y(t)$ as

$$(k+1)Y(k+1) = F(Y(k)), \tag{1.74}$$

where $F(\cdot)$ denotes the transformed function of $f(t, y(k))$. From the initial condition Eq. (1.72), it can be obtained that

$$Y(0) = \alpha. \tag{1.75}$$

In the first subdomain, $y(t)$ can be described by $y_0(t)$. From Eqs. (1.74) and (1.75), $y_0(t)$ can be represented in terms of its nth order Taylor polynomial about a, that is

$$y_0(t) = Y_0(0) + Y_0(1)(t-a) + Y_0(2)(t-a)^2 + \ldots + Y_0(n)(t-a)^n, \tag{1.76}$$

where the subscript 0 denotes that the Taylor polynomial is expanded about $t_0 = (a)$. Once the Taylor polynomial is obtained, $y(t_1)$ can be evaluated as

$$\begin{aligned} y(t_1) &\approx y_0(t_1) \\ &= Y_0(0) + Y_0(1)(t-a) + Y_0(2)(t-a)^2 + \ldots + Y_0(n)(t-a)^n \\ &= Y_0(0) + Y_0(1)(h) + Y_0(2)h^2 + Y_0(n)h^n \\ &= \sum_{j=0}^{n} Y_0(j)h^j \end{aligned} \tag{1.77}$$

The final value $y_0(t_1)$ of the first subdomain is the initial value of the second subdomain, i.e., $y_1(t_1) = Y_1(0) = y_0(t_1)$. In a similar manner, $y(t_2)$ can be represented as:

$$\begin{aligned} y(t_2) &\approx y_1(t_1) \\ &= Y_1(0) + Y_1(1)(t - t_1) + Y_1(2)(t - t_1)^2 + \ldots + Y_1(n)(t - t_1)^n \\ &= Y_1(0) + Y_1(1)(h) + Y_1(2)h^2 + Y_1(n)h^n \\ &= \sum_{j=0}^{n} Y_1(j)h^j. \end{aligned} \tag{1.78}$$

Hence, the solution on the grid points (t_{i+1}) can be obtained as follows:

$$\begin{aligned} y(t_{i+1}) &\approx y_i(t_{i+1}) \\ &= Y_i(0) + Y_i(1)(t_{i+1} - t_i) + Y_i(2)(t_{i+1} - t_i)^2 + \ldots + Y_i(n)(t_{i+1} - t_i)^n \\ &= Y_i(0) + Y_i(1)(h) + Y_i(2)h^2 + Y_i(n)h^n \\ &= \sum_{j=0}^{n} Y_i(j)h^j. \end{aligned} \tag{1.79}$$

For illustrative purposes, the procedure for solving the initial-value problem of linear and nonlinear differential equations is demonstrated as follows.

Example A. Consider the nonlinear initial-value problem

$$\dot{y} = -(y+1)(y+3), \ 0 \le t \le 3, \quad \text{with } y(0) = -2. \tag{1.80}$$

Let $N = 10$ and $h = 0.3$. Taking the differential transformation of Eq. (1.80), it can be obtained that

$$\begin{aligned} &Y_i(k+1) = -[(Y_i(k) + \delta(k)) \otimes (Y_i + 3\delta(k))]/(k+1) = \\ &-[Y_i(k) \otimes Y_i(k) + 4Y_i(k) + 3\delta(k)]/(k+1) \end{aligned} \tag{1.81}$$

With $Y_0(0) = -2$ the approximation of $y(t)$ on the grid point can be obtained by Eq. (1.81). The exact solution of this problem is$y(t) = -3 + (1 + e^{-2t})^{-1}$.

B. Differential Transformation Method With Varying Grid Size

The adaptive technique of the DTM can improve the efficiency of computation by the use of varying grid size techniques in integration approximation. The precondition is that complications from these methods do not influence the nature of the system. The truncation errors of computation can be estimated according to different grid sizes and without approximation of the higher derivatives of the function. These methods are called adaptive because they adapt the number and position of the grid points used, such that the truncation error is bounded within a specified bound [79]. Using the ideal difference equation to solve the initial-value problem of Eq. (1.71), the approximated solution is

$$w_{i+1} = w_i + h_i\Phi(t_i, w_i, h_i) \quad \text{for } i = 0, 1, \ldots, n. \tag{1.82}$$

where Φ is the incremental function of t_i, w_i and h_i. Theoretically, if a tolerance e is given, the global error would not exceed e for any grid points. Thus,

$$|y(t_i) - w_i| \leq \varepsilon \quad \text{for } i = 0, 1, \ldots, n. \tag{1.83}$$

From error estimates of the Taylor transformation method, if sufficient differentiability conditions are satisfied, an nth-order Taylor transformation method will have local error $O(h^{n+1})$ and global error $O(h^n)$. The local error can be predicted by changing the order of the Taylor series expansion. Through this method, one can choose a proper grid size according to some criterions, such that the global error is constrained within a specified bound [79].

Example B. Consider the system of the initial-value problem as follows:

$$\dot{u}_1(t) = 9u_1(t) + 24u_2(t) + 5\cos t - \frac{1}{3}\sin t, \quad \text{with } u_1(0) = \frac{4}{3},$$

$$\dot{u}_2(t) = -24u_1(t) - 51u_2(t) - 9\cos t + \frac{1}{3}\sin t, \quad \text{with } u_2(0) = \frac{2}{3}, \tag{1.84}$$

The exact solutions of Eq. (1.84) are

$$u_1(t) = 2e^{-3t} - e^{-39t} + \frac{1}{3}\cos t,$$

$$u_2(t) = -e^{-3t} + 2e^{-39t} - \frac{1}{3}\cos t.$$

The differential equation of the system, for $t \in [t_i, t_{i+1}]$, can be represented as

$$\dot{u}_1(t^*) = 9u_1(t^*) + 24u_2(t) + \left(5\cos t_i - \frac{1}{3}\sin t_i\right)\cos t^* - \left(5\sin t_i + \frac{1}{3}\cos t_i\right)\sin t^*,$$
$$\dot{u}_2(t^*) = -24u_1(t^*) - 51u_2(t) + \left(-9\cos t_i + \frac{1}{3}\sin t_i\right)\cos t^* + \left(9\sin t_i - \frac{1}{3}\cos t_i\right)\sin t^*, \tag{1.85}$$

where $t^* = t - t_i$. Taking the differential transformation of Eq. (1.85), the corresponding transformed equation can be obtained as

$$\begin{aligned}
&(k+1)U_{1i}(k+1) \\
&= 9U_{1i}(k) + 24U_{2i}(k) + \frac{1}{k!}\left(5\cos t_i - \frac{1}{3}\sin t_i\right)\cos\left(\frac{\Pi}{2}k\right) \\
&-\frac{1}{k!}\left(5\sin t_i + \frac{1}{3}\cos t_i\right)\sin\left(\frac{\Pi}{2}k\right), \\
&(k+1)U_{2i}(k+1) \\
&= -24U_{1i}(k) - 51U_{2i}(k) + \frac{1}{k!}\left(-\cos t_i + \frac{1}{3}\sin t_i\right)\cos\left(\frac{\Pi}{2}k\right) \\
&\frac{1}{k!}\left(9\sin t_i - \frac{1}{3}\cos t_i\right)\sin\left(\frac{\Pi}{2}k\right)
\end{aligned} \tag{1.86}$$

With $U_{10}(0) = \frac{4}{3}$ and $U_{20}(0) = \frac{2}{3}$, the approximation of $y(t)$ on adaptive grid points can be obtained by Eq. (1.86). More details about the accuracy and errors of examples can be found in [79].

1.6 TWO-DIMENSIONAL DIFFERENTIAL TRANSFORMATION METHOD FOR PARTIAL DIFFERENTIAL EQUATIONS

In this section, two-dimensional differential transform method of solution of the initial-value problem for partial differential equations (PDEs) has been studied. The basic definitions and fundamental theorems 1−13 of the two-dimensional transform are defined in [80] as follows:

$$W(k,h) = \frac{1}{k!h!}\left[\frac{\partial^{k+h} w(x,y)}{\partial x^k \partial y^h}\right]_{(0,0)}, \tag{1.87}$$

where $w(x,y)$ is the original function, and $W(k, h)$ is the transformed function. Letters represent the original and transformed functions respectively. The differential inverse transform of $W(k, h)$ is defined as

$$w(x, y) = \sum_{k=0}^{\infty} \sum_{h=0}^{\infty} W(k, h) x^k y^h w\ddot{o}x;\ y\ddot{o}1/4 \tag{1.88}$$

and from Eqs. (1.87) and (1.88), the following can be concluded

$$w(x, y) = \sum_{k=0}^{\infty} \sum_{h=0}^{\infty} \frac{1}{k!h!} \left[\frac{\partial^{k+h} w(x, y)}{\partial x^k \partial y^h} \right]_{(0,0)} x^k y^h \tag{1.89}$$

Theorem 1. If $w(x, y) = u(x, y) \pm v(x, y)$, then $W(k, h) = U(k, h) \pm V(k, h)$.

Theorem 2. If $w(x, y) = \lambda u(x, y)$, then $W(k, h) = \lambda U(k, h)$. Here, λ is a constant.

Theorem 3. If $w(x, y) = \partial u(x, y)/\partial x$, then $W(k, h) = (k + 1)U(k + 1, h)$.

Theorem 4. If $w(x, y) = \partial u(x, y)/\partial y$, then $W(k, h) = (k + 1)U(k, h + 1)$.

Theorem 5. If $w(x, y) = \partial^{r+s} u(x, y)/\partial x^r \partial y^s$, then

$$W(k,h) = (k+1)(k+2)\ldots(k+r)(h+1)(h+2)\ldots(h+s)\lambda U(k+r, h+s).$$

Theorem 6. If$w(x,y) = u(x,y)v(x,y)$, then

$$W(k, h) = \sum_{r=0}^{k} \sum_{s=0}^{h} U(r, h-s) V(k-r, s). \tag{1.90}$$

Theorem 7. If $w(x,y) = x^m y^n$,

$$W(k, h) = \delta(k-m, h-n) = \delta(k-m)\delta(h-n). \tag{1.91}$$

where

$$\delta(k-m) = \begin{cases} 1, k = m, \&, h = n \\ 0, \text{ otherwise} \end{cases}$$

Theorem 8. If

$$W(x, y) = \left[\frac{\partial u(x, y)}{\partial x} \frac{\partial v(x, y)}{\partial x} \right],$$

then

$$W(k, h) = \sum_{r=0}^{k} \sum_{s=0}^{h} (r+1)(k-r+1) U(r+1, h-s) V(k-r+1, s).$$

From the above definition,

$$W(0,0) = \left[\frac{\partial u(x,y)}{\partial x}\frac{\partial v(x,y)}{\partial x}\right]_{(0,0)} = U(1,0)V(1,0) \tag{1.92}$$

$$W(1,0) = \frac{1}{1!0!}\frac{\partial}{\partial x}\left[\frac{\partial u(x,y)}{\partial x}\frac{\partial v(x,y)}{\partial x}\right]_{(0,0)}$$

$$= \frac{1}{1!0!}\left[\frac{\partial^2 u(x,y)}{\partial x^2}\frac{\partial v(x,y)}{\partial x} + \frac{\partial u(x,y)}{\partial x}\frac{\partial^2 v(x,y)}{\partial x^2}\right]_{(0,0)} \tag{1.93}$$

$$= 2U(2,0)V(1,0) + 2U(1,0)V(2,0),$$

$$W(2,0) = \frac{1}{2!0!}\frac{\partial^2}{\partial x^2}\left[\frac{\partial u(x,y)}{\partial x}\frac{\partial v(x,y)}{\partial x}\right]_{(0,0)} \tag{1.94}$$

$$= 3U(3,0)V(1,0) + 4U(2,0)V(2,0) + 3U(1,0)V(3,0),$$

$$W(0,1) = U(1,1)V(1,0) + U(1,0)V(1,1), \tag{1.95}$$

$$W(1,1) = 2U(2,1)V(1,0) + 2U(2,0)V(1,1) + 2U(1,1)V(2,0) \\ +2U(1,0)V(2,1), \tag{1.96}$$

$$W(2,2) = 3U(3,2)V(1,0) + 3U(3,1)V(1,1) + 3U(3,0)V(1,2) \\ +4U(2,2)V(2,0) + 4U(2,1)V(2,1) + 4U(2,0)V(2,2) \\ +3U(1,2)V(3,0) + 3U(1,1)V(3,1) + 3U(1,0)V(3,2). \tag{1.97}$$

Theorem 9. If

$$w(x,y) = \frac{\partial u(x,y)}{\partial y}\frac{\partial v(x,y)}{\partial y},$$

then

$$W(k,h) = \sum_{r=0}^{k}\sum_{s=0}^{h}(k-r+1)(h-s+1)U(k-r+1)V(r,h-s+1).$$

From the above definition,

$$W(0,0) = \left[\frac{\partial u(x,y)}{\partial x}\frac{\partial v(x,y)}{\partial y}\right]_{(0,0)} = U(1,0)V(0,1), \tag{1.98}$$

$$W(1,0) = \frac{1}{1!0!}\frac{\partial}{\partial x}\left[\frac{\partial u(x,y)}{\partial y}\frac{\partial v(x,y)}{\partial y}\right]_{(0,0)}$$

$$= \frac{1}{1!0!}\left[\frac{\partial^2 u(x,y)}{\partial x \partial y}\frac{\partial v(x,y)}{\partial y} + \frac{\partial u(x,y)}{\partial y}\frac{\partial^2 v(x,y)}{\partial x \partial y}\right]_{(0,0)} \quad (1.99)$$

$$U(1,1)V(0,1) + U(0,1)V(1,1),$$

$$W(2,0) = \frac{1}{2!0!}\frac{\partial^2}{\partial x^2}\left[\frac{\partial u(x,y)}{\partial x}\frac{\partial v(x,y)}{\partial x}\right]_{(0,0)} \quad (1.100)$$

$$= U(0,1)V(2,1) + U(1,1)V(1,1) + U(2,1)V(0,1),$$

$$W(0,1) = 2U(0,2)V(0,1) + 2U(0,1)V(0,2), \quad (1.101)$$

$$W(1,1) = 2U(0,2)V(1,1) + 2U(0,1)V(1,2) + 2U(1,2)V(0,1)$$
$$+2U(1,1)V(0,2), \quad (1.102)$$

$$W(2,2) = 3U(0,3)V(2,1) + 4U(0,2)V(2,2) + 3U(0,1)V(2,3)$$
$$+3U(1,3)V(1,1) + 4U(1,2)V(1,2) + 3U(1,1)V(1,3)$$
$$+3U(2,3)V(0,1) + 4U(2,2)V(0,2) + 3U(2,1)V(0,3). \quad (1.103)$$

Theorem 10. If

$$w(x,y) = \frac{\partial u(x,y)}{\partial x}\frac{\partial v(x,y)}{\partial y},$$

then

$$W(k,h) = \sum_{r=0}^{k}\sum_{s=0}^{h}(k-r+1)(h-s+1)U(k-r+1,s)V(r,h-s+1).$$

From the above definition,

$$W(0,0) = \frac{\partial}{\partial x}\left[\frac{\partial u(x,y)}{\partial x}\frac{\partial v(x,y)}{\partial y}\right]_{(0,0)} = U(1,0)V(0,1), \quad (1.104)$$

$$W(1,0) = \frac{1}{1!0!}\frac{\partial}{\partial x}\left[\frac{\partial u(x,y)}{\partial x}\frac{\partial v(x,y)}{\partial y}\right]_{(0,0)}$$

$$= \frac{1}{1!0!}\left[\frac{\partial^2 u(x,y)}{\partial x^2}\frac{\partial v(x,y)}{\partial y} + \frac{\partial u(x,y)}{\partial x}\frac{\partial^2 v(x,y)}{\partial x \partial y}\right]_{(0,0)} \quad (1.105)$$

$$= 2U(2,0)V(0,1) + U(1,0)V(1,1),$$

$$W(2,0)=\frac{1}{2!0!}\frac{\partial^2}{\partial x^2}\left[\frac{\partial u(x,y)}{\partial x}\frac{\partial v(x,y)}{\partial y}\right]_{(0,0)} \tag{1.106}$$

$$=3U(3,0)V(1,0)+2U(2,0)V(1,1)+U(1,0)V(2,1), \tag{1.107}$$

$$W(0,1)=2U(1,0)V(0,2)+2U(1,1)V(0,1), \tag{1.108}$$

$$W(1,1)=4U(2,0)V(0,2)+2U(2,1)V(0,1)+2U(1,0)V(1,2)$$
$$+U(1,1)V(1,1), \tag{1.109}$$

$$W(2,2)=9U(3,0)V(0,3)+6U(3,1)V(0,2)+3U(3,2)V(0,1)$$
$$+6U(2,0)V(1,3)+4U(2,1)V(1,2)+2U(2,2)V(1,1)$$
$$+3U(1,0)V(2,3)+2U(1,1)V(2,2)+U(1,2)V(2,1). \tag{1.110}$$

Theorem 11. If $w(x,y)=u(x,y)v(x,y)\omega(x,y)$, then

$$W(k,h)=\sum_{r=0}^{k}\sum_{t=0}^{k-r}\sum_{s=0}^{h}\sum_{p=0}^{h-s}U(r,h-s-p)V(t,s)\Omega(k-r-t,P).$$

From the definition of transform,

$$W(0,0)=[u(x,y)v(x,y)\omega(x,y)]_{(0,0)}=U(0,0)V(0,0)\Omega(0,0), \tag{1.111}$$

$$W(1,0)=\frac{1}{1!0!}\frac{\partial}{\partial x}[u(x,y)v(x,y)\omega(x,y)]_{(0,0)}$$
$$=U(0,0)V(0,0)\Omega(1,0)+U(0,0)V(1,0)\Omega(0,0) \tag{1.112}$$
$$+U(1,0)V(0,0)\Omega(0,0),$$

$$W(2,0)=\frac{1}{2!0!}\frac{\partial^2}{\partial x^2}[u(x,y)v(x,y)\omega(x,y)]_{(0,0)}$$
$$=U(0,0)V(0,0)\Omega(2,0)+U(0,0)V(1,0)\Omega(1,0) \tag{1.113}$$
$$+U(0,0)V(2,0)\Omega(0,0)+U(1,0)V(0,0)\Omega(1,0)$$
$$+U(1,0)V(1,0)\Omega(0,0)+U(2,0)V(0,0)\Omega(0,0),$$

$$W(0,1)=U(0,1)V(0,0)\Omega(0,0)+U(0,0)V(0,1)\Omega(0,0)$$
$$+U(0,0)V(0,0)\Omega(0,1), \tag{1.114}$$

$$
\begin{aligned}
W(0,2) &= U(0,2)V(0,0)\Omega(0,0) + U(0,1)V(0,0)\Omega(0,1) \\
&+U(0,0)V(0,0)\Omega(0,2) + U(0,1)V(0,1)\Omega(0,0) \\
&+U(0,0)V(0,1)\Omega(0,1) + U(0,0)V(0,2)\Omega(0,0),
\end{aligned} \tag{1.115}
$$

$$
\begin{aligned}
W(1,1) &= U(0,1)V(0,0)\Omega(1,0) + U(0,0)V(0,0)\Omega(1,1) \\
&+U(0,0)V(0,1)\Omega(1,0) + U(0,1)V(1,0)\Omega(0,0) \\
&+U(0,0)V(1,0)\Omega(0,1) + U(0,0)V(1,1)\Omega(0,0) \\
&+U(1,1)V(0,0)\Omega(0,0) + U(1,0)V(0,0)\Omega(0,1) \\
&+U(1,0)V(0,1)\Omega(0,0).
\end{aligned} \tag{1.116}
$$

Theorem 12. If

$$W(x,y) = u(x,y)\frac{\partial u(x,y)}{\partial x}\frac{\partial v(x,y)}{\partial x},$$

then

$$
\begin{aligned}
W(k,h) &= \sum_{r=0}^{k}\sum_{t=0}^{k-r}\sum_{s=0}^{h}\sum_{p=0}^{h-s}(t+1)(k-r-t+1) \\
&\times U(r,h-s-p)V(t+1,s)\Omega(k-r-t+1,P).
\end{aligned}
$$

From the definition of transform,

$$W(0,0) = \left[u(x,y)\frac{\partial v(x,y)}{\partial x}\frac{\partial \omega(x,y)}{\partial x}\right]_{(0,0)} = +U(0,0)V(1,0)\Omega(1,0), \tag{1.117}$$

$$
\begin{aligned}
W(1,0) &= \frac{1}{1!0!}\frac{\partial}{\partial x}\left[u(x,y)\frac{\partial v(x,y)}{\partial x}\frac{\partial \omega(x,y)}{\partial x}\right]_{(0,0)} \\
&= U(1,0)V(1,0)\Omega(1,0) + 2U(0,0)V(2,0)\Omega(1,0) + 2U(0,0)V(1,0)\Omega(2,0),
\end{aligned} \tag{1.118}
$$

$$
\begin{aligned}
W(2,0) &= \frac{1}{2!0!}\frac{\partial^2}{\partial x^2}\left[u(x,y)\frac{\partial v(x,y)}{\partial x}\frac{\partial \omega(x,y)}{\partial x}\right]_{(0,0)} \\
&= 3U(0,0)V(1,0)\Omega(3,0) + 4U(0,0)V(2,0)\Omega(2,0) \\
&+U3(0,0)V(3,0)\Omega(1,0) + 2U(1,0)V(1,0)\Omega(2,0) \\
&+2U(1,0)V(2,0)\Omega(1,0) + U(2,0)V(1,0)\Omega(1,0),
\end{aligned} \tag{1.119}
$$

$$
\begin{aligned}
W(0,1) &= U(0,1)V(1,0)\Omega(1,0) + U(0,0)V(1,0)\Omega(1,1) \\
&U(0,0)V(1,1)\Omega(1,0),
\end{aligned} \tag{1.120}
$$

$$\begin{aligned}W(0,2) &= U(0,2)V(0,0)\Omega(0,0) + U(0,1)V(0,0)\Omega(0,1)\\ &+U(0,0)V(0,0)\Omega(0,2) + U(0,1)V(0,1)\Omega(0,0)\\ &+U(0,0)V(0,1)\Omega(0,1) + U(0,0)V(0,2)\Omega(0,0),\end{aligned} \tag{1.121}$$

$$\begin{aligned}W(1,1) &= 2U(0,1)V(1,0)\Omega(2,0) + U(0,0)V(1,0)\Omega(2,1)\\ &+U(0,0)V(1,1)\Omega(2,0) + 2U(0,1)V(2,0)\Omega(1,0)\\ &+U(0,0)V(2,0)\Omega(1,1) + U(0,0)V(2,1)\Omega(1,0)\\ &+U(1,1)V(1,0)\Omega(1,0) + U(1,0)V(1,0)\Omega(1,1)\\ &+U(0,1)V(1,1)\Omega(1,0)\end{aligned} \tag{1.122}$$

Theorem 13. If

$$W(x,y) = u(x,y)v(x,y)\frac{\partial^2\omega(x,y)}{\partial x^2},$$

then

$$W(k,h) = \sum_{r=0}^{k}\sum_{t=0}^{k-r}\sum_{s=0}^{h}\sum_{p=0}^{h-s}(K-r-t+1)(k-r-t+1)$$
$$\times U(r,h-s-p)V(t,s)\Omega(k-r-t+2,p).$$

From the definition of transform,

$$W(0,0) = \left[u(x,y)v(x,y)\frac{\partial^2\omega(x,y)}{\partial x^2}\right]_{(0,0)} = U(0,0)V(0,0)\Omega(2,0), \tag{1.123}$$

$$\begin{aligned}W(1,0) &= \frac{1}{1!0!}\frac{\partial}{\partial x}\left[u(x,y)v(x,y)\frac{\partial^2\omega(x,y)}{\partial x^2}\right]_{(0,0)}\\ &= 6U(0,0)V(0,0)\Omega(3,0) + 2U(0,0)V(1,0)\Omega(2,0)\\ &+2U(1,0)V(0,0)\Omega(2,0),\end{aligned} \tag{1.124}$$

$$\begin{aligned}W(2,0) &= \frac{1}{2!0!}\frac{\partial^2}{\partial x^2}\left[u(x,y)v(x,y)\frac{\partial^2\omega(x,y)}{\partial x^2}\right]_{(0,0)}\\ &= 12U(0,0)V(0,0)\Omega(4,0) + 6U(0,0)V(1,0)\Omega(3,0)\\ &+2U(0,0)V(2,0)\Omega(2,0) + 6U(1,0)V(0,0)\Omega(3,0)\\ &+2U(1,0)V(1,0)\Omega(2,0) + 2U(2,0)V(0,0)\Omega(2,0),\end{aligned} \tag{1.125}$$

$$W(0,1) = 2U(0,1)V(0,0)\Omega(2,0) + 2U(0,0)V(0,0)\Omega(2,1) \\ +2U(0,0)V(0,1)\Omega(2,0), \tag{1.126}$$

$$W(0,2) = 2U(0,2)V(0,0)\Omega(2,0) + 2U(0,1)V(0,0)\Omega(2,1) \\ +2U(0,0)V(0,0)\Omega(2,1) + 2U(0,1)V(0,1)\Omega(2,0) \\ +2U(0,0)V(0,1)\Omega(2,1) + 2U(0,0)V(0,2)\Omega(2,0), \tag{1.127}$$

$$W(1,1) = U(0,1)V(0,0)\Omega(1,0) + U(0,0)V(0,0)\Omega(1,1) \\ +U(0,0)V(0,1)\Omega(1,0) + U(0,1)V(1,0)\Omega(0,0) \\ +U(0,0)V(1,0)\Omega(0,1) + U(0,0)V(1,1)\Omega(0,0) \\ +U(1,1)V(0,0)\Omega(0,0) + U(1,0)V(0,0)\Omega(0,1) \\ +U(1,0)V(0,1)\Omega(0,0), \tag{1.128}$$

Example A. Consider the following linear PDE with the initial condition by Eqs. (1.130) and (1.131),

$$\frac{\partial^2 u}{\partial x^2} - 3\frac{\partial^2 u}{\partial x \partial t} + \frac{\partial^2 u}{\partial t^2} = 0, \tag{1.129}$$

$$\text{I: C1} \quad (x,0) = x^2, \tag{1.130}$$

$$\text{I: C2} \quad \frac{\partial u(x,0)}{\partial t} = e^x. \tag{1.131}$$

Taking two-dimensional transform of Eq. (1.129), we obtain

$$(k+1)(k+2)U(k+2,h) - 3(k+1)(h+1)U(k+1,h+1) \\ -4(k+1)(k+2)U(k+2,h) = 0, \tag{1.132}$$

from Eq. (1.130),

$$U(0,0) = 0 \tag{1.133}$$

$$U(1,0) = 0 \tag{1.134}$$

$$U(2,0) = 1 \tag{1.135}$$

$$U(\mathrm{i},0) = 0, i = 3,4,\ldots,n, \tag{1.136}$$

and from Eq. (1.131),

$$U(0,1) = 1, \tag{1.137}$$

$$U(1,1) = 1, \tag{1.138}$$

$$U(2,1) = \frac{2}{2!}, \tag{1.139}$$

$$U(i,1) = \frac{2}{n!}, i = 3, 4, \ldots, n. \tag{1.140}$$

Substituting Eqs. (1.133)–(1.140) into Eq. (1.132), and by recursive method, the results corresponding to $n \to \infty$ are listed as

$$U(0,2) = -\frac{1}{8}, \tag{1.141}$$

$$U(0,2) = \frac{13}{96}, \tag{1.142}$$

$$U(1,1) = 1, \tag{1.143}$$

$$U(2,1) = \frac{1}{2}. \tag{1.144}$$

Substituting all $U(k,h)$ into Eq. (1.88), we obtain the series solution form as follows:

$$u(x,t) = t - \frac{1}{8}t^2 + x^2 + xt + \frac{1}{8}x^2 t + \ldots \tag{1.145}$$

and the analytical solution of the problem is given as follows for comparison:

$$u(x,t) = \frac{4}{5}\left(e^{(x+(t/4))} - e^{(x-t)}\right) + x^2 + \frac{1}{4}t^2 \tag{1.146}$$

Example B. Consider the following wave equation [81]:

$$\frac{\partial^2 w(x,t)}{\partial t^2} - 4\frac{\partial^2 w(x,t)}{\partial x^2} = 0, \quad 0 \le x \le 1, \ 0 < t \tag{1.147}$$

with the boundary conditions

$$w(0,t) = w(1,t) = 0, \quad 0 < t \tag{1.148}$$

and initial conditions

$$\begin{aligned} &w(0,t) = \sin(\pi x), \quad 0 \le x \le 1, \\ &\frac{\partial w(x,0)}{\partial t} = 0, \quad 0 \le x \le 1. \end{aligned} \tag{1.149}$$

Taking the differential transform of Eq. (1.147), then

$$(k+2)(k+1)W(i,k+2) = 4(i+2)(i+1)W(i+2,k). \tag{1.150}$$

From the initial condition given by Eq. (1.149),

$$w(x,0) = \sum_{i=0}^{\infty} W(i,0)x^i = \sin(\pi x) = \sum_{i=1,3,\ldots}^{\infty} \frac{(-1)^{(i-1)/2}}{i!}\pi^i x^i \tag{1.151}$$

the corresponding spectra can be obtained as follows:

$$W(i,0) = \begin{cases} 0, & \text{for } i \text{ is even,} \\ \dfrac{(-1)^{(i-1)/2}}{i!}\pi^{i}, & \text{for } i \text{ is odd} \end{cases} \tag{1.152}$$

and from Eq. (1.149) it can be obtained that

$$\frac{\partial w(x,0)}{\partial t} = \sum_{i=0}^{\infty} W(i,1)x^i = 0. \tag{1.153}$$

Hence,

$$w(i,1) = 0. \tag{1.154}$$

Substituting Eqs. (1.153) and (1.154) to Eq. (1.150), all spectra can be found as

$$W(i,k) = \begin{cases} 0, & \text{for } i \text{ is even or } k \text{ is odd,} \\ \dfrac{2^k(-1)^{(i+k-1)/2}}{i!k!}\pi^{i+k}, & \text{for } i \text{ is odd or } k \text{ is even.} \end{cases} \tag{1.155}$$

Therefore, the closed form of the solution can be easily written as

$$\begin{aligned} w(x,t) &= \sum_{i=0}^{\infty}\sum_{k=0}^{\infty} W(i,k)x^i t^k = \sum_{i=0}^{\infty}\sum_{k=0}^{\infty} \frac{2^k}{k!i!}(-1)^{(i+k-1)/2}\pi^{i+k}x^i t^k \\ &= \left(\sum_{i=1,3,\ldots}^{\infty} \frac{1}{i!}(-1)^{(i-1)/2}(\pi x)^i\right)\left(\sum_{k=0,2,\ldots}^{\infty} \frac{1}{k!}(-1)^{(k)/2}(2\pi t)^k\right) \\ &= \sin(\pi x)\cos(2\pi t). \end{aligned} \tag{1.156}$$

As seen in Fig. 1.7, when the solution of the PDE is calculated by using DTM, it is clearly appears that the boundary conditions are provided, as the approximate solutions remain close to the exact solutions.

Example C. Consider the one-dimensional unsteady heat conduction problem as follows:

$$u_{xx} = 4u_t \tag{1.157}$$

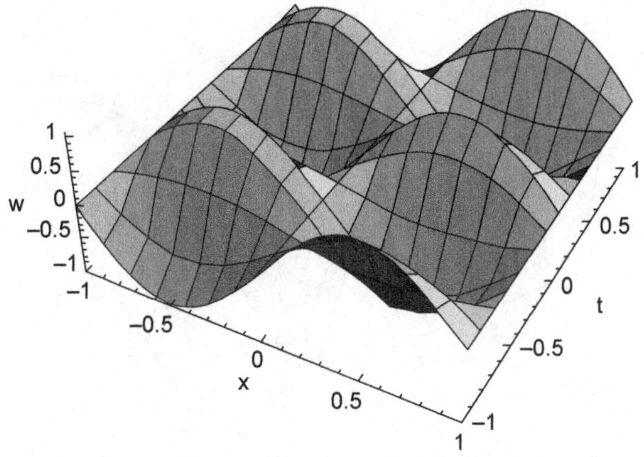

Figure 1.7 The analytic solution of Example 1.

with the initial and boundary conditions are

$$u(0,t) = 0, \quad u(2,t) = 0, \quad u(x,0) = 2\sin\frac{\pi x}{2}. \tag{1.158}$$

Taking the differential transform of Eq. (1.157), then

$$(k+1)(k+2)U(k+2,h) = 4(h+1)U(k,h+1)$$

and also

$$U(k,h+1) = \frac{(k+1)(k+2)}{4(h+1)}U(k+2,h) \tag{1.159}$$

Substituting $U(k,h)$ in DTM principle, it can be obtained that the closed form of the solution is,

$$\begin{aligned} u(x,t) &= 2\left(x - \frac{1}{3!}x^3 + \frac{1}{5!}x^5 - \frac{1}{7!}x^7 + \frac{1}{9!}x^9 - \ldots..\right) \\ &-t\left(\frac{1.3}{1!}x - \frac{2.5}{3!}x^3 + \frac{3.7}{5!}x^5 - \frac{4.9}{7!}x^7 + \frac{5.11}{9!}x^9 - \ldots..\right) \\ &+t^2\left(\frac{3.5}{2!2}x - \frac{3.5.7}{3!4}x^3 + \frac{3.7.9}{5!2}x^5 - \frac{4.9.11}{7!2}x^7 + \ldots..\right) + \ldots. \end{aligned} \tag{1.160}$$

The analytic solution of this equation is

$$u(x,t) = 2\sin\frac{\pi x}{2}\cdot e^{-\pi^2 t/16} \tag{1.161}$$

and the related analytic graph is given by Fig. 1.8.

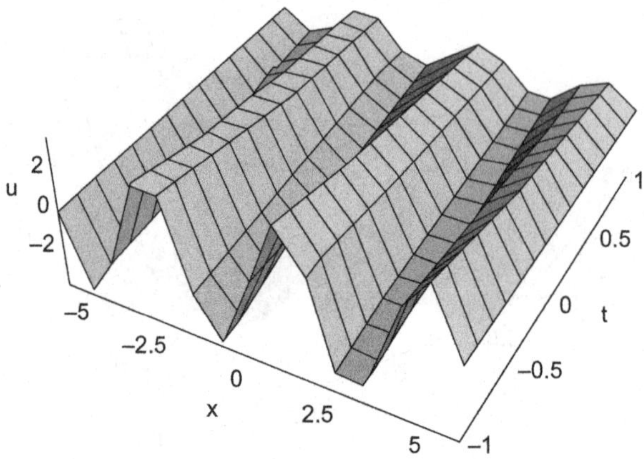

Figure 1.8 The approximate solution of Example C.

1.7 DIFFERENTIAL TRANSFORMATION METHOD—PADÉ APPROXIMATION

A Padé approximant is the ratio of two polynomials constructed from the coefficients of the Taylor series expansion of a function $u(x)$. The $[L/M]$ Pade approximants to a function $y(x)$ are given by [31,82].

$$\left[\frac{L}{M}\right] = \frac{P_L(x)}{Q_M(x)} \tag{1.162}$$

where $P_L(x)$ is polynomial of the degree of at most L, and $Q_M(x)$ is a polynomial of the degree of at most M. The formal power series

$$y(x) = \sum_{i=0}^{\infty} a_i x^i \tag{1.163}$$

and

$$y(x) - \frac{P_L(x)}{Q_M(x)} = O(x^{L+M+1}) \tag{1.164}$$

determine the coefficients of $P_L(x)$ and $Q_M(x)$ by the equation. Since we can clearly multiply the numerator and denominator by a constant and leave $[L/M]$ unchanged, we imposed the normalization condition which is

$$Q_M(0) = 1.0 \tag{1.165}$$

Finally, $P_L(x)$ and $Q_M(x)$ need to include noncommon factors. If the coefficient of $P_L(x)$ and $Q_M(x)$ are written as

$$\begin{cases} P_L(x) = p_0 + p_1 x + p_2 x^2 + \ldots + p_L x^L, \\ Q_M(x) = q_0 + q_1 x + q_2 x^2 + \ldots + q_M x^M, \end{cases} \tag{1.166}$$

and using Eqs. (1.165) and (1.166), we may multiply (14) by $Q_M(x)$, which linearizes the coefficient equations. Eq. (1.164) can be presented in more detail as

$$\begin{cases} a_{L+1} + a_L q_1 + \ldots + a_{L-M+1} q_M = 0, \\ a_{L+2} + a_{L+1} q_1 + \ldots + a_{L-M+2} q_M = 0, \\ \vdots \\ a_{L+M} + a_{L+M-1} q_1 + \ldots + a_L q_M = 0, \end{cases} \tag{1.167}$$

$$\begin{cases} a_0 = p_0, \\ a_1 + a_0 q_1 = p_1, \\ a_2 + a_1 q_1 + a_0 q_2 = p_2, \\ \vdots \\ a_L + a_{L-1} q_1 + \ldots + a_0 q_L = p_L \end{cases} \tag{1.168}$$

To solve these equations, we start with Eq. (1.167), which is a set of linear equations for all the unknown $q's$. Once the $q's$ are known, then Eq. (1.168) gives an explicit formula for the unknown $p's$, which complete the solution. If Eqs. (1.167) and (1.168) are nonsingular, then we can solve them directly and obtain Eq. (1.169), where Eq. (1.169) holds, and if the lower index on a sum exceeds the upper, the sum is replaced by zero:

$$\left[\frac{L}{M}\right] = \frac{\det \begin{bmatrix} a_{L-M+1} & a_{L-M+2} & \cdots & a_{L+1} \\ \vdots & \vdots & \ddots & \vdots \\ a_L & a_{L+1} & \cdots & a_{L+M} \\ \sum\limits_{j=M}^{L} a_{j-M} x^j & \sum\limits_{j=M-1}^{L} a_{j-M+1} x^j & \cdots & \sum\limits_{j=0}^{L} a_j x^j \end{bmatrix}}{\det \begin{bmatrix} a_{L-M+1} & a_{L-M+2} & \cdots & a_{L+1} \\ \vdots & \vdots & \ddots & \vdots \\ a_L & a_{L+1} & \cdots & a_{L+M} \\ x^M & x^{M-1} & \cdots & 1 \end{bmatrix}} \tag{1.169}$$

To obtain a diagonal Padé approximants of a different order such as [2/2], [4/4], or [6/6], the symbolic calculus software Maple is used.

Example A. We considered the heat transfer analysis in the unsteady two-dimensional squeezing nanofluid flow between the infinite parallel plates (Fig. 1.9). The two plates are placed at $z = \pm \ell(1 - \alpha t)^{1/2} = \pm h(t)$. For $\alpha > 0$, the two plates are squeezed until they touch $t = 1/\alpha$, and for $\alpha < 0$

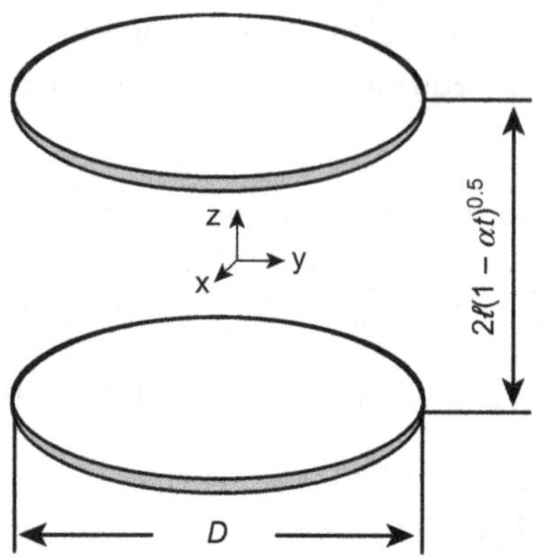

Figure 1.9 Nanofluid between parallel plates.

the two plates are separated. The viscous dissipation effect, the generation of heat due to friction caused by shear in the flow, is retained. This effect is quite important in the case when the fluid is largely viscous or flowing at a high speed. This behavior occurs at high Eckert number (>>1). Further the symmetric nature of the flow is adopted. The fluid is a water-based nanofluid containing Cu (copper) nanoparticles. The nanofluid is a two component mixture with the following assumptions: incompressible; no chemical reaction; negligible viscous dissipation; negligible radiative heat transfer; nanosolid-particles; and the base fluid are in thermal equilibrium and no slip occurs between them. The thermo-physical properties of the nanofluid are given in Table 1.4. The governing equations for momentum and energy in unsteady two-dimensional flow of a nanofluid are [31]:

$$\frac{\partial u}{\partial x}+\frac{\partial v}{\partial y}=0, \tag{1.170}$$

$$\rho_{nf}\left(\frac{\partial u}{\partial t}+u\frac{\partial u}{\partial v}+v\frac{\partial u}{\partial y}\right)=-\frac{\partial p}{\partial x}+\mu_{nf}\left(\frac{\partial^2 u}{\partial x^2}+\frac{\partial^2 u}{\partial y^2}\right), \tag{1.171}$$

Table 1.4 Thermo-Physical Properties of Water and Nanoparticles

	ρ(kg/m^3)	C_p(j/kgk)	k(W/m·k)
Pure water	997.1	4179	0.613
Copper (Cu)	8933	385	401

$$\rho_{nf}\left(\frac{\partial v}{\partial t}+u\frac{\partial v}{\partial v}+v\frac{\partial v}{\partial y}\right)=-\frac{\partial p}{\partial y}+\mu_{nf}\left(\frac{\partial^2 v}{\partial x^2}+\frac{\partial^2 v}{\partial y^2}\right), \tag{1.172}$$

$$\frac{\partial T}{\partial t}+u\frac{\partial T}{\partial x}+v\frac{\partial T}{\partial y}=\frac{k_{nf}}{(\rho C_p)_{nf}}\left(\frac{\partial^2 T}{\partial x^2}+\frac{\partial^2 T}{\partial y^2}\right)$$
$$+\frac{\mu_{nf}}{(\rho C_p)_{nf}}\left(4\left(\frac{\partial u}{\partial x}\right)^2+\left(\frac{\partial u}{\partial x}+\frac{\partial u}{\partial y}\right)^2\right), \tag{1.173}$$

Here u and v are the velocities in the x- and y-directions respectively, T is the temperature, P is the pressure, effective density (ρ_{nf}), the effective dynamic viscosity (μ_{nf}), the effective heat capacity $(\rho C_p)_{nf}$, and the effective thermal conductivity k_{nf} of the nanofluid are defined as [31]:

$$\rho_{nf}=(1-\phi)\rho_f+\phi\rho_s,\quad \mu_{nf}=\frac{\mu_f}{(1-\phi)^{2.5}},\quad (\rho C_p)_{nf}=(1-\phi)(\rho C_p)_f+\phi(\rho C_p)_s$$
$$\frac{k_{nf}}{k_f}=\frac{k_s+2k_f-2\phi(k_f-k_s)}{k_s+2k_f+2\phi(k_f-k_s)} \tag{1.174}$$

The relevant boundary conditions are:

$$\begin{aligned} v&=v_w=dh/dt,\quad T=T_H\quad at\ y=h(t),\\ v&=\partial u/\partial y=\partial T/\partial y=0\quad at\ y=0. \end{aligned} \tag{1.175}$$

We introduce these parameters:

$$\eta=\frac{y}{\left[l(1-\alpha t)^{1/2}\right]},\quad u=\frac{\alpha x}{[2(1-\alpha t)]}f'(\eta),$$
$$v=-\frac{\alpha l}{\left[2(1-\alpha t)^{1/2}\right]}f(\eta),\quad \theta=\frac{T}{T_H}, \tag{1.176}$$
$$A_1=(1-\phi)+\phi\frac{\rho_s}{\rho_f}.$$

Substituting the above variables into Eqs. (1.171) and (1.172), and then eliminating the pressure gradient from the resulting equations give:

$$f^{iv} - S\ A_1(1-\phi)^{2.5}\left(\eta f''' + 3f'' + f'f'' - ff'''\right) = 0, \tag{1.177}$$

Using Eq. (1.176), Eqs. (1.172) and (1.173) reduce to the following differential equations:

$$\theta'' + \mathrm{Pr}S\left(\frac{A_2}{A_3}\right)(f\theta' - \eta\theta') + \frac{\mathrm{Pr}\ Ec}{A_3(1-\phi)^{2.5}}\left(f''^2 + 4\delta^2 f'^2\right) = 0, \tag{1.178}$$

Here A_2 and A_3 are constants given by:

$$A_2 = (1-\phi) + \phi\frac{(\rho C_p)_s}{(\rho C_p)_f}, \quad A_3 = \frac{k_{nf}}{k_f} = \frac{k_s + 2\,k_f - 2\,\phi(k_f - k_s)}{k_s + 2\,k_f + 2\,\phi(k_f - k_s)} \tag{1.179}$$

With these boundary conditions:

$$\begin{aligned} &f(0) = 0, \quad f''(0) = 0, \\ &f(1) = 1, \quad f'(1) = 0, \\ &\theta'(0) = 0, \quad \theta(1) = 1. \end{aligned} \tag{1.180}$$

where S is the squeeze number, Pr is the Prandtl number, and Ec is the Eckert number, which are defined as:

$$S = \frac{\alpha l^2}{2\upsilon_f}, \quad \mathrm{Pr} = \frac{\mu_f(\rho C_p)_f}{\rho_f k_f}, \quad Ec = \frac{\rho_f}{(\rho C_p)_f}\left(\frac{\alpha x}{2(1-\alpha t)}\right)^2, \quad \delta = \frac{l}{x}, \tag{1.181}$$

Physical quantities of interest are the skin fraction coefficient and Nusselt number, which are defined as:

$$Cf = \frac{\mu_{nf}\left(\frac{\partial u}{\partial y}\right)_{y=h(t)}}{\rho_{nf} v_w^2}, \quad Nu = \frac{-lk_{nf}\left(\frac{\partial T}{\partial y}\right)_{y=h(t)}}{kT_H}. \tag{1.182}$$

In terms of Eq. (1.176), we obtain

$$\begin{aligned} &C_f^* = l^2/x^2(1-\alpha t)\mathrm{Re}_x C_f = A_1(1-\phi)^{2.5} f''(1), \\ &Nu^* = \sqrt{1-\alpha t}Nu = -A_3\,\theta'(1). \end{aligned} \tag{1.183}$$

Now DTM into governing equations has been applied. Taking the differential transforms of Eqs. (1.177)–(1.179) with respect to χ and considering $H = 1$ gives:

$$\begin{aligned}
&(k+1)(k+2)(k+3)(k+4)F[k+4] + SA_1(1-\phi)^{2.5}\\
&\quad\sum_{m=0}^{k}(\Delta[k-m-1](m+1)(m+2)(m+3)F[m+3])\\
&-3S(k+1)(k+2)F[k+2] - SA_1(1-\phi)^{2.5}\\
&\quad\sum_{m=0}^{k}((k-m+1)F[k-m+1](m+1)(m+2)F[m+2])\\
&+SA_1(1-\phi)^{2.5}\sum_{m=0}^{k}(F[k-m](m+1)(m+2)(m+3)F[m+3]) = 0,\\
&\Delta[m] = \begin{cases} 1 & m = 1 \\ 0 & m \neq 1 \end{cases}
\end{aligned} \tag{1.184}$$

$$F[0] = 0,\ \ F[1] = a_1,\ \ F[2] = 0,\ \ F[3] = a_2 \tag{1.185}$$

$$\begin{aligned}
&(k+1)(k+2)\Theta[k+2] + \Pr\cdot S\cdot\left(\frac{A_2}{A_3}\right)\sum_{m=0}^{k}(F[k-m](m+1)\Theta[m+1])\\
&-\Pr.S.\left(\frac{A_2}{A_3}\right)\sum_{m=0}^{k}(\Delta[k-m](m+1)\Theta[m+1])\\
&+\frac{\Pr Ec}{A_3(1-\phi)^{2.5}}\sum_{m=0}^{k}((k-m+1)(k-m+2)F[k-m+2](m+1)(m+2)F[m+2])\\
&+4\frac{\Pr Ec}{A_3(1-\phi)^{2.5}}\delta^2\sum_{m=0}^{k}((k-m+1)F[k-m+1](m+1)F[m+1]),\\
&\Delta[m] = \begin{cases} 1 & m = 1 \\ 0 & m \neq 1 \end{cases}
\end{aligned} \tag{1.186}$$

$$\Theta[0] = a_3,\ \Theta[1] = 0 \tag{1.187}$$

where $F[k]$ and $\Theta[k]$ are the differential transforms of $f(\eta)$, $\theta(\eta)$ and a_1, a_2, a_3 are constants, which can be obtained through boundary condition. This problem can be solved as followed:

$$F[0]=0,\ \ F[1]=a_1,\ \ F[2]=0,\ \ F[3]=a_2, F[4]=0$$
$$F[5]=\frac{3}{20}S\ A_1(1-\phi)^{2.5}a_2+\frac{1}{20}S\ A_1(1-\phi)^{2.5}a_1a_2+\frac{1}{20}a_1a_2,\ldots \tag{1.188}$$

$$\begin{aligned}
&\Theta[0]=a_3,\ \ \Theta[1]=0, \Theta[2]=-2\frac{\Pr Ec}{A_3(1-\phi)^{2.5}}\delta^2a_1^2,\\
&\Theta[3]=0,\\
&\Theta[4]=\frac{1}{3}\frac{\Pr Ec}{A_3(1-\phi)^{2.5}}\Pr S\left(\frac{A_2}{A_3}\right)a_1^3\,\delta^2-3\frac{\Pr Ec}{A_3(1-\phi)^{2.5}}\,a_2^2-2\frac{\Pr Ec}{A_3(1-\phi)^{2.5}}a_1a_2,\\
&\Theta[5]=0,\ldots
\end{aligned} \tag{1.189}$$

The above process is continuous. By substituting Eqs. (1.188) and (1.189) into the main equation based on DTM, it can be obtained that the closed form of the solutions is:

$$\begin{aligned}
F(\eta)=a_1\eta+a_2\eta^3+\Bigg(&\frac{3}{20}SA_1(1-\phi)^{2.5}a_2+\frac{1}{20}SA_1(1-\phi)^{2.5}a_1a_2\\
&+\frac{1}{20}a_1a_2\Bigg)\eta^4+\ldots
\end{aligned} \tag{1.190}$$

$$\begin{aligned}
\theta(\eta)=a_3+&\left(-2\frac{\Pr Ec}{A_3(1-\phi)^{2.5}}\delta^2a_1^2\right)\eta^2\\
&+\Bigg(\frac{1}{3}a_1^3\frac{\Pr Ec}{A_3(1-\phi)^{2.5}}\Pr S\left(\frac{A_2}{A_3}\right)\delta^2-3\Pr Ec\frac{\Pr Ec}{A_3(1-\phi)^{2.5}}a_2^2\\
&-2\Pr Ec\frac{\Pr Ec}{A_3(1-\phi)^{2.5}}a_1a_2\Bigg)\eta^4+\ldots
\end{aligned} \tag{1.191}$$

by substituting the boundary condition from Eq. (1.180) into Eqs. (1.190) and (1.191), in point $\eta=1$, it can be obtained the values of a_1, a_2, a_3. By substituting obtained a_1, a_2, a_3 into Eqs. (1.190) and (1.161), it can be obtained the expression of $F(\eta)$ and $\Theta(\eta)$. For example, for Cu-water nanofluid, when $\Pr = 6.2$. $Ec = 0.05$, $\delta = 0.1$, $S = 0.1$, and $\varphi = 0.01$ following equations will be obtained:

$$f(\eta) = 1.4870\eta - 0.47373\eta^3 - 0.01368\eta^5 + 0.0001428\eta^6 + 0.0002479\eta^7 - 3.174 \times 10^{-7}\eta^8 \quad (1.192)$$

$$\theta(\eta) = 1.227 - 0.0135\eta^2 - 0.20073\eta^4 - 0.00935\eta^6 + 0.000176\eta^7 - 0.00374\eta^8 \quad (1.193)$$

by applying Padé approximation to Eqs. (1.192) and (1.193) (for Padé [6,6] accuracy), we have,

$$\text{Padé }[6,6](f(\eta)) = \frac{1.487026\eta - 0.024727\eta^2 - 0.454146\eta^3 + 0.007668\eta^4 - 0.0197056\eta^5 + 0.00043511\eta^6}{0.9999 - 0.016628\eta + 0.0131716\eta^2 - 0.00014\eta^3 + 0.000146\eta^4 - 0.00000133\eta^5 + 0.00000264\eta^6} \quad (1.194)$$

$$\text{Padé }[6,6](\theta(\eta)) = \frac{1.22716 + 0.011948\eta - 0.2471506\eta^2 + 0.0004207\eta^3 - 0.21051\eta^4 - 0.002414\eta^5 + 0.03393\eta^6}{0.9999 + 0.009736\eta - 0.1903901\eta^2 + 0.0004\eta^3 - 0.0100589\eta^4 - 0.00037\eta^5 + 0.004023\eta^6} \quad (1.195)$$

Figs. 1.10 and 1.11 show the results of DTM and DTM−Padé [6,6], respectively for solving Eqs. (1.177) and (1.178) in different Eckert and

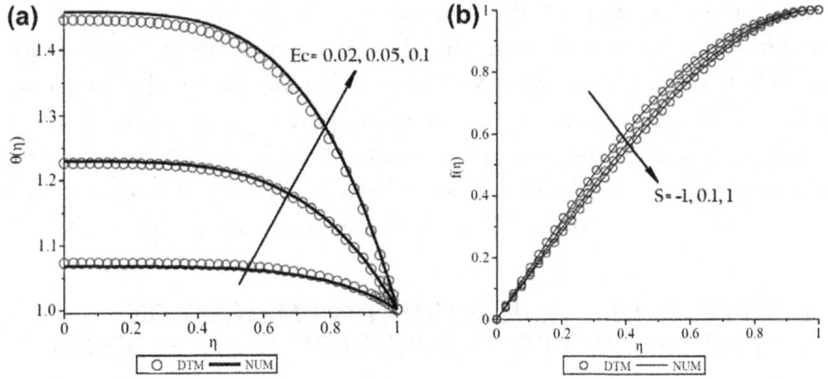

Figure 1.10 Comparison of Differential Transformation Method and numerical results for (a) $\theta(\eta)$ when $\Pr = 6.2$, $S = 0.1$, $\delta = 0.1$, and $\varphi = 0.01$ (b) $f(\eta)$ when $\Pr = 6.2$, $Ec = 0.05$, $\delta = 0.1$, and $\varphi = 0.01$.

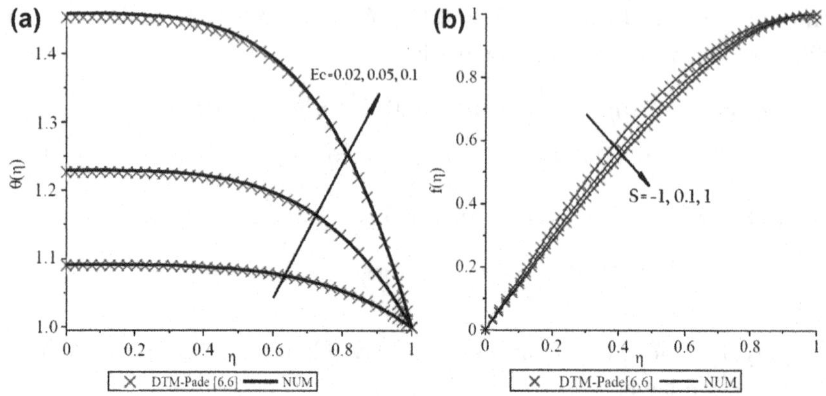

Figure 1.11 Comparison of Differential Transformation Method—Padé [6,6] and numerical results for (a) $\theta(\eta)$ when $\Pr = 6.2$, $S = 0.1$, $\delta = 0.1$, and $\varphi = 0.01$ (b) $f(\eta)$ when $\Pr = 6.2$, $Ec = 0.05$, $\delta = 0.1$, and $\phi = 0.01$.

squeeze numbers. As seen in these figures, DTM and DTM—Padé have good agreement with numerical method in wide range of Ec and S numbers, also Table 1.5, which is designed for an especial case of these two figures ($\Pr = 6.2$, $Ec = 0.05$, $\delta = 0.1$, $S = 0.1$, $\varphi = 0.01$), confirms that DTM—Padé has an excellent congruity with the numerical procedure and is more reliable than DTM. Also it can be seen that by increasing the accuracy of Padé from [3,3] to [6,6], DTM results have more agreement with numerical results. Fig. 1.12 shows the effect of squeeze number (S) on nondimensional temperature and velocity profiles respectively. When two plates move together, thermal boundary layer thickness increases as the absolute magnitude of the squeeze number enhances. The positive and negative squeeze numbers have different effects on the velocity profile. For the case of squeezing flow, the velocity increases due to an increase in the absolute value of squeeze number when $\eta < 0.5$, while it decreases for $\eta > 0.5$. In other words, when S is minus, α is minus too and by decreasing its magnitude, its behavior is like squeezing flow so temperature profile is decreased. The same treatment is observed in velocity profile. By increasing in positive values for S number, space between two plates decreased and consequently velocity profile near the lower plate is decreased.

1.8 DIFFERENTIAL TRANSFORMATION METHOD ON SINGULAR TWO-POINT BOUNDARY VALUE PROBLEM

We consider the singular two-point boundary value problem (BVP) [83].

$$y''(x) + \frac{1}{x}y'(x) + q(x)y(x) = r(x), \quad 0 < x \le 1 \tag{1.196}$$

Table 1.5 Comparison of Differential Transformation Method and Differential Transformation Method—Pade Results for Cu-Water Nanofluid When Pr = 6.2, $Ec = 0.05$, $\delta = 0.1$, $S = 0.1$, and $\varphi = 0.01$

	$\theta(\eta)$				$f(\eta)$			
η	**NUM**	**DTM**	**DTM—Pade [3,3]**	**DTM—Pade [6,6]**	**NUM**	**DTM**	**DTM—Pade [3,3]**	**DTM—Pade [6,6]**
0.0	1.2293076	1.227169	1.227169	1.227169	0.00000	0.00000	0.0000	0.00000
0.1	1.2291512	1.227014	1.227010	1.227014	0.148563042	0.1482287	0.1482287	0.1482287
0.2	1.2284338	1.226307	1.225837	1.226307	0.294239491	0.2936109	0.2936109	0.2936109
0.3	1.2264082	1.224320	1.230774	1.224320	0.434130652	0.4332839	0.4332838	0.4332839
0.4	1.2218221	1.219828	1.228738	1.219828	0.565313773	0.5643523	0.5643513	0.5643523
0.5	1.2129029	1.211086	1.228413	1.211086	0.684830361	0.6838729	0.6838679	0.6838729
0.6	1.1973334	1.195796	1.228287	1.195796	0.789674916	0.788838	0.7888208	0.7888388
0.7	1.1722092	1.171051	1.228223	1.171051	0.876784230	0.8761652	0.8761122	0.8761652
0.8	1.1339720	1.133258	1.228185	1.133258	0.943027415	0.9426750	0.9425400	0.9426750
0.9	1.0783086	1.078024	1.228161	1.078027	0.985196818	0.9850864	0.9847780	0.9850864
1.0	1.0000000	0.999999	1.228144	1.000001	1.00000	0.9999999	0.9993538	1.000001

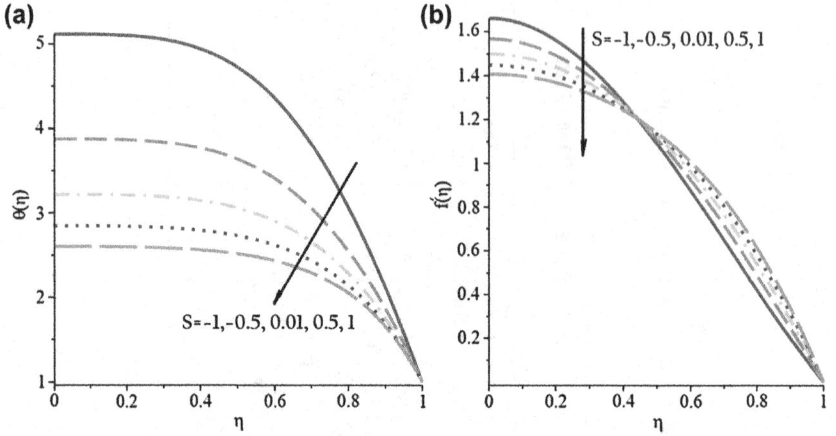

Figure 1.12 Effect of squeeze number (S) on (a) Temperature profile and (b) Velocity profile when $Pr = 6.2$, $Ec = 0.5$, $\delta = 0.1$, and $\phi = 0.06$.

subject to the boundary conditions

$$\begin{cases} y(0) = a_1, & y(1) = b_1, \quad \text{or} \\ y'(0) = a_2, & y(1) = b_2 \end{cases} \tag{1.197}$$

where $q(x)$ and $r(x)$ are continuous functions on (0,1); and a_1, a_2, b_1, and b_2 are real constants. These problems generally arise frequently in many areas of science and engineering, for example, fluid mechanics, quantum mechanics, optimal control, chemical reactor theory, aerodynamics, reaction–diffusion process, geophysics, etc.

Example A. Consider the following singular two-point BVP

$$y''(x) + \frac{1}{x}y'(x) + y(x) = \frac{5}{4} + \frac{x^2}{16}, \quad 0 < x \le 1 \tag{1.198}$$

subject to the boundary conditions

$$y'(0) = 0, \quad y(1) = \frac{17}{16} \tag{1.199}$$

The exact solution of this problem is $y(x) = 1 + \frac{x^2}{16}$. The transformed version of Eq. (1.198) is

$$\sum_{l=0}^{k} \delta(l-1)(k-l+1)(k-l+2)Y(k-l+2) + (k+1)Y(k+1) + \\ + \sum_{l=0}^{k} \delta(l-1)Y(k-l) - \frac{5}{4}\delta(k-1) - \frac{1}{16}\delta(k-3) = 0 \tag{1.200}$$

The transformed boundary conditions are

$$(k+1)Y(k+1) = 0,$$
$$\sum_{k=0}^{N} Y(k) = \frac{17}{16} \tag{1.201}$$

Using Eqs. (1.200) and (1.201) by taking $N = 6$, the following system for $k = 0, \ldots, 5$ is obtained:

$$\begin{aligned}
&Y(1) = 0\\
&4Y(2) + Y(0) = \frac{5}{4}\\
&9Y(3) + Y(1) = 0\\
&16Y(4) + Y(2) = \frac{1}{16}\\
&25Y(5) + Y(3) = 0\\
&36Y(6) + Y(4) = 0
\end{aligned} \tag{1.202}$$

Solving the above system and using the inverse transformation rule, we get the following series solution

$$y(x) = 1 + \frac{x^2}{16} \tag{1.203}$$

Note that for $N > 6$ we evaluate the same solution, which is the exact solution of Eq. (1.198) with the boundary conditions Eq. (1.199).

Example B. Consider the following singular two-point BVP

$$\left(1 - \frac{x}{2}\right)y''(x) + \frac{3}{2}\left(\frac{1}{x} - 1\right)y'(x) + \left(\frac{x}{2} - 1\right)y(x)$$
$$= 5 - \frac{29x}{2} + \frac{13x^2}{2} + \frac{3x^3}{2} - \frac{x^4}{2} \tag{1.204}$$

with following boundary conditions

$$y(0) = 0, \quad y(1) = 0 \tag{1.205}$$

The exact solution of this problem is $y(x) = x^2 - x^3$. The transformed version of Eq. (1.204) is

$$\sum_{l=0}^{k}\delta(l-1)(k-l+1)(k-l+2)Y(k-l+2)$$

$$-\frac{1}{2}\sum_{l=0}^{k}\delta(l-1)(k-l+1)(k-l+2)Y(k-l+2)+\frac{3}{2}(k+1)Y(k+1)+$$

$$-\frac{3}{2}\sum_{l=0}^{k}\delta(l-1)(k-l+1)Y(k-l+1)+\frac{1}{2}\sum_{l=0}^{k}\delta(l-2)Y(k-l)$$

$$-\sum_{l=0}^{k}\delta(l-1)Y(k-l)-5\delta(k-1)-\frac{29}{2}\delta(k-2)$$

$$-\frac{13}{2}(k-3)-\frac{3}{2}\delta(k-4)+\frac{\delta(k-5)}{2}=0 \tag{1.206}$$

The transformed boundary conditions are

$$Y(0)=0,\quad \sum_{k=0}^{N}Y(k)=0 \tag{1.207}$$

Using Eqs. (1.204) and (1.205) by taking $N=6$, the following system for $k=0,\ldots,5$ is obtained:

$$\begin{aligned}
&Y(1)=0\\
&-\frac{3}{2}Y(1)+5Y(2)+Y(0)=5\\
&-Y(1)+\frac{21}{2}Y(3)-4Y(2)+\frac{1}{2}Y(0)=-\frac{29}{2}\\
&18Y(4)-\frac{15}{2}Y(3)-Y(2)+\frac{1}{2}Y(1)=\frac{13}{2}\\
&\frac{55}{2}Y(5)-12Y(4)+\frac{1}{2}Y(2)-Y(3)=\frac{3}{2}\\
&39Y(6)-\frac{35}{2}Y(5)+\frac{1}{2}Y(3)-Y(4)=-\frac{1}{2}
\end{aligned} \tag{1.208}$$

REFERENCES

[1] Hatami M, Ganji DD. Optimization of the longitudinal fins with different geometries for increasing the heat transfer. In: ISER- 12th international conference on recent trends in engineering and technology (ICRTET-2015), Kuala Lumpur, Malaysia November 8, 2015; November 2015.

[2] Jalili Palandi S, Hatami M, Ghasemi SE, Ganji DD. Temperature distribution of convective fin with temperature-dependent internal heat generation and thermal conductivity using DTM. In: International conference on nonlinear modeling & optimization; August 2012.

[3] Jalili Palandi S, Ghasemi SE, Hatami M, Ganji DD. Efficient analytical approaches for motion of a spherical solid particle in plane couette fluid flow. In: International conference on nonlinear modeling & optimization; August 2012.

[4] Gorji M, Hatami M, Hasanpour A, Ganji DD. Nonlinear thermal analysis of solar air heater for the purpose of energy saving. January 2012. http://dx.doi.org/10.5829/idosi.ijee.2012.03.04.010.

[5] Ghasemi SE, Hatami M, Mehdizadeh Ahangar GHR, Ganji DD. Electrohydrodynamic flow analysis in a circular cylindrical conduit using Least Square Method. Journal of Electrostatics January 2013;72(1). http://dx.doi.org/10.1016/j.elstat.2013.11.005.

[6] Hatami M, Ganji DD. Investigation of refrigeration efficiency for fully wet circular porous fins with variable sections by combined heat and mass transfer analysis. International Journal of Refrigeration January 2013;40. http://dx.doi.org/10.1016/j.ijrefrig.2013.11.002.

[7] Khavaji A, Ganji DD, Roshan N, Moheimani R, Hatami M, Hasanpour A. Slope variation effect on large deflection of compliant beam using analytical approach. Structural Engineering & Mechanics November 2012;44(3). http://dx.doi.org/10.12989/sem.2012.44.3.405.

[8] Hatami M, Ganji DD. Heat transfer and flow analysis for SA-TiO_2 non-Newtonian nanofluid passing through the porous media between two coaxial cylinders. Journal of Molecular Liquids December 2013;188:155–61. http://dx.doi.org/10.1016/j.molliq.2013.10.009.

[9] Hatami M, Hatami J, Ganji DD. Computer simulation of MHD blood conveying gold nanoparticles as a third grade non-Newtonian nanofluid in a hollow porous vessel. Computer Methods and Programs in Biomedicine November 2013;113(2). http://dx.doi.org/10.1016/j.cmpb.2013.11.001.

[10] Hatami M, Nouri R, Ganji DD. Forced convection analysis for MHD Al2O3-water nanofluid flow over a horizontal plate. Journal of Molecular Liquids November 2013;187:294–301. http://dx.doi.org/10.1016/j.molliq.2013.08.008.

[11] Ghasemi SE, Hatami M, Ganji DD. Analytical thermal analysis of air-heating solar collectors. Journal of Mechanical Science and Technology November 2013;27(11). http://dx.doi.org/10.1007/s12206-013-0878-0.

[12] Hatami M, Hasanpour A, Ganji DD. Heat transfer study through porous fins (Si_3N_4 and AL) with temperature-dependent heat generation. Energy Conversion and Management October 2013;74:9–16. http://dx.doi.org/10.1016/j.enconman.2013.04.034.

[13] Ganji DD, Gorji M, Hatami M, Hasanpour A, Khademzadeh N. Propulsion and launching analysis of variable-mass rockets by analytical methods September 2013;2(3):225–33. http://dx.doi.org/10.1016/j.jppr.2013.07.006.

[14] Hatami M, Ganji DD, Boubaker K. Temperature variations analysis for Condensed matter micro- and nanoparticles combustion burning in gaseous oxidizing media by DTM and BPES September 2013;2013(24). http://dx.doi.org/10.1155/2013/129571.

[15] Sheikholeslami M, Hatami M, Ganji DD. Analytical investigation of MHD nanofluid flow in a semi-porous channel. Powder Technology September 2013;246:327–36. http://dx.doi.org/10.1016/j.powtec.2013.05.030.
[16] Gorji M, Ganji DD, Hatami M, Hasanpour A, Hossein AK. Launching process for variable-mass rockets by series solution. June 2013. http://dx.doi.org/10.18005/JMET0101003.
[17] Hatami M, Ganji DD. Thermal performance of circular convective-radiative porous fins with different section shapes and materials. Energy Conversion and Management December 2013;76:185–93. http://dx.doi.org/10.1016/j.enconman.2013.07.040.
[18] Hatami M, Sheikholeslami M, Ganji DD. Laminar flow and heat transfer of nanofluid between contracting and rotating disks by least square method. Powder Technology February 2014;253:769–79. http://dx.doi.org/10.1016/j.powtec.2013.12.053.
[19] Hatami M, Ganji DD. Thermal and flow analysis of microchannel heat sink (MCHS) cooled by Cu-water nanofluid using porous media approach and least square method. Energy Conversion and Management February 2014;78:347–58. http://dx.doi.org/10.1016/j.enconman.2013.10.063.
[20] Hatami M, Ganji DD. Heat transfer and nanofluid flow in suction and blowing process between parallel disks in presence of variable magnetic field. Journal of Molecular Liquids February 2014;190:159–68. http://dx.doi.org/10.1016/j.molliq.2013.11.005.
[21] Sheikholeslami M, Hatami M, Ganji DD. Nanofluid flow and heat transfer in a rotating system in the presence of a magnetic field. Journal of Molecular Liquids February 2014;190:112–20. http://dx.doi.org/10.1016/j.molliq.2013.11.002.
[22] Hatami M, Ganji DD. Motion of a spherical particle in a fluid forced vortex by DQM and DTM. Particuology March 2014;16. http://dx.doi.org/10.1016/j.partic.2014.01.001.
[23] Hatami M, Domairry G. Transient vertically motion of a soluble particle in a Newtonian fluid media. Powder Technology February 2014;253:481–5. http://dx.doi.org/10.1016/j.powtec.2013.12.015.
[24] Hatami M, Ganji DD. Natural convection of sodium alginate (SA) non-Newtonian nanofluid flow between two vertical flat plates by analytical and numerical methods. Case Studies in Thermal Engineering March 2014;2:14–22. http://dx.doi.org/10.1016/j.csite.2013.11.001.
[25] Nouri R, Ganji DD, Hatami M. Unsteady sedimentation analysis of spherical particles in Newtonian fluid media using analytical methods June 2014;3(2). http://dx.doi.org/10.1016/j.jppr.2014.05.003.
[26] Hatami M, Ganji DD. Thermal behavior of longitudinal convective–radiative porous fins with different section shapes and ceramic materials (SiC and Si_3N_4). Ceramics International June 2014;40(5):6765–75. http://dx.doi.org/10.1016/j.ceramint.2013.11.140.
[27] Sheikholeslami M, Hatami M, Ganji DD. Micropolar fluid flow and heat transfer in a permeable channel using analytical method. Journal of Molecular Liquids June 2014;194:30–6. http://dx.doi.org/10.1016/j.molliq.2014.01.005.
[28] Hatami M, Ganji DD. MHD nanofluid flow analysis in divergent and convergent channels using WRMs and numerical method. International Journal of Numerical Methods for Heat and Fluid Flow May 2014;24(5). http://dx.doi.org/10.1108/HFF-01-2013-0010.
[29] Hatami M, Ganji DD. Motion of a spherical particle on a rotating parabola using Lagrangian and high accuracy Multi-step Differential Transformation Method. Powder Technology May 2014;258:94–8. http://dx.doi.org/10.1016/j.powtec.2014.03.007.
[30] Ahmadi AR, Zahmatkesh A, Hatami M, Ganji DD. A comprehensive analysis of the flow and heat transfer for a nanofluid over an unsteady stretching flat plate. Powder Technology May 2014;258:125–33. http://dx.doi.org/10.1016/j.powtec.2014.03.021.

[31] Domairry G, Hatami M. Squeezing Cu-water nanofluid flow analysis between parallel plates by DTM-Padé Method. Journal of Molecular Liquids May 2014;193:37–44. http://dx.doi.org/10.1016/j.molliq.2013.12.034.
[32] Ganji DD, Hatami M. Three weighted residual methods based on Jeffery-Hamel flow. International Journal of Numerical Methods for Heat Fluid Flow June 2014;24(3). http://dx.doi.org/10.1108/HFF-06-2012-0137.
[33] Hatami M, Sheikholeslami M, Hosseini M, Ganji DD. Analytical investigation of MHD nanofluidflow in non-parallel walls. Journal of Molecular Liquids June 2014;194:251–9. http://dx.doi.org/10.1016/j.molliq.2014.03.002.
[34] Hatami M, Sheikholeslami M, Ganji DD. Nanofluid flow and heat transfer in an asymmetric porous channel with expanding or contracting wall. Journal of Molecular Liquids July 2014;195:230–9. http://dx.doi.org/10.1016/j.molliq.2014.02.024.
[35] Hatami M, Sheikholeslami M, Domairry G. High accuracy analysis for motion of a spherical particle in plane Couette fluid flow by Multi-step Differential Transformation Method. Powder Technology July 2014;260:59–67. http://dx.doi.org/10.1016/j.powtec.2014.02.057.
[36] Hatami M, Mehdizadeh Ahangar GHR, Ganji DD, Boubaker K. Refrigeration efficiency analysis for fully wet semi-spherical porous fins. Energy Conversion and Management August 2014;84:533–40. http://dx.doi.org/10.1016/j.enconman.2014.05.007.
[37] Hatami M, Hosseinzadeh Kh, Domairry G, Behnamfar MT. Numerical study of MHD two-phase Couette flow analysis for fluid-particle suspension between moving parallel plates. Journal of the Taiwan Institute of Chemical Engineers September 2014;45(5). http://dx.doi.org/10.1016/j.jtice.2014.05.018.
[38] Ghasemi SE, Hatami M, Ganji DD. Thermal analysis of convective fin with temperature-dependent thermal conductivity and heat generation. Case Studies in Thermal Engineering November 2014;4. http://dx.doi.org/10.1016/j.csite.2014.05.002.
[39] Hatami M, Ganji DD, Jafaryar M, Farkhadnia F. Transient combustion analysis for Iron micro-particles in a gaseous media by weighted residual methods (WRMs). Case Studies in Thermal Engineering November 2014;4. http://dx.doi.org/10.1016/j.csite.2014.06.003.
[40] Mosayebidorcheh S, Sheikholeslami M, Hatami M, Ganji DD. Analysis of turbulent MHD Couette nanofluid flow and heat transfer using hybrid DTM–FDM. Particuology November 2014. http://dx.doi.org/10.1016/j.partic.2014.07.004S.
[41] Ghasemi E, Valipour P, Hatami M, Ganji DD. Heat transfer study on solid and porous convective fins with temperature-dependent heat generation using efficient analytical method. Journal of Central South University of Technology December 2014;21(12). http://dx.doi.org/10.1007/s11771-014-2465-7.
[42] Fakour M, Vahabzadeh A, Ganji DD, Hatami M. Analytical study of micropolar fluid flow and heat transfer in a channel with permeable walls. Journal of Molecular Liquids January 2015;204. http://dx.doi.org/10.1016/j.molliq.2015.01.040.
[43] Sheikholeslami M, Hatami M, Domairry G. Numerical simulation of two phase unsteady nanofluid flow and heat transfer between parallel plates in presence of time dependent magnetic field. Journal of the Taiwan Institute of Chemical Engineers January 2015;46:43–50. http://dx.doi.org/10.1016/j.jtice.2014.09.025.
[44] Hatami M, Hatami J, Jafaryar M, Domairry G. Differential transformation method for Newtonian and Non-Newtonian fluids flow analysis: comparison with HPM and numerical solution. Journal of the Brazilian Society of Mechanical Sciences and Engineering February 2015. http://dx.doi.org/10.1007/s40430-014-0275-3.

[45] Rahimi-Gorji M, Pourmehran O, Hatami M, Ganji DD. Statistical optimization of microchannel heat sink (MCHS) geometry cooled by different nanofluids using RSM analysis. European Physical Journal Plus February 2015;130(2). http://dx.doi.org/10.1140/epjp/i2015-15022-8.
[46] Hatami M, Ghasemi SE, Sarokolaie AK, Ganji DD. Study on Blood flow containing nanoparticles trough porous arteries in presence of magnetic field using analytical methods. Physica E Low-dimensional Systems and Nanostructures March 2015;70. http://dx.doi.org/10.1016/j.physe.2015.03.002.
[47] Dogonchi AS, Hatami M, Hosseinzadeh K, Domairry G. Non-spherical particles sedimentation in an incompressible Newtonian medium by Pade approximation. Powder Technology 2015;278:248–56.
[48] Dogonchi AS, Hatami M, Domairry G. Motion analysis of a spherical solid particle in plane Couette Newtonian fluid flow. Powder Technology April 2015;274. http://dx.doi.org/10.1016/j.powtec.2015.01.018.
[49] Pourmehran O, Rahimi-Gorji M, Hatami M, Sahebi SAR, Domairry G. Numerical optimization of microchannel heat sink (MCHS) performance cooled by KKL based nanofluids in saturated porous medium. Journal of the Taiwan Institute of Chemical Engineers May 2015;55. http://dx.doi.org/10.1016/j.jtice.2015.04.016.
[50] Atouei SA, Hosseinzadeh Kh, Hatami M, Ghasemi SE, Sahebi SAR, Ganji DD. Heat transfer study on convective–radiative semi-spherical fins with temperature-dependent properties and heat generation using efficient computational methods. Applied Thermal Engineering June 2015;89. http://dx.doi.org/10.1016/j.applthermaleng.2015.05.084.
[51] Dogonchi AS, Hatami M, Hosseinzadeh K, Domairry G. Non-spherical particles sedimentation in an incompressible Newtonian medium by Padé approximation. Powder Technology July 2015;278. http://dx.doi.org/10.1016/j.powtec.2015.03.036.
[52] Mosayebidorcheh S, Hatami M, Mosayebidorcheh T, Ganji DD. Effect of periodic body acceleration and pulsatile pressure gradient pressure on non-Newtonian blood flow in arteries. Journal of the Brazilian Society of Mechanical Sciences and Engineering August 2015;38(3). http://dx.doi.org/10.1007/s40430-015-0404-7.
[53] Mosayebidorcheh S, Hatami M, Ganji DD, Mosayebidorcheh T, Mirmohammadsadeghi SM. Investigation of transient MHD Couette flow and heat transfer of Dusty fluid with temperature-dependent properties. Journal of Applied Fluid Mechanics September 2015;8(4).
[54] Ghasemi SE, Hatami M, Hatami J, Sahebi SAR, Ganji DD. An efficient approach to study the pulsatile blood flow in femoral and coronary arteries by Differential Quadrature Method. Physica A: Statistical Mechanics and its Applications October 2015;443. http://dx.doi.org/10.1016/j.physa.2015.09.039.
[55] Hatami M, Ghasemi SE, Sahebi SAR, Mosayebidorcheh S, Ganji DD, Hatami J. Investigation of third-grade non-Newtonian blood flow in arteries under periodic body acceleration using multi-step differential transformation method. Applied Mathematics and Mechanics November 2015;36(11). http://dx.doi.org/10.1007/s10483-015-1995-7.
[56] Mosayebidorcheh S, Hatami M, Mosayebidorcheh T, Ganji DD. Optimization analysis of convective–radiative longitudinal fins with temperature-dependent properties and different section shapes and materials. Energy Conversion and Management November 2015;106. http://dx.doi.org/10.1016/j.enconman.2015.10.067.
[57] Ghasemi SE, Zolfagharian A, Hatami M, Ganji DD. Analytical thermal study on nonlinear fundamental heat transfer cases using a novel computational technique. Applied Thermal Engineering December 2015;98:88–97. http://dx.doi.org/10.1016/j.applthermaleng.2015.11.120.

[58] Ghasemi SE, Vatani M, Hatami M, Ganji DD. Analytical and numerical investigation of nanoparticle effect on peristaltic fluid flow in drug delivery systems. Journal of Molecular Liquids March 2016;215:88–97. http://dx.doi.org/10.1016/j.molliq.2015.12.001.
[59] Rashidi MM. The modified differential transform method for solving MHD boundary-layer equations. Computer Physics Communications 2009;180(11):2210–7.
[60] Rashidi MM, Laraqi N, Sadri SM. A novel analytical solution of mixed convection about an inclined flat plate embedded in a porous medium using the DTM-Padé. International Journal of Thermal Sciences 2010;49(12):2405–12.
[61] Rashidi MM, Mohimanian Pour SA. A novel analytical solution of heat transfer of a micropolar fluid through a porous medium with radiation by DTM-Padé. Heat Transfer—Asian Research 2010;39(8):575–89.
[62] Rashidi MM, Parsa AB, Bég OA, Shamekhi L, Sadri SM, Bég TA. Parametric analysis of entropy generation in magneto-hemodynamic flow in a semi-porous channel with OHAM and DTM. Applied Bionics and Biomechanics 2014;11(1–2):47–60.
[63] Rashidi MM, Keimanesh M. Using differential transform method and padé approximant for solving mhd flow in a laminar liquid film from a horizontal stretching surface. Mathematical Problems in Engineering 2010;2010.
[64] Erfani E, Rashidi MM, Parsa AB. The modified differential transform method for solving off-centered stagnation flow toward a rotating disc. International Journal of Computational Methods 2010;7(04):655–70.
[65] Rashidi MM, Laraqi N, Basiri Parsa A. Analytical modeling of heat convection in magnetized micropolar fluid by using modified differential transform method. Heat Transfer—Asian Research 2011;40(3):187–204.
[66] Rashidi MM, Hayat T, Basiri Parsa A. Solving of boundary-layer equations with transpiration effects, governance on a vertical permeable cylinder using modified differential transform method. Heat Transfer—Asian Research 2011;40(8):677–92.
[67] Rashidi MM, Rahimzadeh N, Ferdows M, Uddin MJ, Bég OA. Group theory and differential transform analysis of mixed convective heat and mass transfer from a horizontal surface with chemical reaction effects. Chemical Engineering Communications 2012;199(8):1012–43.
[68] Rashidi MM, Hayat T, Keimanesh T, Yousefian H. A study on heat transfer in a second-grade fluid through a porous medium with the modified differential transform method. Heat Transfer—Asian Research 2013;42(1):31–45.
[69] Rashidi MM, Chamkh AJ, Keimanesh M. Application of multi-step differential transform method on flow of a second-grade fluid over a stretching or Shrinking Sheet. American Journal of Computational Mathematics 2011;6:119–28.
[70] Rashidi MM, Abelman S, Freidooni Mehr N. Entropy generation in steady MHD flow due to a rotating porous disk in a nanofluid. International Journal of Heat and Mass Transfer 2013;62:515–25.
[71] Keimanesh M, Rashidi MM, Chamkha AJ, Jafari R. Study of a third grade non-Newtonian fluid flow between two parallel plates using the multi-step differential transform method. Computers & Mathematics with Applications 2011;62(8):2871–91.
[72] Rashidi MM, Mohimanian Pour SA, Laraqi N. A semi-analytical solution of micro-Polar flow in a porous channel with mass Injection by using differential transform method. Nonlinear Analysis Modelling and Control 2010;15(3):341–50 [ISI].
[73] Rashidi MM, Shahmohamadi H, Domairry G. Variational iteration method for solving three-dimensional Navier–Stokes equations of flow between two stretchable disks. Numerical Methods for Partial Differential Equations 2011;27(2):292–301.
[74] Rashidi MM, Erfani E. The modified differential transform method for investigating nano boundary-layers over stretching surfaces. International Journal of Numerical Methods for Heat and Fluid Flow 2011;21(7):864–83.

[75] Basiri Parsa A, Rashidi MM, Anwar Bég O, Sadri SM. Semi-Computational simulation of magneto-hemodynamic flow in a semi-porous channel using optimal homotopy and differential transform methods. Computers in Biology and Medicine 2013;43(9): 1142–53.
[76] Rashidi MM, Sadri SM. Solution of the laminar viscous flow in a semi-porous channel in the presence of a uniform magnetic field by using the differential transform method. International Journal of Contemporary Mathematical Sciences 2010;5(15):711–20.
[77] Zhou K. Differential transformation and its applications for electrical circuits. Wuhan, China: Huazhong University Press; 1986.
[78] Mosayebidorcheh S, Sheikholeslami M, Hatami M, Ganji DD. Analysis of turbulent MHD Couette nanofluid flow and heat transfer using hybrid DTM–FDM. Particuology 2016;26:95–101.
[79] Jang M-J, Chen C-L, Liy Y-C. On solving the initial-value problems using the differential transformation method. Applied Mathematics and Computation 2000;115(2):145–60.
[80] Ayaz F. On the two-dimensional differential transform method. Applied Mathematics and Computation 2003;143(2):361–74.
[81] Bildik N, Konuralp A, Bek FO, Küçükarslan S. Solution of different type of the partial differential equation by differential transform method and Adomian's decomposition method. Applied Mathematics and Computation 2006;172(1):551–67.
[82] Bervillier C. Status of the differential transformation method. Applied Mathematics and Computation 2012;218(20):10158–70.
[83] Ravi Kanth ASV, Aruna K. Solution of singular two-point boundary value problems using differential transformation method. Physics Letters A 2008;372:4671–3.

CHAPTER 2

Differential Transformation Method in Advance

2.1 INTRODUCTION

Many phenomena in viscoelasticity, fluid mechanics, biology, chemistry, acoustics, control theory, psychology, and other areas of science can be successfully modeled by the use of fractional-order derivatives. That is because of the fact that, a realistic modeling of a physical phenomenon having dependence not only at the time instant, but also the previous time history can be successfully achieved by using fractional calculus. In previous chapter the principle of Differential Transformation Method (DTM) was introduced, which can be applied in most of mechanical engineering problems [1]. But in some mechanical problems such as eigenvalue problem, higher-order initial problems, fractional integro-differential equations, etc., the governing equations are some complicated and cannot be solved by the traditional DTM. This chapter introduces DTM for advance problems and contains the following sections:

2.1 Introduction
2.2 Differential Transformation Method for Higher-Order Initial Value Problems
2.3 Fractional Differential Transform Method
2.4 Differential Transformation Method for Integro-Differential Equation
2.5 Differential Transformation Method for Eigenvalue Problems
2.6 Two-Dimensional Differential Transformation Method for Fractional Order Partial Differential Equations
2.7 Reduced Differential Transform Method
2.8 Modified Differential Transformation Method

2.2 DIFFERENTIAL TRANSFORMATION METHOD FOR HIGHER-ORDER INITIAL VALUE PROBLEMS

A. Consider the Second-Order Initial Value Problem [2]

$$y''(t) - 2y'(t) + 2y(t) = \exp(2t)\sin(t), \quad 0 \le t \le 1 \tag{2.1}$$

Differential Transformation Method for Mechanical Engineering Problems
ISBN 978-0-12-805190-0
http://dx.doi.org/10.1016/B978-0-12-805190-0.00002-4

with initial conditions

$$y(0) = -0.4 \tag{2.2}$$

$$y'(0) = -0.6. \tag{2.3}$$

As above, with $u_1(t) = y(t)$ and $u_2(t) = y'(t)$. Eq. (2.1) transformed into the system of the first-order differential equation

$$u_1'(t) = u_2(t), \tag{2.4}$$

$$u_2' = \exp(2t)\sin(t) + 2u_2(t) \tag{2.5}$$

and the initial conditions Eqs. (2.2) and (2.3) become

$$u_1(0) = -0.4 \tag{2.6}$$

$$u_2(0) = -0.6 \tag{2.7}$$

Let $h = 0.1$ and $N = 10$. The differential equations of the system Eqs. (2.4) and (2.5) between t_i and t_{i+1} can be represented as

$$u_1'(t^*) = u_2(t^*) \tag{2.8}$$

$$u_2'(t^*) = \exp(2t^* + 2t_i)\sin(t^* + t_i) + 2u_1(t^*) + 2u_2(t^*), \tag{2.9}$$

where $t^* = t - t_i$. Taking the differential transformation of Eqs. (2.8) and (2.9), respectively, we get

$$U_{1i}(k+1) = \frac{1}{(k+1)} U_{2i}(k) \tag{2.10}$$

and

$$U_{2i}(k+1) = \frac{1}{(k+1)} \left\{ \exp(2t_i)\cos(t_i) \sum_{l=0}^{k} \frac{2^l}{(l!)(k-l)!} \sin\left(\frac{\pi}{2}\right)(k-l) + \exp(2t_i)\sin(t_i) \sum_{l=0}^{k} \frac{2^l}{(l!)(k-l)!} \cos\left(\frac{\pi}{2}\right)(k-l) - 2u_{1i}(k) + 2u_{2i}(k) \right\}, \tag{2.11}$$

with

$$U_{10}(0) = -0.4, \quad U_{20}(0) = -0.6, \tag{2.12}$$

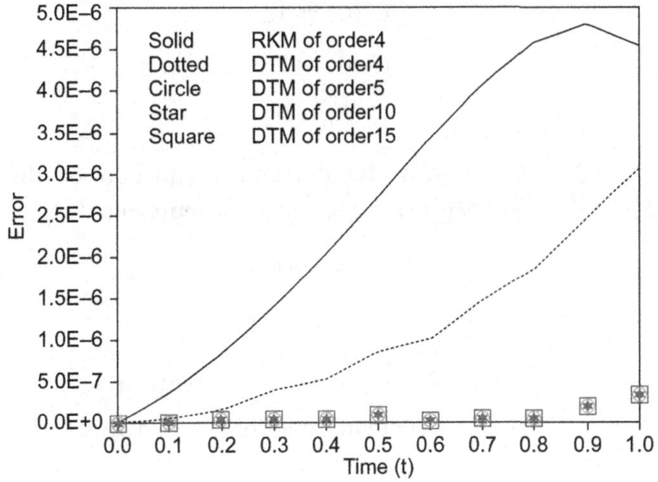

Figure 2.1 Computational errors corresponding to different order of Differential Transformation Method for case A.

The approximation of $u_1(t) = y(t)$ on the grid points can be obtained from Eqs. (2.10) and (2.8). The actual solution of Eqs. (2.1)–(2.3) is

$$y(t) = u_1(t) = 0.2\exp(2t) + \sin(t), \tag{2.13}$$

Fig. 2.1, shows errors involved with different order of DTM, along with the result obtained by the Runge–Kutta fourth-order method.

B. Consider the Third-Order Initial Value Problem

$$y'''(t) + 2y''(t) - y'(t) - 2y(t) = \exp(t), \quad 0 \le t \le 3 \tag{2.14}$$

with initial conditions

$$y(0) = 1. \tag{2.15}$$

$$y'(0) = 2 \tag{2.16}$$

$$y''(0) = 0 \tag{2.17}$$

With $U_1(t) = y(t)$, $U_2(t) = y'(t)$, and $U_3(t) = y''(t)$. Eq. (2.14) transformed into the system of the first-order differential equation

$$u_1'(t) = u_2(t). \tag{2.18}$$

$$u_2'(t) = u_3(t), \tag{2.19}$$

$$u_3'(t) = -2u_3(t) + u_2(t) + 2u_1(t) + \exp(t) \tag{2.20}$$

$$U_1(0) = 1 \tag{2.21}$$

$$U_2(0) = 2 \tag{2.22}$$

$$U_3(0) = 0 \tag{2.23}$$

Let $h = 0{:}2$ and $N = 15$. The differential equation of the system Eqs. (2.18)–(2.20) between t_i and $t_i + 1$ can be represented as

$$u_1'\left(t^*\right) = u_2\left(t^*\right). \tag{2.24}$$

$$u_2'\left(t^*\right) = u_3\left(t^*\right) \tag{2.25}$$

$$u_3'\left(t^*\right) = -2u_3\left(t^*\right) + u_2\left(t^*\right) + 2u_1\left(t^*\right) + \exp\left(t^* + t_i\right), \tag{2.26}$$

where $t^* = t - t_i$ Taking the differential transformation of Eqs. (2.24)–(2.26), respectively, we get

$$u_{1i}(k+1) = \frac{1}{(k+1)}u_{2i}(k) \tag{2.27}$$

$$u_{3i}(k+1) = \frac{1}{(k+1)}\left\{-\ 2u_{3i}(k) + u_{2i}(k) + 2u_{1i}(k) + \frac{1}{k!}\exp(t_i)\right\}, \tag{2.28}$$

and

$$u_{2i}(k+1) = \frac{1}{(k+1)}u_{3i}(k) \tag{2.29}$$

with

$$\begin{aligned} u_{10}(0) &= 0, \\ u_{20}(0) &= 0, \\ u_{30}(0) &= 0. \end{aligned} \tag{2.30}$$

The approximation of $u_{1t} = y(t)$ on the grid points can be obtained from Eq. (2.27). The actual solution of Eqs. (2.1)–(2.3) is

$$y(t) = u_{1t} = \frac{43}{36}\exp(t) + \frac{1}{4}\exp(-t) - \frac{4}{9}\exp(-2t) + \frac{1}{6}[t\exp(t)]. \tag{2.31}$$

$$y(t) = u_{1t} = \frac{43}{36}\exp(t) + \frac{1}{4}\exp(-t) - \frac{4}{9}\exp(-2t) + \frac{1}{6}[t\exp(t)]. \tag{2.32}$$

Fig. 2.2 shows errors involved with different order of DTM, along with the result obtained by the Runge–Kutta fourth-order method. As indicated in Figs. 2.1 and 2.2, the computational error decreases as the order of

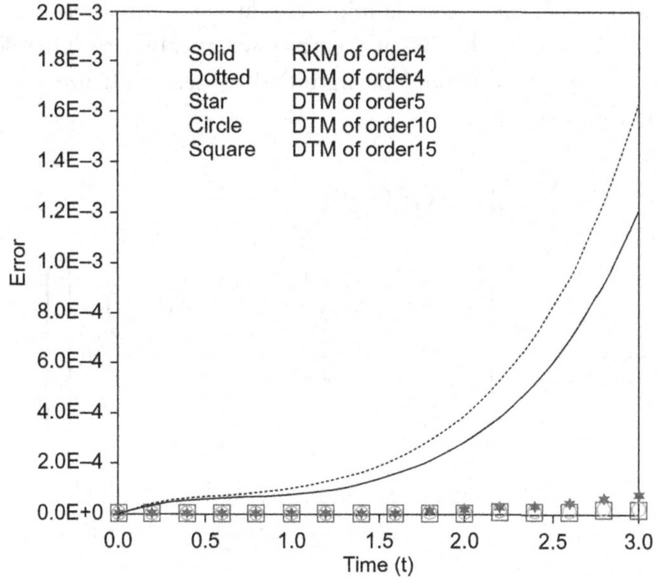

Figure 2.2 Computational errors corresponding to different order of Differential Transformation Method for case B.

Taylor series increases. The order of computational error corresponding to the Runge–Kutta method and the DTM of the same order is the same.

2.3 FRACTIONAL DIFFERENTIAL TRANSFORM METHOD

There are several approaches to the generalization of the notion of differentiation to fractional orders. The fractional differentiation in Riemann–Liouville sense is defined by [3].

$$D_{x_0}^{q} f(x) = \frac{1}{\Gamma(m-q)} \frac{d^m}{dx^m} \left[\int_{x_0}^{x} \frac{f(t)}{(x-t)^{1+q-m}} dt \right] \tag{2.33}$$

for $m - 1 \leq q < m$, $m \in Z^+$, $x > x_0$. Let us expand the analytical and continuous function $f(x)$ in terms of a fractional power series as follows:

$$f(x) = \sum_{k=0}^{\infty} f(k)(x - x_0)^{\frac{k}{\alpha}} \tag{2.34}$$

where a is the order of fraction and $F(k)$ is the fractional differential transform off(x). Concerning the practical applications encountered in various branches of science, the fractional initial conditions are frequently not available, and it may not be clear what their physical meaning is. Therefore, the definition in Eq. (2.33) should be modified to deal with integer ordered initial conditions as follows:

$$D_{x_0}^{q}\left[f(x)-\sum_{k=0}^{m-1}\frac{1}{k!}(x-x_0)^k f^{(k)}(0)\right] = \frac{1}{\Gamma(m-q)}\frac{d^m}{dx^m}\left\{\int_0^x\left[\frac{f(t)-\sum_{k=0}^{m-1}\frac{1}{k!}(t-x_0)^k f^{(k)}(0)}{(x-t)^{1+q-m}}dt\right]\right\} \quad (2.35)$$

Since the initial conditions are implemented to the integer order derivatives, the transformation of the initial conditions is defined as follows:

$$F(k)=\begin{cases}\text{if } \frac{k}{\alpha}\in z^+ \rightarrow \frac{1}{\left(\frac{k}{\alpha}\right)!}\left[\frac{d^{\frac{k}{\alpha}}f(x)}{dx^{\frac{k}{\alpha}}}\right]_{x=x_0} \\ \text{if } \frac{k}{\alpha}\notin z^+ \rightarrow 0\end{cases} \quad \text{for } k=0,1,2,\ldots.(n\alpha-1) \quad (2.36)$$

where, n is the order of fractional differential equation considered. Using Eqs. (2.33) and (2.34), the theorems of fractional differential transform method (FDTM), are introduced below.

Theorem 1. If $f(x) = g(x) \pm h(x)$, then $F(k) = G(k) \pm H(k)$.

Theorem 2. If $f(x) = g(x)h(x)$, then $F(K) = \sum_{t=0}^{K} G(l)H(k-1)$

Theorem 3. If $f(x) = g_1(x)g_2(x),\ldots,g_{n-1}(x)g_n(x)$, then

$$F(k)=\sum_{k_{n-1}}^{k}\sum_{k_{n-2}}^{k_{n-1}}\cdots\sum_{k_2=0}^{k_3}\sum_{k_1=0}^{k_2}G_1(K_1)G_2(K_2-K_1)\ldots G_{n-1}(k_{n-1}-k_{n-2})G_N(K-K_{N-1})$$

Theorem 4. If $f(x) = (x - x_0)^p$, then $F(k) = \delta(k - \alpha p)$, where,
$$\delta(k)=\begin{cases}1\rightarrow \text{if}, k=0\\ 0\rightarrow \text{if}, k\neq 0\end{cases}$$

Theorem 5. If $f(x) = D^q_{x_0}[g(x)]$, then $F(k) = \frac{\Gamma\left(q+1+\frac{k}{\alpha}\right)}{\Gamma\left(1+\frac{k}{\alpha}\right)} G(K+\alpha q)$

Theorem 6. For the production of fractional derivatives in the most general form, if $f(x) = \frac{d^{q_1}}{dx^{q_1}}[g_1(x)]\frac{d^{q2}}{dx^{q2}}[g_2(x)]..\frac{d^{q_{n-1}}}{dx^{q_{n-1}}}[g_{n-1}(x)]$ $\frac{d^{q_n}}{dx^{q_n}}[g_n(x)]$, then:

$$F(K) = \sum_{k_{n-1}=0}^{k}\sum_{k_{n-2}=0}^{k_{n-1}}\cdots\sum_{k_2=0}^{k_3}\sum_{k_1=0}^{k_2}\frac{\Gamma(q_1+1+k_1/\alpha)}{\Gamma(1+k_1/\alpha)}\,\frac{\Gamma(q_2+1+k_2-k_1/\alpha)}{\Gamma(1+k_2-k_1/\alpha)}$$

$$\cdots\frac{\Gamma(q_{n-1}+1+k_{n-1}-k_{n-2}/\alpha)}{\Gamma(1+k_2-k_1/\alpha)}\times\frac{\Gamma(q_n+1+(k-k_{n-1})/\alpha)}{\Gamma(1+(k-k_{n-1})/\alpha)}$$

$$G_1(K_1+\alpha q_1)G_2(K_2-K_1+\alpha q_2)\ldots G_{N-1}(K_{N-1}-K_{N-2}+\alpha q_{n-1})$$

$$\times\, G_{N-1}(K-K_{N-1}+\alpha q_n)$$

where, $\alpha q_i \in z^+$, for, $i = 1, 2, 3,\ldots,n$.

Proofs of all above theorems are presented in Ref. [3]. Following some examples are discussed.

Example A. Now, let us consider the Bagley–Torvik equation that governs the motion of a rigid plate immersed in a Newtonian fluid.

$$A\frac{d^2x}{dt^2} + B\frac{d^{\frac{3}{2}}x}{dx^{\frac{3}{2}}} + cx = f(t) \tag{2.37}$$

It is considered the case $f(t) = C(1 + t)$, $A = 1$; $B = 1$; and $C = 1$, with the following boundary conditions:

$$x(0) = 1,\ \text{and},\ x'(0) = 1 \tag{2.38}$$

Selecting the order of fraction as alpha = 2, the boundary conditions are transformed by using Eq. (2.36) as follows:

$$x(0) = 1,\ x(1) = 0,\ x(2) = 1,\ \text{and}\ x(3) = 0 \tag{2.39}$$

By using theorems 4 and 5, the transform of Eq. (2.37) leads to the following recurrence relation:

$$X(k+4) = -\frac{\Gamma(5/2+k/2)X(k+3) + \Gamma(1+k/2)[X(k)-\delta(k-2)-\delta(k)]}{\Gamma(3+k/2)} \tag{2.40}$$

Using Eqs. (2.39) and (2.40), $X(k)$ is evaluated up to a certain number of terms and then using the inverse transformation rule, $x(t)$ is evaluated as follows:

$$X(t) = 1 + t \tag{2.41}$$

Example B. Consider the composite fractional oscillation equation [3]

$$\frac{d^2u}{dt^2} - \alpha\frac{d^\alpha u}{dt^\alpha} - bu = 8 \rightarrow t > 0, 0 < \alpha \leq 2 \tag{2.42}$$

with the initial conditions

$$U(0) = 0 \text{ and } u'(0) = 0 \tag{2.43}$$

Taking $a = b = -1$ and using theorems 4 and 5, Eq. (2.42) can be transformed as follows:

$$u(k + 2\beta) = -\frac{\Gamma(\alpha + 1 + k/\beta)u(k + \beta\alpha) + \Gamma(1 + k/\beta)[u(k) - 8\delta(k)]}{\Gamma(3 + k/\beta)} \tag{2.44}$$

where b is the unknown value of the fraction. The conditions in Eq. (2.43) can be transformed by using Eq. (2.36) as follows:

$$u(k) = 0, \quad \text{for } k = 0, 1, \ldots, 2\beta - 1 \tag{2.45}$$

Example C. We consider the equation [3].

$$D^{2.2}x(t) + 1.3D^{1.5}x(t) + 2.6x(t) = \sin(2t) \tag{2.46}$$

with the initial conditions

$$x(0) = x'(0) = x''(0) = 0 \tag{2.47}$$

By choosing $a = 10$ and using theorem 4, Eq. (2.46) can be transformed as follows:

$$x(k + 22) = \frac{\Gamma(1 + 0.1k)[s(k) - 2.6x(k)] - 1.3\Gamma(2.5 + 0.1k)x(k + 15)}{\Gamma(3.2 + 0.1k)} \tag{2.48}$$

where $S(k)$ is the fractional differential transform of $\sin(2t)$ that can be evaluated using Eq. (2.34) as

$$s(k) = \sum_{i=0}^{\infty} \frac{(-1)^i 2^{2i+1}}{(2i + 1)!}\delta[k - 10(2i + 1)] \tag{2.49}$$

The conditions in Eq. (2.47) can be transformed by using Eq. (2.36) as

$$X(k) = 0 \quad k = 0, 1, 2, \ldots 21 \tag{2.50}$$

Using the inverse transformation rule in Eq. (2.34) the following series solution is obtained:

$$\begin{aligned} x(t) = &\frac{28561}{3600000}t^6 + \frac{2}{\Gamma(21/6)}t^{16/5} - \frac{13}{5\Gamma(49/10)}t^{39/10} \\ &+ \frac{169}{50\Gamma(28/5)}t^{23/5} - \frac{8}{\Gamma(21/5)}t^{26/5} - \frac{2197}{500\Gamma(63/10)}t^{53/10} \\ &- \frac{26}{5\Gamma(32/5)}t^{27/5} + \frac{52}{5\Gamma(69/10)}t^{59/10} + \ldots \end{aligned} \tag{2.51}$$

By using a mathematical software package, $x(t)$ is evaluated up to $N = 1000$ and plotted in Fig. 2.3.

Example D. Lastly, the fractional Ricatti equation is considered that is frequently encountered in optimal control problems

$$\frac{d^\beta y}{dt^\beta} = 2y - y^2 + 1, 0 < \beta \le 1 \tag{2.52}$$

with the initial condition

$$Y(0) = 0 \tag{2.53}$$

by using theorems 2, 4, and 5 will be as follows:

$$Y(k + \alpha\beta) = \frac{\Gamma(1 + k/\alpha)}{\Gamma(\beta + 1 + k/\alpha)}\left[2Y(k) - \sum_{k_1=0}^{k} Y(k_1)Y(k - k_1) + \delta(k)\right] \tag{2.54}$$

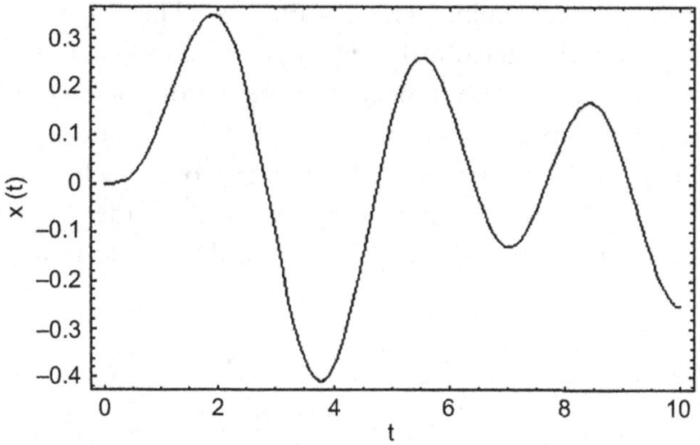

Figure 2.3 Time response of $x(t)$ for $N = 1000$ terms.

The condition in Eq. (2.53) is transformed by using Eq. (2.36) as follows:

$$y(k) = 0, \text{for}, k = 0, 1, \ldots., \alpha\beta - 1 \tag{2.55}$$

For the values of $b = 1/2$ and $a = 2$, $k = 6y(t)$ is obtained as follows:

$$y(t) = \frac{2}{\sqrt{\pi}}t^{1/2} + 2t + \frac{16(\pi - 1)}{3\pi^{3/2}}t^{3/2} + \frac{\pi - 4}{\pi}t^2 - \frac{32(3\pi^2 + 44\pi - 32)}{45\pi^{5/2}}t^{5/2} + \left(\frac{128}{9\pi^2} - \frac{71}{9\pi} - \frac{37}{4}\right)t^3 + \cdots \tag{2.56}$$

2.4 DIFFERENTIAL TRANSFORMATION METHOD FOR INTEGRO-DIFFERENTIAL EQUATION

Consider the fractional-order integro-differential equation of the form [4].

$$D^q y(x) = f(x) + \int_a^x k_1(x, t)y(t)dt + \int_a^b k_2(x, t)y(t)dt \tag{2.57}$$

With the nonlocal boundary conditions,

$$\sum_{j=1}^{m}\left(\gamma_{ij}y^{(j-1)}(a) + \eta_{ij}y^{(j-1)}(b)\right) + \lambda_i \int_a^b H_i(t)y(t)dt = d_i, \quad i = 1, 2, \ldots, m \tag{2.58}$$

where D^q denotes a differential operator with fractional order q, $f(x)$ and $k_i(x, t)$ $(i = 1, 2)$ are holomorphic functions, $H_i(t)$ is a continuous function. γ_{ij}, η_{ij}, λ_i, and $d_i(i = 1, 2,\ldots,m)$ are constants, and $y(x)$ is a function of class C (a class of functions that are piecewise continuous on $J' = (0, \infty)$ and integrable on any finite subinterval $J = [0, \infty)$). There are various types of definition for the fractional derivative of order $q > 0$; the most commonly used definitions among various definitions of fractional derivatives of order $q > 0$ are the Riemann–Liouville and Caputo formulas, ones which use fractional integrations and derivatives of the whole order. The difference between the two definitions is in the order of evaluation. Riemann–Liouville fractional integration of order q is defined as

$$J_{x_0}^q f(x) = \frac{1}{\Gamma(q)}\int_{x_0}^x (x - t)^{q-1}f(t)dt, \quad q > 0, \quad x > 0 \tag{2.59}$$

The following equations define Riemann–Liouville and Caputo fractional derivatives of order q, respectively:

$$D_{x_0}^{q} f(x) = \frac{d^m}{dx^m}\left[J_{x_0}^{m-q} f(x)\right] \tag{2.60}$$

$$D_{*x_0}^{q} f(x) = J_{x_0}^{m-q}\left[\frac{d^m}{dx^m} f(x)\right] \tag{2.61}$$

where $m-1 \leq q < m$ and $m \in N$. From Eqs. (2.59) and (2.60), we have

$$D_{x_0}^{q} f(x) = \frac{1}{\Gamma(m-q)}\frac{d^m}{dx^m}\int_{x_0}^{x}(x-t)^{m-q-1} f(t)\,dt, \quad x > x_0 \tag{2.62}$$

Consider the fractional-order integro-differential Eq. (2.57) with the nonlocal boundary conditions (2.58), where we assumed that the functions $k_i(x, t)$, $(i = 1, 2)$ and $f(x)$ are holomorphic and $y(x)$ is a function of class C; hence $k_i(x,t)$, $(i = 1,2)$ can be approximated by separable functions. Therefore one can write $k_i(x,t) = \sum_{j=0}^{n} u_{ij}(x)v_{ij}(t), \quad (i = 1, 2)$ and then

$$\int_a^b k_i(x,t)y(t)dt = \sum_{j=0}^{n}\int_a^b u_{ij}(x)v(t)y(t)dt = \sum_{j=0}^{n} u_{ij}(x)\int_a^b v_{ij}(t)y(t)dt \tag{2.63}$$

By using FDTM, Eq. (2.57) is transformed to the following recurrence relation:

$$Y(k+\alpha q) = \frac{\Gamma\left(1+\frac{k}{\alpha}\right)}{\Gamma\left(1+q+\frac{k}{\alpha}\right)}\left(F(k) + R(k) + \sum_{j=0}^{n} U_{2j}(k)\int_a^b v_{2j}(t)y(y)dt\right),$$
$$k = 0, 1, \ldots, N \tag{2.64}$$

where N is an arbitrary natural number, $R(k)$ is the transformation of $\int_\alpha^x k_1(x,t)y(t)dt$, and $Y(0)=y(\alpha)$, $Y(\alpha)=y'(\alpha)$, $\dots Y\left((m-1)\alpha = \frac{y^{(m-1)}(\alpha)}{(m-1)!}\right)$ are the unknowns to be determined. If we set,

$$\beta_j = \int_a^b v_{2j}(t)y(t)dt \tag{2.65}$$

then Eq. (2.64) implies that

$$Y(k+\alpha q) = L(k)\left(A(k) + \sum_{j=0}^{n} U_{2j}(k)\beta_j\right), \quad k = 0, 1, \dots, N \tag{2.66}$$

where

$$L(k) = \frac{\Gamma\left(1+\frac{k}{\alpha}\right)}{\Gamma\left(1+q+\frac{k}{\alpha}\right)}, \quad A(k) = F(k) + R(k) \tag{2.67}$$

By using inverse transformation on both sides of Eq. (2.66), $y(x)$ is obtained as

$$\begin{aligned} y(x) &= \sum_{k=0}^{\alpha q+N} Y(k)(x-a)^{\frac{k}{\alpha}} \\ &= \sum_{k=0}^{m-1} \frac{Y^{(k)}}{k!}(x-a)^k + \sum_{k=0}^{N}\left(A(k) + \sum_{j=0}^{n} U_{2j}(k)\beta_j\right)L(k)(x-a)^{\frac{\alpha q+k}{\alpha}} \end{aligned} \tag{2.68}$$

Substituting $y(x)$ in the boundary conditions, we obtain, for $i = 1, 2,\dots,m$

$$y(a)\left[\gamma_{i1}+\eta_{i1}+\lambda_i\int_a^b H_i(t)dt\right]+y'(a)\left[\gamma_{i2}+(b-a)\eta_{i1}+\eta_{i2}+\lambda_i\int_a^b H_i(t)(t-a)dt\right]+\ldots$$

$$+y^{(m-1)}(a)\left[\gamma_{im}+\eta_{i1}\frac{(b-a)^{(m-1)}}{(m-1)!}+\eta_{i2}\frac{(b-a)^{(m-2)}}{(m-2)!}+\eta_{i3}\frac{(b-a)^{(m-3)}}{(m-3)!}+\ldots\right.$$

$$\left.+\eta_{im-1}(b-a)+\eta_{im}+\lambda_i\int_a^b H_i(t)\frac{(t-a)^{(m-1)}}{(m-1)!}dt\right]$$

$$+\beta_0\left[\eta_{i1}\sum_{k=0}^{N}U_{20}(k)L(k)(b-a)^{\frac{\alpha q+k}{\alpha}}+\eta_{i2}\sum_{k=0}^{N}\left(\frac{\alpha q+k}{\alpha}\right)U_{20}(k)L(k)(b-a)^{\frac{\alpha q+k}{\alpha}-1}+\ldots\right.$$

$$+\eta_{im}\sum_{k=0}^{N}\left(\frac{\alpha q+k}{\alpha}\right)\left(\frac{\alpha q+k}{\alpha}-1\right)\left(\frac{\alpha q+k}{\alpha}-m+2\right)U_{20}(k)L(k)(b-a)^{\frac{\alpha q+k}{\alpha}-m+1}$$

$$\left.+\lambda_i\int_a^b H_i(t)\sum_{k=0}^{N}U_{20}(k)L(k)(t-a)^{\frac{\alpha q+k}{\alpha}}dt\right]$$

$$+\beta_1\left[\eta_{i1}\sum_{k=0}^{N}U_{21}(k)L(k)(b-a)^{\frac{\alpha q+k}{\alpha}}+\eta_{i2}\sum_{k=0}^{N}\left(\frac{\alpha q+k}{\alpha}\right)U_{21}(k)L(k)(b-a)^{\frac{\alpha q+k}{\alpha}-1}+\ldots\right.$$

$$+\eta_{im}\sum_{k=0}^{N}\left(\frac{\alpha q+k}{\alpha}\right)\left(\frac{\alpha q+k}{\alpha}-1\right)\ldots\left(\frac{\alpha q+k}{\alpha}-m+2\right)U_{21}(k)L(k)(b-a)^{\frac{\alpha q+k}{\alpha}-m+1} \tag{2.69}$$

$$\left.+\lambda_i\int_a^b H_i(t)\sum_{k=0}^{N}U_{21}(k)L(k)(t-a)^{\frac{\alpha q+k}{\alpha}}dt\right]+\ldots$$

$$+\beta_n\left[\eta_{i1}\sum_{k=0}^{N}U_{2n}(k)L(k)(b-a)^{\frac{\alpha q+k}{\alpha}}+\eta_{i2}\sum_{k=0}^{N}\left(\frac{\alpha q+k}{\alpha}\right)U_{2n}(k)L(k)(b-a)^{\frac{\alpha q+k}{\alpha}-1}+\ldots\right.$$

$$+\eta_{im}\sum_{k=0}^{N}\left(\frac{\alpha q+k}{\alpha}\right)\left(\frac{\alpha q+k}{\alpha}-1\right)\ldots\left(\frac{\alpha q+k}{\alpha}-m-2\right)U_{2n}(k)L(k)(b-a)^{\frac{\alpha q+k}{\alpha}-m+1}$$

$$\left.+\lambda_i\int_a^b H_i(t)\sum_{k=0}^{N}U_{2n}(k)L(k)(t-a)^{\frac{\alpha q+k}{\alpha}}dt\right]$$

$$=-\left(\eta_{i1}\sum_{k=0}^{N}L(k)A(k)(b-a)^{\frac{\alpha q+k}{\alpha}}+\eta_{i2}\sum_{k=0}^{N}\left(\frac{\alpha q+k}{\alpha}\right)L(k)A(k)(b-a)^{\frac{\alpha q+k}{\alpha}-1}+\ldots\right.$$

$$+\eta_{im}\sum_{k=0}^{N}\left(\frac{\alpha q+k}{\alpha}\right)\left(\frac{\alpha q+k}{\alpha}-1\right)..\left(\frac{\alpha q+k}{\alpha}-m+2\right)L(k)A(k)(b-a)^{\frac{\alpha q+k}{\alpha}-m+1}$$

$$\left.+\lambda_i\int_a^b H_i(t)\sum_{k=0}^{N}A(k)L(k)(t-a)^{\frac{\alpha q+k}{\alpha}}\right]+d_i$$

On the other hand; $\beta_j=\int_a^b v_{2i}(t)y(t)dt$. Substituting $y(x)$ from Eq. (2.68) and rearranging the terms, for $j=0,1,2,\ldots,n$ yields

$$\beta_j\left(\int_a^b v_{2j}(t)\sum_{k=0}^{N}U_{2j}(k)L(k)(t-a)^{\frac{\alpha q+k}{\alpha}}dt-1\right)$$

$$+\sum_{\substack{l=0\\ l\neq j}}^{N}\left(\beta_l\int_a^b v_{2j}(t)\sum_{k=0}^{N}L(k)U_{21}(k)(t-a)^{\frac{\alpha q+k}{\alpha}}dt\right) \tag{2.70}$$

$$=-\int_a^b v_{2j}(t)\left(\sum_{k=0}^{m-1}\frac{y^{(k)}(a)}{k!}(t-a)^k+\sum_{k=0}^{N}A(k)L(k)(t-a)^{\frac{\alpha q+k}{\alpha}}\right)dt$$

Example A. Consider the following linear fractional integro-differential equation with the given nonlocal condition

$$D^{\frac{1}{2}}y(x)=-x^2\frac{e^x}{3}y(x)-\frac{1}{2}x^2+\frac{1}{\Gamma\left(\frac{3}{2}\right)}x^{\frac{1}{2}}+e^x\int_0^x ty(t)dt+\int_0^1 x^2y(t)dt \tag{2.71}$$

$$y(0)+y(1)-3\int_0^1 ty(t)dt=0 \tag{2.72}$$

where the order of fraction is $\alpha=2$. If we set $\beta_0=\int_0^1 y(t)dt$, then Eq. (2.71) is transformed to

$$Y(k+1)=\frac{\Gamma\left(1+\frac{k}{2}\right)}{\Gamma\left(\frac{3}{2}+\frac{k}{2}\right)}\left(-\frac{1}{3}\sum_{n=0}^{k}\sum_{j=0}^{n}\delta(j-4)E(n-j)Y(j-n)\right.$$

$$-\frac{1}{2}\delta(k-4)+\frac{1}{\Gamma\left(\frac{3}{2}\right)}\delta(k-1)$$

$$\left.+2\sum_{i=2}^{k}\frac{1}{j}\sum_{m=0}^{i-2}\delta(m-2)Y(j-m-2)E(k-j)+\delta(k-4)\beta_0\right).$$

where $E(k)$ denotes the transformation of $y(x)$ that can be expressed by using Section 2.3 as follows:

$$E(k)=\begin{cases}0, & \frac{k}{\alpha}\notin Z^+\\ \frac{1}{\left(\frac{k}{\alpha}\right)!}, & \frac{k}{\alpha}\in Z^+\end{cases} \tag{2.73}$$

$$\text{Thus } L(k) = \frac{\Gamma\left(1+\frac{k}{2}\right)}{\Gamma\left(\frac{3}{2}+\frac{k}{2}\right)}, \quad U_{20} = (k) = \delta(k-4),$$

and

$$A(k) = -\frac{1}{3}\sum_{n=0}^{k}\sum_{j=0}^{n}\delta(j-4)E(n-j)Y(k-n) - \frac{1}{2}\delta(k-4)$$

$$+\frac{1}{\Gamma\left(\frac{3}{2}\right)}\delta(k-1)$$

$$+2\sum_{i=2}^{k}\frac{1}{j}\sum_{m=0}^{j-2}\delta(m-2)Y(j-m-2) - E(k-j).$$

Now, we solve the system

$$\begin{aligned} &\frac{1}{2}y(0) + \frac{2}{3\Gamma\left(\frac{7}{2}\right)}\beta_0 + \sum_{k=0}^{N}\frac{k-1}{5+k}L(k)A(k) = 0 \\ &y(0) + \left(\frac{4}{7\Gamma\left(\frac{7}{2}\right)} - 1\right)\beta_0 + \sum_{k=0}^{N}\frac{2}{3+k}L(k)A(k) = 0 \end{aligned} \tag{2.74}$$

and obtain the following approximation for $Y(0) = y(0)$ and β_0. For $N = 20$: $Y(0) = y(0) = -0.96658 \times 10^{-6}$ and $\beta_0 = 0.5 + 4.2 \times 10^{-6}$. The exact values of $y(0)$ and β_0 are 0 and 0.5, respectively, and the exact solution is $y(x) = x$. For $N = 20$, the approximate solution is

$$\begin{aligned} y(x) =& -0.96658 \times 10^{-6} + x + 0.24306 \times 10^{-5}x^{\frac{5}{2}} - 0.831 \times 10^{-7}x^{\frac{7}{2}} - 0.36933 \times 10^{-7}x^{\frac{9}{2}} \\ &- 0.16178 \times 10^{-6}x^{5} - 0.11192 \times 10^{-7}x^{\frac{11}{2}} - 0.10295 \times 10^{-6}x^{6} - 0.25827 \times 10^{-8}x^{\frac{13}{2}} \\ &+ 0.157 \times 10^{-6}x^{7} + 0.37576 \times 10^{-7}x^{\frac{15}{2}} - 0.10389 \times 10^{-7}x^{8} + 0.64758 \times 10^{-7}x^{\frac{17}{2}} \\ &+ 0.89992 \times 10^{-8}x^{9} + 0.13509 \times 10^{-7}x^{\frac{19}{2}} + 0.73501 \times 10^{-7}x^{10} + \ldots. \end{aligned} \tag{2.75}$$

Example B. Consider the following linear fractional integro-differential equation with the given nonlocal condition

$$D^{\frac{1}{3}}y(x) = \frac{3}{2}\frac{x^{\frac{3}{2}}}{\Gamma\left(\frac{3}{2}\right)} - 1 + e^{x^2} - x^2e^{x^2} + \int_0^x x^2e^{xt}y(t)dt \tag{2.76}$$

$$y(0) + 2y(1) + 3\int_0^1 ty(t)dt = 3 \tag{2.77}$$

with exact solution $y(x) = x$. We have

$$k(x,t) \approx \sum_{i=0}^{M}\frac{x^{2+i}t^i}{i!} \tag{2.78}$$

By using FDTM, Eq. (2.76) is transformed to the following recurrence relation

$$\begin{aligned} Y(k+1) = \frac{\Gamma\left(1+\frac{k}{3}\right)}{\Gamma\left(\frac{4}{3}+\frac{k}{3}\right)} &\left(\frac{3}{2\Gamma\left(\frac{2}{3}\right)}\delta(k-2) - \delta(k) - E_1(k) - \sum_{k_1=0}^{k}\delta(k_1-6)\right. \\ &\times E_1(k-k_1) + 3\sum_{k_1=3}^{k}\frac{1}{k_1}Y(k_1-3)\delta(k-k_1-6) + 3\sum_{k_1=3}^{k}\sum_{k_2=0}^{k_1-3} \\ &\left.\times\sum_{k_3=1}^{M}\frac{1}{k_1}Y(k_1-k_2-3)\left(\frac{1}{k_3!}\delta(k_2-3k_3)\delta(k-k_1-3(k_3-2))\right)\right) \end{aligned} \tag{2.79}$$

where $E_1(k)$ denotes to the transformation of e^{x^2} that can be expressed by using Section 2.3 as follows:

$$E_1(k) = \begin{cases} 0, & \frac{k}{2\alpha} \notin Z^+ \\ \frac{1}{\left(\frac{k}{2\alpha}\right)!}, & \frac{k}{2\alpha} \in Z^+ \end{cases} \tag{2.80}$$

Thus we have $L(k) = \dfrac{\Gamma\left(1+\frac{k}{3}\right)}{\Gamma\left(\frac{4}{3}+\frac{k}{3}\right)}$ and

$$A(k) = \left(\frac{3}{2\Gamma\left(\frac{3}{2}\right)} \delta(k-2) - \delta(k) + E_1(k) - \sum_{k_1=0}^{k} \delta(k_1 - 6) E_1(k - k_1) \right.$$
$$+3 \sum_{k_1=3}^{k} \frac{1}{k_1} Y(k_1 - 3)\delta(k - k_1 - 6) + 3 \sum_{k_1=3}^{k} \sum_{k_2=0}^{k_1-3} \sum_{k_3=1}^{M}$$
$$\left. \times \frac{1}{k_1} Y(k_1 - k_2 - 3) \left(\frac{1}{k_3!} \delta(k_2 - 3k_3)\delta(k - k_1 - 3(k_3 + 2)) \right) \right) \quad (2.81)$$

By solving the following equation we get the value of $Y(0) = y(0)$:

$$\frac{9}{2} y(0) + \sum_{k=0}^{N} \left(\frac{2k + 23}{7 + k} \right) L(k) A(k) = 3 \quad (2.82)$$

For $N = 30$ and $M = 3$, we have $y(0) = 0.14247 \times 10^{-4}$. For $N = 20$, the approximate solution is

$$y(x) = 1.4247 \times 10^{-5} + 9.9997 \times 10^{-1} x + 0.92308 \times 10^{-5} x^{\frac{10}{3}} - 0.89711 \times 10^{-5} x^{\frac{13}{3}} \quad (2.83)$$

2.5 DIFFERENTIAL TRANSFORMATION METHOD FOR EIGENVALUE PROBLEMS

Following Strum–Liouville problem is used to illustrate how to solve eigenvalues and eigenfunctions by DTM [5]

$$\frac{d}{dx}\left[p(x) \frac{dy(x)}{dx} \right] + [q(x) + \lambda w(x)] y(x) = 0 \quad (2.84)$$

Boundary conditions:

$$y(0) + \alpha y'(0) = 0 \quad (2.85)$$

$$y(1) + \beta y'(1) = 0 \quad (2.86)$$

where α and β are constants.

Taking differential transformation of Eq. (2.84) and using DTM principle, we have

$$\sum_{l=0}^{k}(l+1)P(l+1)(k-l+1)Y(k-l+1) + \sum_{l=0}^{k}P(l)(k-l+1)(k-l+2)Y(k-l+1) + \sum_{l=0}^{k}[\lambda W(l)-Q(l)]Y(k-1)=0 \tag{2.87}$$

where $P(k)$, $Q(k)$, $W(k)$, and $Y(k)$ are T-function of $p(x)$, $q(x)$, $w(x)$, and $y(x)$, respectively.

DTM transformation of boundary condition (2.85) becomes

$$Y(0)+\alpha Y(1)=0 \tag{2.88}$$

and boundary condition (2.89) becomes

$$\sum_{k=0}^{n}(1+\beta k)Y(k)=0 \tag{2.89}$$

Let

$$Y(0)=c. \tag{2.90}$$

From Eq. (2.88) we have

$$Y(1)=-\frac{c}{\alpha}. \tag{2.91}$$

Substituting Eqs. (2.90) and (2.91), and $k=0$ into Eq. (2.87), we have

$$Y(2)=\frac{c}{2P(0)}\left[\frac{p(1)}{\alpha}+Q(0)-\lambda W(0)\right] \tag{2.92}$$

Substituting Eqs. (2.90)–(2.92), and $k=1$ into Eq. (2.87), we have

$$Y(3)=\frac{c}{6P(0)}\left\{\frac{2P(2)}{\alpha}-\frac{2(P(1))^2}{\alpha P(0)}\right\}+ \lambda\left[\frac{W(0)}{\alpha}-W(1)+\frac{2P(1)W(0)}{P(0)}\right]-\frac{2P(1)Q(0)}{P(0)}-\frac{Q(0)}{\alpha}+Q(1) \tag{2.93}$$

Following the same recursive procedure, we calculate up to the nth term $Y(n)$, and n is decided by the convergence of the eigenvalue, as described later. Substituting $Y(1) - Y(n)$ into Eq. (2.89), we have

$$c\left[f^{(n)}(\lambda)\right] = 0 \tag{2.94}$$

where $f^{(n)}(\lambda)$ is a polynomial of a corresponding to n. For nontrivial solutions of eigenfunctions, we have $c \neq 0$, and

$$f^{(n)}(\lambda) = 0 \tag{2.95}$$

Solving Eq. (2.95), we get

$$\lambda = \lambda_i^{(n)}, \text{where } i = 1, 2, \ldots \tag{2.96}$$

$\lambda_i^{(n)}$ is the ith estimated eigenvalue corresponding to n, and n is decided by the following equation

$$\left|\lambda_i^{(n)} + \lambda_i^{(n-1)}\right| \leq \varepsilon \tag{2.97}$$

where $\lambda_i^{(n-1)}$ is the ith estimated eigenvalue corresponding to $n - 1$ and ε is a small value, we set. From Eq. (2.24) we have two cases to discuss.

Case 1:

If Eq. (2.97) is satisfied, then $\lambda_i^{(n)}$ is the ith eigenvalue λ_i. Substituting λ_i into $Y(0) - Y(n)$ and using:

$$y_i(x) = \sum_{k=0}^{n} x^k Y_{\lambda_i}(k) \tag{2.98}$$

where $Y_{\lambda_i}(k)$ is $Y(k)$, whose λ is substituted by λ_i, and $y_i(x)$ is the eigen function corresponding to eigenvalue λ_i. For comparison with the analytic solution as shown later, the ith normalized eigenfunction is defined as follows:

$$\widehat{y}_i(x) = \frac{y_i(x)}{\int_0^1 |y_i(x)|dx} \tag{2.99}$$

Case 2:

If Eq. (2.97) is not satisfied, then repeat the following steps until the ith eigenvalue and the ith normalized eigenfunction are found.

Step 1: substituting $n + 1$ for n.

Step 2: following the same procedure as shown in Eqs. (2.94)–(2.99).

At first glance, the method introduced in this section looks very involved in computation; in fact, however, these algebraic computations can be finished very fast by symbolic computational software.

Example A. Consider the following equation,

$$y'' + \lambda y = 0 \tag{2.100}$$

$$y(0) - y'(0) = 0 \tag{2.101}$$

$$y(1) + y'(1) = 0 \tag{2.102}$$

Taking differential transform of Eq. (2.100), we obtain

$$Y(k+2) = -\frac{\lambda Y(k)}{(k+1)(k+2)} \tag{2.103}$$

Boundary condition (2.101) becomes

$$Y(0) - Y(1) = 0 \tag{2.104}$$

Boundary condition (2.102) becomes

$$\sum_{k=0}^{n}(1+k)Y(k) = 0 \tag{2.105}$$

1. Solving the first eigenvalue and eigenfunction. For ease of demonstration, list the computation and result corresponding to $n = 6$.

   Let

$$Y(0) = c \tag{2.106}$$

and from Eq. (2.104), we have

$$Y(1) = c \tag{2.107}$$

Substituting Eqs. (2.106) and (2.107), and $k = 0$ into Eq. (2.103), we have

$$Y(2) = -\frac{c}{2}\lambda \tag{2.108}$$

Following the same recursive procedure, we have

$$Y(3) = -\frac{c}{6}\lambda \tag{2.109}$$

$$Y(4) = -\frac{c}{24}\lambda^2 \tag{2.110}$$

$$Y(5) = -\frac{c}{120}\lambda^2 \tag{2.111}$$

$$Y(6) = -\frac{c}{720}\lambda^3 \tag{2.112}$$

Substituting Eqs. (2.106)–(2.112) into Eq. (2.105), we have

$$f^{(6)}(\lambda) = 3 - \frac{13}{6}\lambda + \frac{31}{120}\lambda^2 - \frac{7}{720}\lambda^3 = 0 \tag{2.113}$$

Solving Eq. (2.113), we have

$$\lambda = 1.71, 12.43 \pm 5.08i. \tag{2.114}$$

Take real root

$$\lambda_1^{(6)} = 1.71 \tag{2.115}$$

When $n = 5$, by the same way, we have

$$\lambda_1^{(5)} = 1.75 \tag{2.116}$$

From Eqs. (2.115) and (2.116), we have

$$\left|\lambda_1^{(6)} - \lambda_1^{(5)}\right| = 0.04 \leq \boldsymbol{\varepsilon} \tag{2.117}$$

where $\boldsymbol{\varepsilon}$ is a small value we set. From Eq. (2.117) we have $\lambda_1 = 1.71$ and λ_1 is the first eigenvalue. Substituting λ_1 into $Y(0) - Y(6)$ and DTM principle, we obtain the first eigenfunction.

$$\begin{aligned} y_1(x) = (&1 + x - 0.855x^2 - 0.285x^3 \\ &+ 0.121838x^4 + 0.0243675x^5 - 0.00694474x^6)c \end{aligned} \tag{2.118}$$

By Eq. (2.99), the first normalized eigenfunction is shown as follows

$$\begin{aligned} \widehat{y}_1(x) = 0.853835(&1 + x - 0.855x^2 - 0.285x^3 + 0.121838x^4 \\ &+ 0.0243675x^5 - 0.00694474x^6) \end{aligned} \tag{2.119}$$

we have the first eigenvalue and eigenfunction as follows

$$\lambda_1^{(a)} = 1.71 \tag{2.120}$$

$$y_1^{(a)}(x) = 1.30767\cos(1.30767x) + \sin(1.30767x) \tag{2.121}$$

After normalization, Eq. (2.121) becomes

$$\widehat{y}_1^{(a)}(x) = 0.652999[1.30767\cos(1.30767x) + \sin(1.30767x)] \quad (2.122)$$

Besides $\lambda_1 = \lambda_1^{(a)}$, the calculated results from Eq. (2.119) are compared closely with the analytic results from Eq. (2.122) as shown in Fig. 2.4.

2. Solving the second eigenvalue and eigenfunction: List the result corresponding to $n = 12$.

Following the same procedure as shown earlier, we solve the equation $f^{(12)}(\lambda) = 0$, take the real root and get

$$\lambda_1^{(12)} = 1.71 \quad (2.123)$$

$$\lambda_2^{(12)} = 13.49 \quad (2.124)$$

Note that $\lambda_1^{(12)} = \lambda_1^{(6)}$. Due to $\left|\lambda_2^{(12)} - \lambda_2^{(11)}\right| \leq \varepsilon$, we have the second eigenvalue $\lambda_2 = 13.49$. Following the same procedure shown above, we get the second normalized eigenfunction as follows,

$$\begin{aligned}\widehat{y}_2(x) = 1.40347(&1 + x - 6.745x^2 - 2.24833x^3 + 7.5825x^4 \\ &+ 1.5165x^5 - 3.4096x^6 - 0.487086x^7 + 0.821348x^8 \\ &+ 0.0912609x^9 - 0.123111x^{10} - 0.0111919x^{11} + 0.0125816x^{12})\end{aligned} \quad (2.125)$$

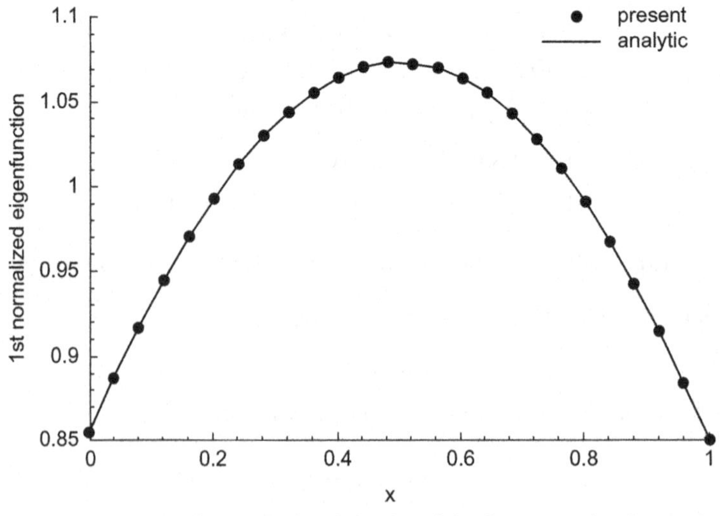

Figure 2.4 Comparison of the calculated results of the first normalized eigenfunction with the analytic results after normalization.

By the analytic method, we have the second eigenvalue and normalized eigenfunction as follows,

$$\lambda_2^{(a)} = 13.49 \tag{2.126}$$

$$\widehat{y}_2^{(a)}(x) = 0.382121[3.67287\cos(3.67287x) + \sin(3.67287x)] \tag{2.127}$$

Besides $\lambda_2 = \lambda_2^{(a)}$, the calculated results from Eq. (2.125) are compared closely with the analytic results from Eq. (2.127) as shown in Fig. 2.5.

3. Solving the third eigenvalue and eigenfunction: List the result corresponding to $n = 22$. Following the same procedure as shown earlier, we solve the equation $f^{(22)}(\lambda) = 0$, take the real root and get

$$\lambda_1^{(22)} = 1.71 \tag{2.128}$$

$$\lambda_2^{(22)} = 13.49 \tag{2.129}$$

$$\lambda_3^{(22)} = 43.36 \tag{2.130}$$

Note that $\lambda_1^{(22)} = \lambda_1^{(12)} = \lambda_1^{(6)}$ and $\lambda_2^{(22)} = \lambda_2^{(12)}$. Due to $\left|\lambda_3^{(22)} - \lambda_3^{(21)}\right| \le \varepsilon$, we have the third eigenvalue $\lambda_3 = 43.36$. Following the same procedure shown earlier, we have the third normalized eigenfunction as follows,

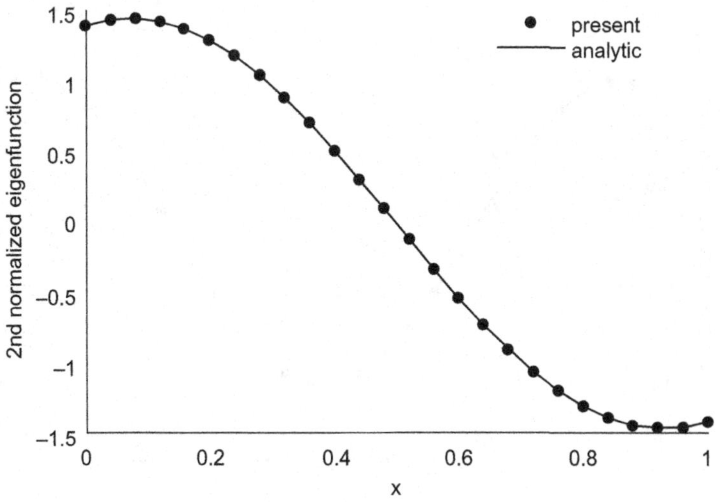

Figure 2.5 Comparison of the calculated results of the second normalized eigenfunction with the analytic results after normalization.

$$\begin{aligned}\widehat{y}_3(x) = 1.51383(1 &+ x - 21.68x^2 - 7.22667x^3 + 78.3371x^4 \\ &+ 15.6674x^5 - 113.223x^6 - 16.1747x^7 + 87.6671x^8 \\ &+ 9.74073x^9 - 42.2361x^{10} - 3.83964x^{11} + 13.8739x^{12} \\ &+ 1.06722x^{13} - 3.30534x^{14} - 0.220356x^{15} + 0.597165x^{16} \\ &+ 0.351274x^{17} - 0.0846179x^{18} - 0.00445358x^{19} + 0.00965535x^{20} \\ &+ 0.000459779x^{21} - 0.000906182x^{22})\end{aligned} \tag{2.131}$$

By the analytic method, we have the third eigenvalue and normalized eigenfunction as follows,

$$\lambda_3^{(a)} = 43.36 \tag{2.132}$$

$$\widehat{y}_3^{(a)}(x) = 0.229896[6.58483\cos(6.58483x) + \sin(6.58483x)] \tag{2.133}$$

Besides $\lambda_3 = \lambda_3^{(a)}$, the calculated results from Eq. (2.131) are compared closely with the analytic results from Eq. (2.133) as shown in Fig. 2.6.

The convergence of eigenvalues $\lambda_1 - \lambda_3$ is shown in Fig. 2.7, where λ_1, λ_2, and λ_3 converge to 1.71, 13.49, and 43.36, respectively.

Example B. IH Abdel–Halim [6] applied this method to solve the eigenvalue problem in the shape of

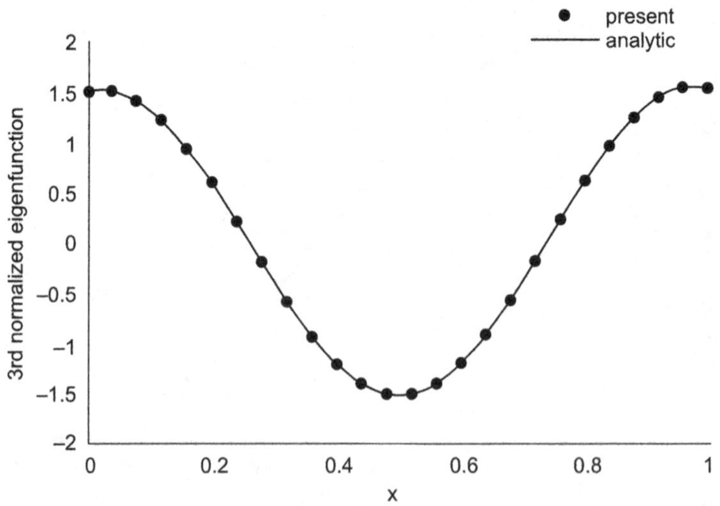

Figure 2.6 Comparison of the calculated results of the third normalized eigenfunction with the analytic results after normalization.

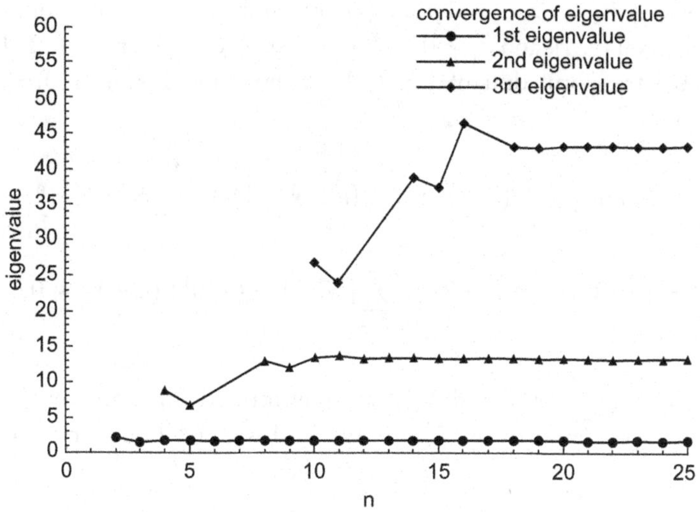

Figure 2.7 The convergence of eigenvalues $\lambda_1 - \lambda_3$.

$$\frac{d}{dy}\left[p(x)\frac{dy}{dx}\right] + [\lambda w(x) - q(x)]y(x) = 0 \tag{2.134}$$

with boundary conditions

$$y(0) + \alpha y'(0) = 0, \tag{2.135}$$

$$y(0) + \beta y'(1) = 0, \tag{2.136}$$

where a and b are constants, and illustrated the results by solving this problem

$$y'' + \lambda x^2 y = 0 \tag{2.137}$$

with boundary conditions

$$y(0) = 0, \quad y(1) = 0 \tag{2.138}$$

Consider the regular Sturm–Liouville eigenvalue problem

$$\frac{d}{dy}\left[p(x)\frac{dy(x)}{dx}\right] + [\lambda r(x) - q(x)]y(x) = 0 \tag{2.139}$$

with boundary conditions

$$\alpha_1 y(0) + \beta_1 y'(0) = 0, \tag{2.140}$$

$$\alpha_2 y(1) + \beta_2 y'(1) = 0, \tag{2.141}$$

where $p(x) > 0$, $r(x) > 0$, and $p(x)$, $r(x)$, $q(x)$ and $p'(x)$ are continuous on the closed interval [0, 1], and $\alpha_i \geq 0$, $b_i \geq 0$, and $\alpha_i + \beta_i \geq 0$ for $i = 1, 2$. Taking differential transformation of Eq. (2.139) and using Eqs. (2.133)–(2.137) we obtain

$$\sum_{\ell=0}^{k}(\ell+1)p(\ell+1)(k-\ell+1)Y(k-\ell+1)+\sum_{l=0}^{k}p(\ell)(k-\ell+1)$$
$$\times(k-\ell+2)Y(k-\ell+2)+\sum_{\ell=0}^{k}[\lambda R(\ell)-Q(\ell)]Y(k-\ell)=0, \tag{2.142}$$

where $P(k)$, $Q(k)$, $R(k)$, and $Y(k)$, are transformed functions of $p(x)$, $r(x)$, $q(x)$, and $y(x)$, respectively. Using Eq. (2.134), the boundary condition (2.140) becomes

$$\alpha_1 Y(0) - \beta_1 Y(1) = 0, \tag{2.143}$$

the boundary condition (2.141) becomes

$$\sum_{k=0}^{n}(\alpha_2+\beta_2 k)Y(k)=0 \tag{2.144}$$

Put

$$y(0) = c \tag{2.145}$$

from Eq. (2.143) we find that

$$Y(1) = \frac{\alpha_1 c}{\beta_1} \tag{2.146}$$

At $k = 0$, and substituting from Eqs. (2.145) and (2.146) into Eq. (2.142), we have

$$Y(2) = -\frac{c}{2p(0)}\left[\frac{\alpha_1 p(1)}{\beta_1}+\lambda R(0)-Q(0)\right]. \tag{2.147}$$

At $k = 1$, and substituting from Eqs. (2.145)–(2.147) into Eq. (2.142), we have

$$Y(3)=\frac{c}{6p(0)}\left\{-\frac{2\alpha_1 p(2)}{\beta_1}+\frac{2\alpha_1 p^2(1)}{P(0)\beta_1}+\lambda\left[\frac{2p(1)R(0)}{\beta_1}-R(1)\right]\right.$$
$$\left.-\frac{2p(1)Q(1)}{p(0)}+\frac{\alpha_1 Q(1)}{\beta_1}+Q(1)\right\} \tag{2.148}$$

Following the same procedure, we calculate up to the nth term $Y(n)$ and substituting from $Y(1)$ to $Y(n)$ into Eq. (2.144), we obtain

$$c\left[f^{(n)}(\lambda)\right] = 0 \tag{2.149}$$

where $f^{(n)}(\lambda)$ is a polynomial of k corresponding to n, and for $c \neq 0$, we have

$$f^{(n)}(\lambda) = 0 \tag{2.150}$$

Solving Eq. (2.150), we $\lambda_i = \lambda_i^n, \quad i = 1, 2, 3, \ldots,$ where $\lambda_i^{(n)}$ is the nth estimated eigenvalue corresponding to n, and n is indicated by

$$\left|\lambda_i^{(n)} - \lambda_i^{(n-1)}\right| \leq \xi, \tag{2.151}$$

where in $\lambda_i^{(n-1)}$ i is the ith estimated eigenvalue corresponding to $n-1$ and ξ is a small value we set and then we have two cases the same as Section 2.5 described.

Example C: Consider the equation of

$$xy''(x) + \lambda^2 y(x) = 0, \tag{2.152}$$

Boundary conditions:

$$y(0) = 0, \tag{2.153}$$

$$y(1) = 0. \tag{2.154}$$

Taking DTM, we obtain

$$\sum_{\ell=0}^{k} \delta(\ell - 1)(k - \ell + 1)(k - \ell + 2)Y(k - \ell + 2) = -\lambda^2 Y(k). \tag{2.155}$$

Using DTM, the boundary condition (2.153) becomes

$$Y(0) = 0. \tag{2.156}$$

and the boundary condition (2.154) becomes

$$\sum_{k=0}^{n} Y(k) = 0. \tag{2.157}$$

(I): Solving the first eigenvalue and eigenfunction: putting

$$Y(1) = c. \tag{2.158}$$

Substituting Eqs. (2.156) and (2.158) at $k = 1$ into Eq. (2.155), we have

$$Y(2) = -\frac{c}{2}\lambda^2. \tag{2.159}$$

Following the same procedure, $Y(3) = Y(7)$ can be solved as follows

$$Y(3) = \frac{c}{12}\lambda^4, \tag{2.160}$$

$$Y(4) = -\frac{c}{144}\lambda^6, \tag{2.161}$$

$$Y(5) = \frac{c}{2880}\lambda^8, \tag{2.162}$$

$$Y(6) = -\frac{c}{86400}\lambda^{10}, \tag{2.163}$$

$$Y(7) = -\frac{c}{3628800}\lambda^{12}. \tag{2.164}$$

Substituting Eqs. (2.156), (2.158)–(2.164) into Eq. (2.157), we have

$$f^{(7)}(\lambda) = 1 - \frac{\lambda^2}{2} + \frac{\lambda^4}{12} - \frac{\lambda^6}{144} + \frac{\lambda^8}{28800} - \frac{\lambda^{10}}{86400} + \frac{\lambda^6}{3628800} = 0. \tag{2.165}$$

Solving Eq. (2.165), and taking the real roots, we have

$$\lambda_1{}^{(7)} = \pm 1.9159. \tag{2.166}$$

When $n = 6$, by the same way, we have

$$\lambda_1^{(6)} = \pm 1.9143. \tag{2.167}$$

From Eqs. (2.166) and (2.167), we have

$$\left|\lambda_1{}^{(7)} - \lambda_1{}^{(6)}\right| = 0.01 \leq \xi. \tag{2.168}$$

From Eq. (2.166) we take $\lambda_1 = 1.92$ as the first eigenvalue. Substituting λ_1 into Eqs. (2.158)–(2.164) and using DTM principle, we obtain the first eigenfunction,

$$\begin{aligned} y_1(x) = {} & (x - 1.8432x^2 + 1.1324x^3 - 0.347892x^4 + 0.0641235x^5 4 \times 10^{-4}x^7 \\ & -0.00787949x^6 + 6.91594 \times 10^{-4}x^7)c. \end{aligned} \tag{2.169}$$

the first normalized eigenfunction is

$$\begin{aligned} \hat{y}_1(x) = {} & 9.19243(x - 1.8432x^2 + 1.1324x^3 - 0.347892x^4 \\ & + 0.0641235x^5 4 \times 10^{-4}x^7). \\ & - 0.00787949x^6 + 6.91594 \times 10^{-4}x^7). \end{aligned} \tag{2.170}$$

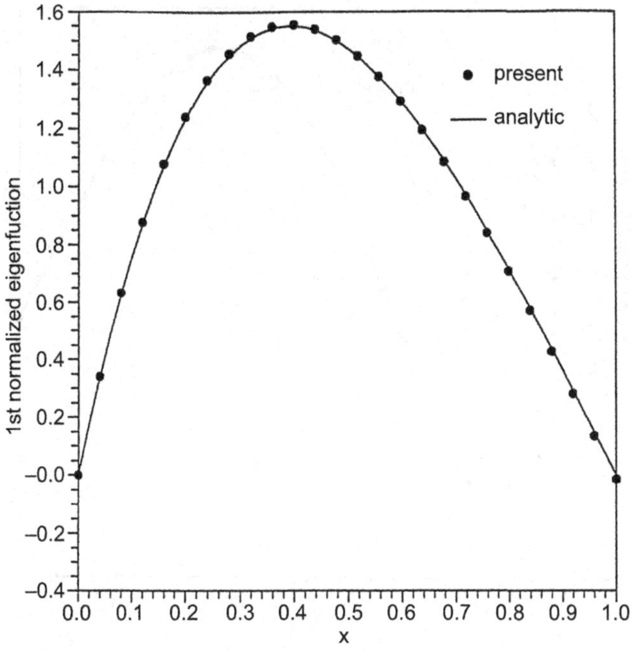

Figure 2.8 Comparison of the calculated results of the first normalized eigenfunction with the analytic results after normalization.

By the analytical method, the first eigenvalue and eigenfunction are

$$\lambda_1{}^{(a)} = 1.916, \tag{2.171}$$

$$y_1{}^{(a)} = \sqrt{x}J_1\left(3.832\sqrt{x}\right). \tag{2.172}$$

After normalization, Eq. (2.172) becomes

$$\hat{y}_1^{(a)}(x) = 4.7589\sqrt{x}J_1\left(3.832\sqrt{x}\right). \tag{2.173}$$

From Eqs. (2.166) and (2.171) we deduce that $\lambda_1 = \lambda_1{}^{(a)}$, and the calculated results from Eq. (2.166) are compared closely with the analytic results from Eq. (2.173) as shown in Fig. 2.8.

(II): Solving the second eigenvalue and eigenfunction: Listing the result corresponding to $n = 11$, and following the same procedure as above, we get $\lambda_2 = 3.507502$ and the second normalized eigenfunction as

$$\begin{aligned}\hat{y}_2(x) = &-41.0133(x - 6.15303x^2 + 12.6199x^3 - 12.9418x^4 + 6.96316x^5 \\ &- 3.26649x^6 + 0.95088x^7 - 0.210321x^8 + 0.0359475x^9 \\ &- 0.00491526x^{10} + 5.49886 \times 10^{-4}x^{11}).\end{aligned} \tag{2.174}$$

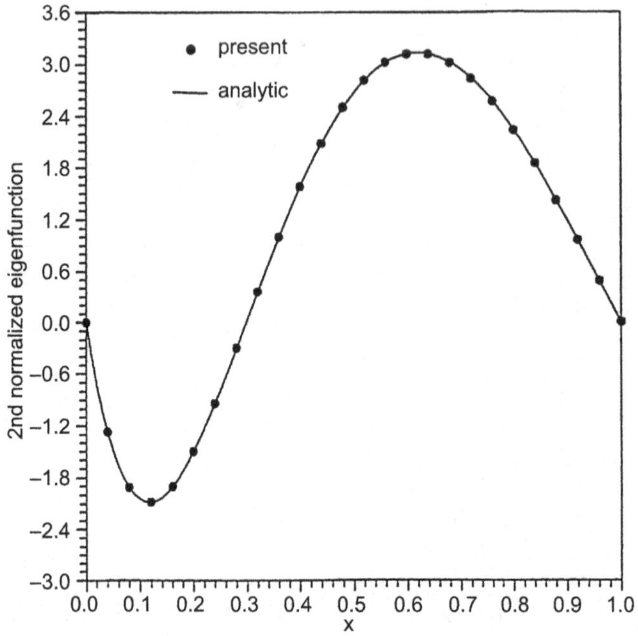

Figure 2.9 Comparison of the calculated results of the second normalized eigenfunction with the analytic results after normalization.

By the analytical method, we have the second eigenvalue and normalized eigenfunction as follows:

$$\lambda_2^{(a)} = 3.508 \tag{2.175}$$

$$\hat{y}^{(a)}(x) = -11.69\sqrt{x}J_1\left(7.016\sqrt{x}\right). \tag{2.176}$$

The calculated results from Eq. (2.174) are compared closely with the analytic results from Eq. (2.176) as shown in Fig. 2.9. Regarding the third normalized eigenfunction, we find that the comparison results are the same from Fig. 2.10.

The convergence of eigenvalues of the problem confirms that λ_1, λ_2, and λ_3 converge to 1.92, 3.51, and 5.09, respectively, as shown in Fig. 2.11.

2.6 TWO-DIMENSIONAL DIFFERENTIAL TRANSFORMATION METHOD FOR FRACTIONAL ORDER PARTIAL DIFFERENTIAL EQUATIONS

There are several definitions of a fractional derivative of order $\alpha > 0$. The Caputo fractional derivative is defined as [7].

$$D_a^{\alpha} f(x) = J_a^{m-\alpha} D^m f(x), \tag{2.177}$$

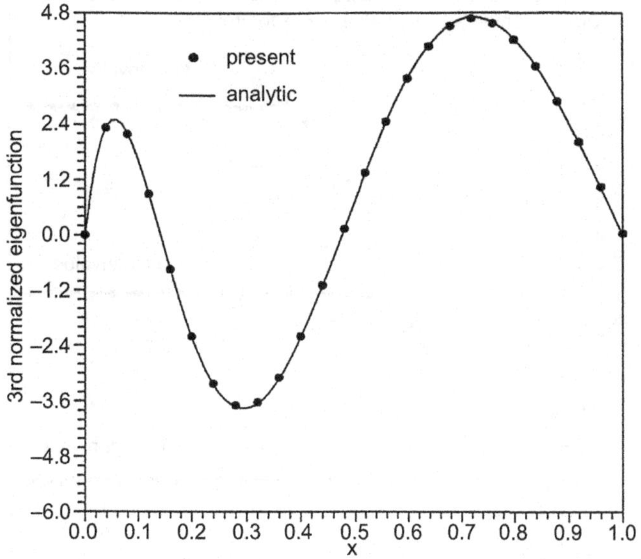

Figure 2.10 Comparison of the calculated results of the third normalized eigenfunction with the analytic results after normalization.

where $m-1<\alpha\leq m$. Here D^m is the usual integer differential operator of order m and J_a^μ is the Riemann–Liouville integral operator of order $\mu>0$, defined by

$$J_a^\mu f(x)=\frac{1}{\Gamma(\mu)}\int_a^x (x-t)^{\mu-1}f(t)dt, \quad x>0. \tag{2.178}$$

The Caputo fractional derivative is considered here because it allows traditional initial and boundary conditions to be included in the formulation of the problem [4].

Consider a function of two variables $u(x, y)$, and suppose that it can be represented as a product of two single-variable functions, i.e., $u(x, y)=f(x)g(y)$. On the basis of the properties of generalized two-dimensional differential transform, the function $u(x, y)$ can be represented as

$$\begin{aligned} u(x,y) &= \sum_{k=0}^{\infty} F_\alpha(k)(x-x_0)^{k\alpha}\sum_{h=0}^{\infty} G_\beta(h)(y-y_0)^{h\beta} \\ &= \sum_{k=0}^{\infty}\sum_{h=0}^{\infty} U_{\alpha,\beta}(k,h)(x-x_0)^{k\alpha}(y-y_0)^{h\beta}, \end{aligned} \tag{2.179}$$

where $0<\alpha,\beta\leq 1$, $U_{\alpha,\beta}(k, h)=F_\alpha(k)G_\beta(h)$ is called the spectrum of $u(x, y)$. The generalized two-dimensional differential transform of the function $u(x, y)$ is as follows:

$$U_{\alpha,\beta}(k,h)=\frac{1}{\Gamma(\alpha k+1)\Gamma(\beta h+1)}\left[\left(D_{x0}^{\alpha}\right)^k\left(D_{y0}^{\beta}\right)^h u(x,y)\right]_{(x0,y0)}, \tag{2.180}$$

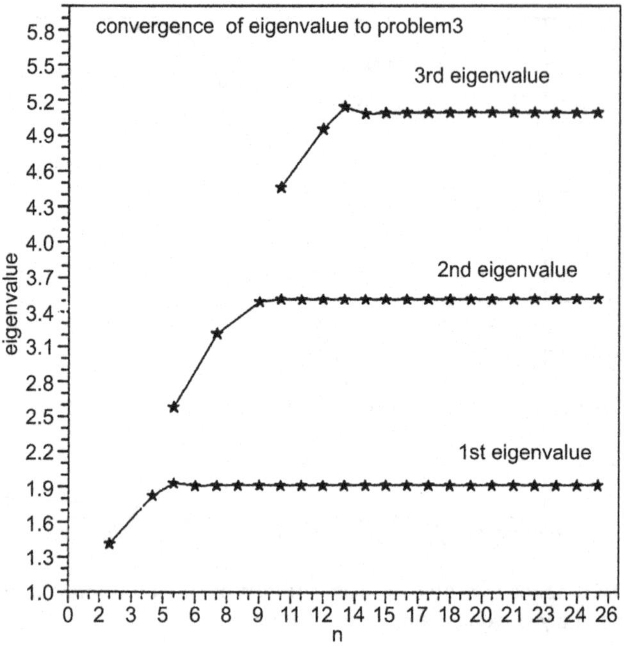

Figure 2.11 Convergence of the eigenvalues $\lambda_1 - \lambda_3$, where λ_1, λ_2 and λ_3 converge to 1.92, 3.51, and 5.09, respectively.

where $\left(D_{x0}^{\alpha}\right)^k = D_{x0}^{\alpha} D_{x0}^{\alpha} \ldots D_{x0}^{\alpha}$, k times. In this work, the lowercase $u(x, y)$ represents the original function while the uppercase $U_{\alpha,\beta}(k, h)$ stands for the transformed function. On the basis of the definitions (I) and (II), we have the following results:

Theorem I. Suppose that $U_{\alpha,\beta}(k, h)$, $V_{\alpha,\beta}(k, h)$, and $W_{\alpha,\beta}(k, h)$ are the differential transformations of the functions $u(x,y)$, $v(x,y)$, and $w(x,y)$, respectively;

$$(a) \text{ if } \; u(x,y) = v(x,y) \pm \omega(x,y), \; \text{ then } \; U_{\alpha,\beta}(k,h) = V_{\alpha,\beta}(k,h) \pm W_{\alpha,\beta}(k,h),$$

$$(b) \text{ if } \; u(x,y) = av(x,y), a \in R, \; \text{ then } \; U_{\alpha,\beta}(k,h) = aV_{\alpha,\beta}(k,h),$$

$$(c) \text{ if } \; u(x,y) = v(x,y)\omega(x,y), \; \text{ then } \; U_{\alpha,\beta}(k,h) = \sum_{r=0}^{k} \sum_{s=0}^{h} V_{\alpha,\beta}(r,h-s) W_{\alpha,\beta}(k-r,s),$$

$$(d) \text{ if } \; u(x,y) = (x-x_0)^{n\alpha}(y-y_0)^{m\beta}, \; \text{ then } \; U_{\alpha,\beta}(k,h) = \delta(k-n)\delta(h-m),$$

$$(e) \text{ if } \; u(x,y) = D_{x0}^{\alpha} v(x,y), 0 < \alpha \le 1, \; \text{ then } \; U_{\alpha,\beta}(k,h) = \frac{\Gamma(\alpha(k+1)+1)}{\Gamma(\alpha k+1)} U_{\alpha,\beta}(k+1,h).$$

Theorem II. If $u(x, y) = f(x)g(y)$ and the function $f(x) = x\_h(x)$, where $\lambda > -1$, $h(x)$ has the generalized Taylor series expansion $h(x) = \sum_{n=0}^{\infty} a_n (x - x0)^{\alpha k}$, and (*a*) $\beta < \lambda + 1$ and α is arbitrary, or

(*b*) $\beta < \lambda + 1$, α is arbitrary and an $= 0$ for $n = 0, 1, \ldots m - 1$, where $-1 < \beta \leq m$.

Then the generalized differential transform becomes

$$U_{\alpha,\beta}(k,h) = \frac{1}{\Gamma(\alpha k + 1)\Gamma(\beta h + 1)}\left[\left(D_{x0}^{\alpha}\right)^k\left(D_{y0}^{\beta}\right)^h u(x,y)\right]_{(x0,y0)}, \tag{2.181}$$

Theorem III. If $v(x, y) = f(x)g(y)$, the function $f(x)$ satisfies the conditions given in Theorem 2.2, and $D_{x0}^{\gamma}v(x,y)$ then

$$U_{\alpha,\beta}(k,h) = \frac{\Gamma(\alpha(k+1)+\gamma)}{\Gamma(\alpha k + 1)} V_{\alpha,\beta}(k + \gamma/\alpha, h) \tag{2.182}$$

Following two examples are presented that demonstrate the performance and efficiency of the generalized DTM for solving linear partial differential equations with time- or space-fractional derivatives.

Example A. Consider the following linear inhomogeneous time-fractional equation:

$$\frac{\partial^\alpha u}{\partial t^\alpha} + x\frac{\partial u}{\partial x} + \frac{\partial^2 x}{\partial x^2} = 2t^\alpha + 2x + 2, U_{\alpha,1}(k,0) = \delta(k-2). \tag{2.183}$$

where $0 < \alpha \leq 1$, subject to the initial condition

$$u(x,0) = x^2. \tag{2.184}$$

Suppose that the solution $u(x,t)$ can be represented as a product of single-valued functions. Selecting $\beta = 1$ and applying the generalized two-dimensional differential transform to both sides of Eq. (2.183), the linear inhomogeneous time-fractional Eq. (2.183) transforms to

$$U_{\alpha,1}(k,h+1) = \frac{\Gamma(\alpha k + 1)}{\Gamma(\alpha(h+1)+1)}\left[\sum_{r=0}^{k}\sum_{s=0}^{h}\delta(r-1)\delta(h-s)(k-r+1)U_{\alpha,1}(k-r+1,s) - (k+1)(k+2)U_{\alpha,1}(k+2,h) + 2\delta(k-2)\delta(h) + 2\delta(k)\delta(h)\right] \tag{2.185}$$

The generalized two-dimensional differential transform of the initial condition Eq. (2.184) is

$$U_{\alpha,1}(k,0) = \delta(k-2). \tag{2.186}$$

Utilizing the recurrence relation Eq. (2.185) and the transformed initial condition Eq. (2.186), we get $U_{\alpha,1}(2,\ 1) = 1$, $U_{\alpha,1}(0,2) = 2\frac{\Gamma(\alpha+1)}{\Gamma(2\alpha+1)}$, and $U_{\alpha,1}(k,h) = 0$ for $k \neq 2$, $h \neq 2$. Therefore, according to above sections, the solution of Eq. (2.183) is given by

$$u(x,t) = x^2 + 2\frac{\Gamma(\alpha+1)}{\Gamma(2\alpha+1)}t^{2\alpha}, \tag{2.187}$$

which is the exact solution of the linear inhomogeneous time-fractional Eq. (2.183).

Example B. Consider the following linear space-fractional telegraph equation

$$\frac{\partial^{1.5}u}{\partial x^{1.5}} = \frac{\partial^2 u}{\partial t^2} + \frac{\partial u}{\partial t} + u, \quad x>0, \tag{2.188}$$

subject to the initial conditions

$$u(0,t) = \exp(-t), \quad u_x(0,t) = \exp(-t)\,. \tag{2.189}$$

Suppose that the solution $u(x,\ t)$ can be represented as a product of single-valued functions, $u(x,\ t) = v(x)w(t)$ where the function $v(x)$ satisfies the conditions given in theorem II. Selecting $\alpha = 1$, $\beta = 0.5$ and applying the generalized two-dimensional differential transform to both sides of Eq. (2.188), the linear space-fractional telegraph Eq. (2.188) transforms to

$$\times\left[(h+1)(h+2)U_{1,1/2}(k,h+2)U_{1,1/2}(k,h+1) + U_{1,1/2}(k,h)\right].$$

$$U_{1,1/2}(k+3,h) = \frac{\Gamma(k/2+1)}{\Gamma(k/2+5/2)} \tag{2.190}$$

The generalized two-dimensional differential transforms of the initial conditions Eq. (2.189) are given by

$$\begin{aligned} U_{1,1/2}(0,h) &= (-1)^h/h!, \\ U_{1,1/2}(1,h) &= 0, \\ U_{1,1/2}(2,h) &= (-1)^h/h!, \end{aligned}$$

Utilizing the recurrence relation Eq. (2.190) and the transformed initial conditions, the first few components of $U_{1,1/2}(k,\ h)$ are calculated and given in Table 2.1 in Ref. [7].

Table 2.1 The Fundamental Operations of Reduced Differential Transform Method

Functional Form	**Transformed Form**
$u(x, t)$	$U_k(x, t) = \frac{1}{k!}\left[\frac{\partial^k}{\partial x^k} u(x, t)\right]_{t=0}$
$w(x,t) = u(x, t) \pm v(x,t)$	$W_k(x) = U_k(x) \pm V_k(x)$
$w(x,t) = \alpha u(x,t)$	$W_k(x) = \alpha U_k(x)$ (α is constant)
$w(x,t) = x^m t^n$	$W_k(x) = x^m \delta(k - n)$
$w(x,t) = x^m t^n u(x,t)$	$W_k(x) = x^m U_{k-n}$
$w(x,t) = u(x,t)v(x,t)$	$W_k(x) = \sum_{r=0}^{k} V_r(x) U_{k-r}(x) = \sum_{r=0}^{k} U_r(x) V_{k-r}(x)$
$w(x, t) = \frac{\partial^k}{\partial t^k} u(x, t)$	$W_k(x) = (k+1)\ldots(k+r)U_{k+r}(x) = \frac{(k+r)!}{k!} U_{k+r}(x)$
$w(x, t) = \frac{\partial}{\partial x} u(x, t)$	$W_k(x) = \frac{\partial}{\partial x} U_k(x)$

Therefore, the approximate solution of the linear space-fractional telegraph Eq. (2.188) can be derived as

$$
\begin{aligned}
u(x, t) = & \left(1 - t + \frac{1}{2!}t - \frac{1}{3!}t^3 + \frac{1}{4!}t^4 - \frac{1}{5!}t^5\right) \\
& + \left(1 - t + \frac{1}{2!}t^2 - \frac{1}{3!}t^3 + \frac{1}{4!}t^4 - \frac{1}{5!}t^5\right) x \\
& + \left(1 - t + \frac{1}{2!}t^2 - \frac{1}{3!}t^3 + \frac{1}{4!}t^4 - \frac{1}{5!}t^5\right) \frac{x^{1.5}}{\Gamma(5/2)} \\
& + \left(1 - t + \frac{1}{2!}t^2 - \frac{1}{3!}t^3 + \frac{1}{4!}t^4 - \frac{1}{5!}t^5\right) \frac{x^{2.5}}{\Gamma(5/2)} + \ldots
\end{aligned}
\tag{2.191}
$$

that is

$$
\begin{aligned}
u(x, t) = \exp(-t) & \left(1 + x + \frac{x^{1.5}}{\Gamma(5/2)} + \frac{x^{2.5}}{\Gamma(7/2)} + \frac{x^3}{\Gamma(4)} \right. \\
& \left. + \frac{x^4}{\Gamma(5)} + \frac{x^{4.5}}{\Gamma(11/2)} + \ldots\right),
\end{aligned}
\tag{2.192}
$$

2.7 REDUCED DIFFERENTIAL TRANSFORM METHOD

The reduced differential transform method (RDTM) was first proposed by the Turkish mathematician Keskin [8] in 2009. It has received much attention since it has applied to solve a wide variety of problems by many authors. In this section after introducing this method, it is applied to solve two examples of NLPDEs [9]. Consider a function of two variables $u(x,t)$ and suppose that it can be represented as a product of two single-variable functions, i.e., $u(x,t) = f(x)g(t)$. Based on the properties of one-dimensional differential transform, the function $u(x,t)$ can be represented as follows [9]:

$$u(x,t) = \left(\sum_{i=0}^{\infty} F(i)x^i\right)\left(\sum_{j=0}^{\infty} G(j)t^j\right) = \sum_{k=0}^{\infty} U_k(x)t^k \tag{2.193}$$

where $U_k(x)$ is called t-dimensional spectrum function of $u(x, t)$. The basic definitions of RDTM are introduced as follows:

Definition I. If function $u(x,t)$ is analytic and differentiated continuously with respect to time t and space x in the domain of interest, then let

$$U_k(x,t) = \frac{1}{k!}\left[\frac{\partial^k}{\partial x^k} u(x,t)\right]_{t=0} \tag{2.194}$$

where the t-dimensional spectrum function $U_k(x)$ is the transformed function. In this paper, the lowercase $u(x, t)$ represents the original function, while the uppercase $U_k(x)$ stands for the transformed function.

Definition II. The differential inverse transform of $U_k(x)$ is defined as follows:

$$u(x,t) = \sum_{k=0}^{\infty} U_k(x)t^k \tag{2.195}$$

Then, combining Eqs. (2.194) and (2.195) we write

$$u(x,t) = \sum_{k=0}^{\infty} \frac{1}{k!}\left[\frac{\partial^k}{\partial t^k} u(x,t)\right]_{t=0} t^k \tag{2.196}$$

From the above definitions, it can be found that the concept of the RDTM is derived from the power series expansion. To illustrate the basic concepts of the RDTM, consider the following nonlinear partial differential equation written in an operator form

$$Lu(x,t) + Ru(x,t) + Nu(x,t) = g(x,t) \tag{2.197}$$

with initial condition

$$u(x,0) = f(x) \tag{2.198}$$

where $L = \frac{\partial}{\partial t}$, R is a linear operator which has partial derivatives, $Nu(x, t)$ is a nonlinear operator and $g(x,t)$ is an inhomogeneous term.

According to the RDTM, we can construct the following iteration formula:

$$(k+1)U_{k+1}(x,t) = G_k(x) - RU_k(x) - NU_k(x) \tag{2.199}$$

where $U_k(x)$, $RU_k(x)$, $NU_k(x)$, and $G_k(x)$ are the transformations of the functions $Lu(x,t)$, $Ru(x,t)$, $Nu(x,t)$, and $g(x,t)$ respectively. From initial condition Eq. (2.198), we write

$$U_0(x) = f(x) \tag{2.200}$$

Substituting Eq. (2.200) into Eq. (2.199) and by straightforward iterative calculation, we get the following $U_k(x)$ values. Then, the inverse transformation of the set of values $\{U_k(x)\}_{k=0}^{n}$ gives the n-terms approximation solution as follows

$$\tilde{u}_n(x,t) = \sum_{k=0}^{n} U_k(x)t^k \tag{2.201}$$

Therefore, the exact solution of the problem is given by

$$u(x,t) = \lim_{n\to\infty} \tilde{u}_n(x,t) \tag{2.202}$$

The fundamental mathematical operations performed by RDTM can be readily obtained and are listed in Table 2.1.

Following, the RDTM is applied to solve two nonlinear partial differential equations (PDEs), namely, generalized Drinfeld–Sokolov (gDS) equations and Kaup–Kupershmidt (KK) equation.

Example A. We consider the generalized Drinfeld–Sokolov (gDS) equations:

$$\begin{aligned} &u_t + u_{xxx} - 6uu_x - 6(v^\alpha)_x = 0 \\ &v_t - 2v_{xxx} + 6uv_x = 0 \end{aligned} \tag{2.203}$$

with initial conditions

$$\begin{aligned} &u(x,0) = \frac{-b^2 - 4k^4}{4k^2} + 2k^2\tanh^2(kx) \\ &v(x,0) = b\tanh(kx) \end{aligned} \tag{2.204}$$

where α is a constant. According to the RDTM and Table 2.1, the differential transform of Eq. (2.203) reads

$$\begin{aligned}(k+1)U_{k+1}(x) &= -\frac{\partial^3}{\partial x^3}U_k(x) + 6A_k(x) + 6B_k(x)\\(k+1)V_{k+1}(x) &= 2\frac{\partial^3}{\partial x^3}V_k(x) - 6C_k(x)\end{aligned} \tag{2.205}$$

where the t-dimensional spectrum functions $U_k(x)$, $V_k(x)$ are the transformed functions. $A_k(x)$, $B_k(x)$, and $C_k(x)$ are transformed form of the nonlinear terms. For the convenience of the reader, the first few nonlinear terms are as follows

$$\begin{aligned}&A_0 = U_0\frac{\partial}{\partial x}U_0, A_1 = U_1\frac{\partial}{\partial x}U_0 + U_0\frac{\partial}{\partial x}U_1,\\&A_2 = U_2\frac{\partial}{\partial x}U_0 + U_1\frac{\partial}{\partial x}U_1 + U_0\frac{\partial}{\partial x}U_2,\\&A_3 = U_3\frac{\partial}{\partial x}U_0 + U_2\frac{\partial}{\partial x}U_1 + U_1\frac{\partial}{\partial x}U_2 + U_0\frac{\partial}{\partial x}U_3\end{aligned} \tag{2.206}$$

$$\begin{aligned}&B_0 = \frac{\partial}{\partial x}\left(V_0^{\alpha}\right), B_1 = \frac{\partial}{\partial x}\left(\alpha V_0^{\alpha-1}V_1\right),\\&B_2 = \frac{\partial}{\partial x}\left(\frac{\alpha(\alpha-1)}{2}V_0^{\alpha-2}V_1^2 + \alpha V_0^{\alpha-1}V_2\right),\\&B_3 = \frac{\partial}{\partial x}\left(\alpha V_0^{\alpha-1}V_3 + \alpha(\alpha-1)V_0^{\alpha-2}V_1V_2\right.\\&\qquad\left.+\frac{1}{6}\alpha(\alpha-1)(\alpha-2)V_0^{\alpha-3}V_1^3\right)\end{aligned} \tag{2.207}$$

$$\begin{aligned}&C_0 = U_0\frac{\partial}{\partial x}V_0, C_1 = U_1\frac{\partial}{\partial x}V_0 + U_0\frac{\partial}{\partial x}V_1,\\&C_2 = U_2\frac{\partial}{\partial x}V_0 + U_1\frac{\partial}{\partial x}V_1 + U_0\frac{\partial}{\partial x}V_2,\\&C_3 = U_3\frac{\partial}{\partial x}V_0 + U_2\frac{\partial}{\partial x}V_1 + U_1\frac{\partial}{\partial x}V_2 + U_0\frac{\partial}{\partial x}V_3\end{aligned} \tag{2.208}$$

from initial conditions Eq. (2.204), we write

$$\begin{aligned}U_0(x) &= \frac{-b^2-4k^4}{4k^2} + 2k^2\tanh^2(kx)\\V_0(x) &= b\tanh(kx)\end{aligned} \tag{2.209}$$

Substituting Eq. (2.209) into Eq. (2.205) (when $\alpha = 2$) and by straightforward iterative steps, we can obtain

$$
\begin{aligned}
U_1(x) &= \frac{2k(4k^4 + 3b^2)\sinh(kx)}{\cosh(kx)^3}, \\
U_2(x) &= -\frac{1}{2}\frac{\left(2\cosh(kx)^2 - 3\right)(4k^4 + 3b^2)^2}{\cosh(kx)^4}, \\
U_3(x) &= \frac{1}{3}\frac{\sinh(kx)\left(\cosh(kx)^2 - 3\right)(4k^4 + 3b^2)^3}{\cosh(kx)^5}, \\
U_4(x) &= -\frac{1}{24}\frac{(4k^4 + 3b^2)^4\left(15 + 2\cosh(kx)^4 - 15\cosh(kx)^2\right)}{\cosh(kx)^6} \\
&\vdots
\end{aligned}
\tag{2.210}
$$

and

$$
\begin{aligned}
V_1(x) &= \frac{1}{2}\frac{b(4k^4 + 3b^2)}{\cosh(kx)^2 k}, \\
V_2(x) &= -\frac{1}{4}\frac{b(4k^4 + 3b^2)^2\sinh(kx)}{\cosh(kx)^3 k^2}, \\
V_3(x) &= \frac{1}{24}\frac{\left(2\cosh(kx)^2 - 3\right)b(4k^4 + 3b^2)^3}{\cosh(kx)^4 k^3}, \\
V_4(x) &= -\frac{1}{48}\frac{\sinh(kx)(4k^4 + 3b^2)^4\left(\cosh(kx)^2 - 3\right)b}{\cosh(kx)^5 k^4}. \\
&\vdots
\end{aligned}
\tag{2.211}
$$

and so on, in the same manner, the rest of components can be obtained by using MAPLE software.

Taking the inverse transformation of the set of values $\{U_k(x)\}_{k=0}^{n}$ and $\{V_k(x)\}_{k=0}^{n}$ gives n-terms approximation solutions as follows

$$
\begin{aligned}
\tilde{u}_n(x,t) &= \sum_{k=0}^{n} U_k(x)t^k = \frac{-b^2 - 4k^4}{4k^2} + 2k^2\tanh^2(kx) + \frac{2k(4k^4 + 3b^2)\sinh(kx)}{\cosh(kx)^3}t + \ldots \\
&+ \frac{1}{n!}\left[\frac{\partial^n}{\partial t^n}\left(\frac{-b^2 - 4k^4}{4k^2} + 2k^2\tanh^2\left(kx + \frac{(4k^4 + 3b^2)}{2k}t\right)\right)\right]_{t=0} t^n
\end{aligned}
\tag{2.212}
$$

$$\tilde{v}_n(x,t) = \sum_{k=0}^{n} V_k(x)t^k = b\tanh(kx) + \frac{1}{2}\frac{b(4k^4+3b^2)}{\cosh(kx)^2 k}t + \ldots$$
$$+\frac{1}{n!}\left[\frac{\partial^n}{\partial t^n}\left(b\tanh\left(kx+\frac{(4k^4+3b^2)}{2k}t\right)\right)\right]_{t=0} t^n \tag{2.213}$$

Therefore, the exact solution of the problem is readily obtained as follows

$$u(x,t) = \lim_{n\to\infty}\tilde{u}_n(x,t) = \frac{-b^2-4k^4}{4k^2} + 2k^2\tanh^2\left(kx+\frac{(4k^4+3b^2)}{2k}t\right)$$
$$v(x,t) = \lim_{n\to\infty}\tilde{v}_n(x,t) = b\tanh\left(kx+\frac{(4k^4+3b^2)}{2k}t\right) \tag{2.214}$$

To examine the accuracy of the RDTM solution, the absolute errors of the 4-terms approximate solutions are plotted in Fig. 2.12.

Example B: Consider the nonlinear Kaup–Kupershmidt (KK) equation:

$$u_t = u_{xxxxx} + 5uu_{xxx} + \frac{25}{2}u_x u_{xx} + 5u^2 u_x \tag{2.215}$$

subject to the initial condition

$$u(x,0) = -2k^2 + \frac{24k^2}{1+e^{kx}} - \frac{24k^2}{(1+e^{kx})^2} \tag{2.216}$$

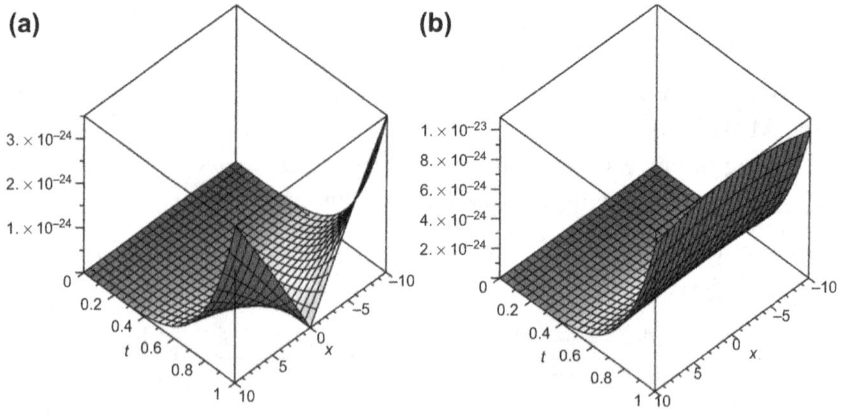

Figure 2.12 The absolute error of (a) $\tilde{u}_4(x,t)$ (b) $\tilde{v}_4(x,t)$ when $k=0.01$ and $b=0.001$.

where k is an arbitrary constant. By taking the differential transform of Eq. (2.215), we obtain

$$(k+1)U_{k+1}(x) = \frac{\partial^5}{\partial x^5}U_k(x) + 5A_k(x) + \frac{25}{2}B_k(x) + 5C_k(x) \tag{2.217}$$

where the t-dimensional spectrum function and $U_k(x)$ is the transformed function. $A_k(x)$, $B_k(x)$, and $C_k(x)$ are the transformed form of the nonlinear terms. For the convenience of the reader, the first few nonlinear terms are as follows

$$\begin{aligned}
A_0 &= U_0\frac{\partial^3}{\partial x^3}U_0,\ A_1 = U_1\frac{\partial^3}{\partial x^3}U_0 + U_0\frac{\partial^3}{\partial x^3}U_1,\\
A_2 &= U_2\frac{\partial^3}{\partial x^3}U_0 + U_1\frac{\partial^3}{\partial x^3}U_1 + U_0\frac{\partial^3}{\partial x^3}U_2,\\
A_3 &= U_3\frac{\partial^3}{\partial x^3}U_0 + U_2\frac{\partial^3}{\partial x^3}U_1 + U_1\frac{\partial^3}{\partial x^3}U_2 + U_0\frac{\partial^3}{\partial x^3}U_3
\end{aligned} \tag{2.218}$$

$$\begin{aligned}
B_0 &= \frac{\partial}{\partial x}U_0\frac{\partial^2}{\partial x^2}U_0,\ B_1 = \frac{\partial}{\partial x}U_0\frac{\partial^2}{\partial x^2}U_1 + \frac{\partial}{\partial x}U_1\frac{\partial^2}{\partial x^2}U_0+,\\
B_2 &= \frac{\partial}{\partial x}U_0\frac{\partial^2}{\partial x^2}U_2 + \frac{\partial}{\partial x}U_1\frac{\partial^2}{\partial x^2}U_1 + \frac{\partial}{\partial x}U_2\frac{\partial^2}{\partial x^2}U_0,\\
B_3 &= \frac{\partial}{\partial x}U_0\frac{\partial^2}{\partial x^2}U_3 + \frac{\partial}{\partial x}U_1\frac{\partial^2}{\partial x^2}U_2 + \frac{\partial}{\partial x}U_2\frac{\partial^2}{\partial x^2}U_1 + \frac{\partial}{\partial x}U_3\frac{\partial^2}{\partial x^2}U_0
\end{aligned} \tag{2.219}$$

$$\begin{aligned}
C_0 &= U_0^2\frac{\partial}{\partial x}U_0,\ C_1 = 2U_0U_1\frac{\partial}{\partial x}U_0 + U_0^2\frac{\partial}{\partial x}U_1,\\
C_2 &= 2U_0U_2\frac{\partial}{\partial x}U_0 + U_1^2\frac{\partial}{\partial x}U_0 + 2U_0U_1\frac{\partial}{\partial x}U_1 + U_0^2\frac{\partial}{\partial x}U_2,\\
C_3 &= 2U_0U_3\frac{\partial}{\partial x}U_0 + 2U_1U_2\frac{\partial}{\partial x}U_0 + U_1^2\frac{\partial}{\partial x}U_1 + 2U_0U_2\frac{\partial}{\partial x}U_1\\
&\quad + 2U_0U_1\frac{\partial}{\partial x}U_2 + U_0^2\frac{\partial}{\partial x}U_3
\end{aligned} \tag{2.220}$$

from the initial conditions Eq. (2.216), we write

$$U_0(x) = -2k^2 + \frac{24k^2}{1+e^{kx}} - \frac{24k^2}{(1+e^{kx})^2} \tag{2.221}$$

Substituting Eq. (2.29) into Eq. (2.25) and by straightforward iterative steps, yields

$$
\begin{aligned}
U_1(x) &= -\frac{264k^7 e^{kx}(-1+e^{kx})}{(1+e^{kx})^3}, \\
U_2(x) &= -\frac{1452k^{12} e^{kx}(4e^{kx}-e^{2kx}-1)}{(1+e^{kx})^4}, \\
U_3(x) &= \frac{5324k^{17} e^{kx}(1-11e^{kx}+11e^{2kx}-e^{3kx})}{(1+e^{kx})^5}, \\
U_4(x) &= \frac{14641k^{22} e^{kx}(1-26e^{kx}+66e^{2kx}-26e^{3kx}+e^{4kx})}{(1+e^{kx})^6} \\
&\vdots
\end{aligned}
\tag{2.222}
$$

and so on.

Taking the inverse transformation of the set of values $\{U_k(x)\}_{k=0}^{n}$ gives n-terms approximation solution as

$$
\begin{aligned}
\tilde{u}_n(x,t) &= \sum_{k=0}^{n} U_k(x)t^k = -2k^2 + \frac{24k^2}{1+e^{kx}} - \frac{24k^2}{(1+e^{kx})^2} - \frac{264k^7 e^{kx}(-1+e^{kx})}{(1+e^{kx})^3}t + \ldots \\
&+ \frac{1}{n!}\left[\frac{\partial^n}{\partial t^n}\left(-2k^2 + \frac{24k^2}{1+e^{kx+11k^5 t}} - \frac{24k^2}{(1+e^{kx+11k^5 t})^2}\right)\right]_{t=0} t^n
\end{aligned}
\tag{2.223}
$$

Therefore, the exact solution of the problem is readily obtained as follows

$$
u(x,t) = \lim_{n\to\infty} \tilde{u}_n(x,t) = -2k^2 + \frac{24k^2}{1+e^{kx+11k^5 t}} - \frac{24k^2}{(1+e^{kx+11k^5 t})^2} \tag{2.224}
$$

2.8 MODIFIED DIFFERENTIAL TRANSFORMATION METHOD

In recent years, glorious developments have been presented by researchers on DTM to improve its accuracy. One of these developments is Modified Differential Transformation Method or MDTM [10]. Let us consider the differential transform for,

$$
u^3(x,t) = \sum_{r=0}^{k}\sum_{q=0}^{k-r}\sum_{s=0}^{h}\sum_{p=0}^{h-s} U_{\alpha,1}(r,h-s-p)U_{\alpha,1}(q,s)U_{\alpha,1}(k-r-q,p) \tag{2.225}
$$

where involves four summations. Thus it is necessary to have a lot of computational work to calculate such differential transform $U_{\alpha,1}(k, h)$ for the large number of (k, h). Since, DTM is based on the Taylor series for all variables. To reduce the complexity in DTM, we introduce the DTM with respect to the specific variable for the function $u(x,t)$ known as the modified version of DTM. Assume that the specific variable is the variable t then, we have the Taylor series expansion of the function $u(x, t)$ at $t = t_0$ as follows,

$$u(x,t) = \sum_{h=0}^{\infty} \frac{1}{\Gamma(\alpha h+1)} \left(\frac{\partial^{\alpha h} u(x,t)}{\partial t^{\alpha h}} \right)_{t=t_0} (t - t_0)^{\alpha h} \tag{2.226}$$

The modified differential transform $U_{\alpha,1}(x, h)$ of $u(x, t)$ with respect to the variable t at t_0 is defined by

$$U_{\alpha,1}(x,h) = \frac{1}{\Gamma(\alpha h+1)} \left(\frac{\partial^{\alpha h} u(x,t)}{\partial t^{\alpha h}} \right)_{t=t_0} \tag{2.227}$$

The modified differential inverse transform $U_{\alpha,1}(x, h)$ with respect to the variable t at t_0 is defined by

$$u(x,t) = \sum_{h=0}^{\infty} U_{\alpha,1}(x,h)(t - t_0)^{\alpha h} \tag{2.228}$$

Since the MDTM results from the Taylor's series of the function with respect to the specific variable, it is expected that the corresponding algebraic equation from the given problem is much simpler than the result obtained by the standard DTM. The fundamental mathematical operations performed by two-dimensional DTM are listed in Tables 2.2 and 2.3.

Example A. Consider the nonlinear fractional Klein–Gordon equation

$$\frac{\partial^{\alpha} u(x,t)}{\partial t^{\alpha}} - \frac{\partial^2 u(x,t)}{\partial x^2} + u^2(x,t) = 0 \tag{2.229}$$

subject to the initial conditions

$$u(x,0) = 1 + \sin x \tag{2.230}$$

DTM transferred is

$$\begin{aligned} &\frac{\Gamma(\alpha(h+1)+1)}{\Gamma(\alpha h+1)} U_{\alpha,1}(k,h+1) - (k+1)(k+2)U_{\alpha,1}(k+2,h) \\ &+ \sum_{r=0}^{k} \sum_{s=0}^{h} U_{\alpha,1}(r,h-s)U_{\alpha,1}(k-r,s) = 0 \end{aligned} \tag{2.231}$$

Table 2.2 The Operations for the Two-Dimensional Differential Transform Method

Original Function	Transformed Function
$w(x,t) = u(x,\ t) \pm v(x,t)$	$W_{\alpha,1}(k,h) = U_{\alpha,1}(k,\ h) \pm V_{\alpha,1}(k,\ h)$
$w(x,t) = \mu u(x,t)$	$W_{\alpha,1}(k,h) = \mu U_{\alpha,1}(k,\ h)$
$w(x,t) = \frac{\partial u(x,t)}{\partial x}$	$W_{\alpha,1}(k,h) = (k+1)U_{\alpha,1}(k+1,\ h)$
$w(x,t) = D^{\alpha}_{*t_0} u(x,t), \quad 0 < \alpha \leq 1$	$W_{\alpha,1}(k,h) = \frac{\Gamma(\alpha(h+1)+1)}{\Gamma(\alpha h+1)} U_{\alpha,1}(k, h+1)$
$w(x,t) = (x - x_0)^m (t - t_0)^{n\alpha}$	$W_{\alpha,1}(k,h) = \delta(k-m,\ h\alpha - n) = \begin{cases} 1 & k=m,\ h=n \\ 0 & \text{otherwise} \end{cases}$
$w(x,t) = u^2(x,t)$	$W_{\alpha,1}(k,h) = \sum_{m=0}^{k} \sum_{n=0}^{h} U_{\alpha,1}(m, h-n) U_{\alpha,1}(k-m, n)$
$w(x,t) = u^3(x,t)$	$W_{\alpha,1}(k,h) = \sum_{r=0}^{k} \sum_{q=0}^{k-r} \sum_{s=0}^{h} \sum_{p=o}^{h-s} U_{\alpha,1}(r, h-s-p) U_{\alpha,1}(q,s) U_{\alpha,1}(k-r-q, p)$

Table 2.3 The Operations for the Modified Differential Transform Method

Original Function	Transformed Function
$w(x,t) = u(x,t) \pm v(x,t)$	$W_{\alpha,1}(x,h) = U_{\alpha,1}(x,h) \pm V_{\alpha,1}(x,h)$
$w(x,t) = \mu u(x,t)$	$W_{\alpha,1}(x,h) = \mu U_{\alpha,1}(x,h)$
$w(x,t) = \frac{\partial u(x,t)}{\partial x}$	$W_{\alpha,1}(x,h) = \frac{\partial U_{\alpha,1}(x,h)}{\partial x}$
$w(x,t) = D^{\alpha}_{*t_0} u(x,t), \quad 0 < \alpha \leq 1$	$W_{\alpha,1}(x,h) = \frac{\Gamma(\alpha(h+1)+1)}{\Gamma(\alpha h+1)} U_{\alpha,1}(x,h+1)$
$w(x,t) = (x-x_0)^m (t-t_0)^{n\alpha}$	$W_{\alpha,1}(x,h) = (x-x_0)^m \delta(h\alpha - n)$
$w(x,t) = u^2(x,t)$	$W_{\alpha,1}(x,h) = \sum_{m=0}^{h} U_{\alpha,1}(x,m) U_{\alpha,1}(x,h-m)$
$w(x,t) = u^3(x,t)$	$W_{\alpha,1}(x,h) = \sum_{m=0}^{h} \sum_{l=0}^{m} U_{\alpha,1}(x,h-m) U_{\alpha,1}(x,l) U_{\alpha,1}(x,m-l)$

The transformed of Eq. (2.230) is

$$U_{\alpha,1}(k,0)=\begin{cases}1, k=0\\ \frac{1}{k!}, k=1,5,9\ldots\\ -\frac{1}{k!}, k=3,7,11\ldots\end{cases} \tag{2.232}$$

By substituting Eq. (2.232) into Eq. (2.231) and using DTM inverse, we have

$$\begin{aligned}u(x,t)&=\left(1+x-\frac{x^3}{3!}+\frac{x^5}{5!}-\frac{x^7}{7!}+\ldots\right)\\&+\left(1+3x+x^2-\frac{x^3}{2}-\frac{x^4}{3}+\ldots\right)\frac{t^\alpha}{\Gamma(\alpha+1)}\\&+\left(11x+12x^2+\frac{x^3}{6}-4x^4+\ldots\right)\frac{t^{2\alpha}}{\Gamma(2\alpha+1)}+\ldots\end{aligned} \tag{2.233}$$

For applying MDTM to Eq. (2.229) we have

$$\begin{aligned}&\frac{\Gamma(\alpha(h+1)+1)}{\Gamma(\alpha h+1)}U_{\alpha,1}(x,h+1)-\frac{\partial^2 U_{\alpha,1}(x,h)}{\partial x^2}\\&+\sum_{m=0}^{h}U_{\alpha,1}(x,m)U_{\alpha,1}(x,h-m)=0\end{aligned} \tag{2.234}$$

The transformed of Eq. (2.230) is

$$U_{\alpha,1}(x,0)=1+\sin x \tag{2.235}$$

Finally after substituting the result is

$$\begin{aligned}u(x,t)&=1+\sin x-\frac{1}{\Gamma(\alpha+1)}(1+3\sin x+\sin^2 x)t^\alpha\\&+\frac{1}{\Gamma(2\alpha+1)}(11\sin x+12\sin^2 x+2\sin^3 x)t^{2\alpha}\end{aligned} \tag{2.236}$$

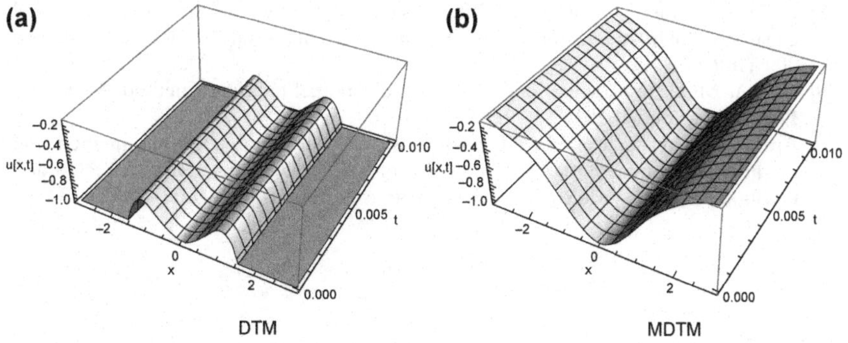

Figure 2.13 Difference between Modified Differential Transformation Method and Differential Transformation Method for *u(x,t)*.

Aruna and Kanth [10] showed that MDTM can improve the results, for example, they showed when the equation is

$$\frac{\partial^{2.5}u(x,t)}{\partial t^{2.5}}-\frac{\partial^{2}u(x,t)}{\partial x^{2}}-u(x,t)+u^{3}(x,t)=0 \tag{2.237}$$

With initial condition of

$$u(x,0)=-\sec hx \tag{2.238}$$

DTM and MDTM have a large difference as shown in Fig. 2.13.

REFERENCES

[1] Joneidi AA, Ganji DD, Babaelahi M. Differential Transformation Method to determine fin efficiency of convective straight fins with temperature dependent thermal conductivity. International Communication in Heat and Mass Transfer 2009;36:757−62.
[2] Hassan IHA-H. Differential transformation technique for solving higher-order initial value problems. Applied Mathematics and Computation 2004;154(2):299−311.
[3] Arikoglu A, Ozkol I. Solution of fractional differential equations by using differential transform method. Chaos, Solitons & Fractals 2007;34(5):1473−81.
[4] Nazari D, Shahmorad S. Application of the fractional differential transform method to fractional-order integro-differential equations with nonlocal boundary conditions. Journal of Computational and Applied Mathematics 2010;234(3):883−91.
[5] Chen C-K, Ho S-H. Application of differential transformation to eigenvalue problems. Applied Mathematics and Computation 1996;79(2):173−88.
[6] Hassan IHA-H. On solving some eigenvalue problems by using a differential transformation. Applied Mathematics and Computation 2002;127(1):1−22.
[7] Odibat Z, Momani S. A generalized differential transform method for linear partial differential equations of fractional order. Applied Mathematics Letters 2008;21(2):194−9.

[8] Keskin Y, Oturanc G. Reduced differential transform method for partial differential equations. International Journal of Nonlinear Sciences and Numerical Simulation 2009;10(6):741–9.
[9] Al-Amr MO. New applications of reduced differential transform method. Alexandria Engineering Journal 2014;53(1):243–7.
[10] Aruna K, Ravi Kanth ASV. Two-dimensional differential transform method and modified differential transform method for solving nonlinear fractional Klein–Gordon equation. National Academy of Science Letters March–April 2014;37(2):163–71.

CHAPTER 3

DTM for Heat Transfer Problems

3.1 INTRODUCTION

Most of the problems in mechanical engineering include the heat transfer phenomena. Industrial engineering, cooling process, oil industry and melting, shaping and deformations, automobile industry and many other process have a heat transfer and researchers need to control it by increase/decrease devices. Fins are the most effective instrument for increasing the rate of heat transfer, as we know they increase the area of heat transfer and cause an increase in the transferred heat volume. A complete review on this topic is presented by Krause et al. [1]. Fins are widely used in the many industrial applications such as air conditioning, refrigeration, automobile, chemical processing equipment, and electrical chips. Also cooling by suitable fluids such as nanofluids is another application of heat transfer discussed in the present chapter, which contains following sections:

3.1 Introduction
3.2 Longitudinal Fins With Constant Profile
3.3 Natural Convection Flow of a Non-Newtonian Nanofluid
3.4 Two-Dimensional Heat Transfer in Longitudinal Rectangular and Convex Parabolic Fins
3.5 Thermal Boundary Layer on Flat Plate
3.6 Falkner–Skan Wedge Flow
3.7 Free Convection Problem

3.2 LONGITUDINAL FINS WITH CONSTANT PROFILE

Consider a longitudinal fin with constant rectangular profile, section area A, length L, perimeter P, thermal conductivity k, and heat generation q^*. The fin is attached to a surface with a constant temperature T_b and losses heat to the surrounding medium with temperature T_∞ through a constant convective heat transfer coefficient h. In the problem we assume that the temperature variation in the transfer direction is negligible, so heat conduction occurs only in the longitudinal direction (x-direction). A schematic of the geometry of described fin and other properties is shown in Fig. 3.1.

Differential Transformation Method for Mechanical Engineering Problems
ISBN 978-0-12-805190-0
http://dx.doi.org/10.1016/B978-0-12-805190-0.00003-6

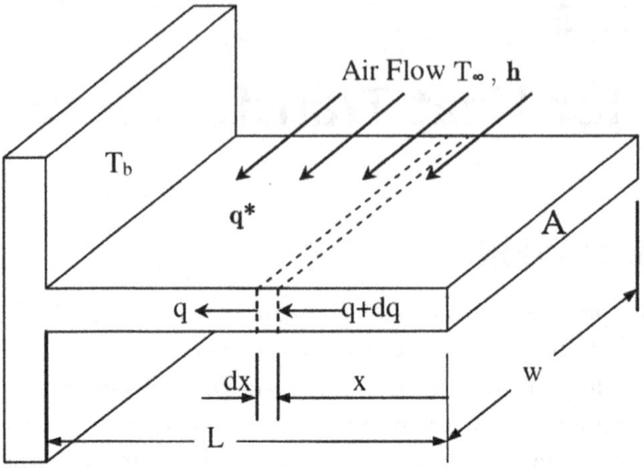

Figure 3.1 Schematic of the fin geometry with heat generation source.

For this problem, the governing differential equation and boundary condition can be written as [2]

$$\frac{d^2T}{dx^2} - \frac{hP}{kA}(T - T_\infty) + \frac{\overset{*}{q}}{k} = 0 \tag{3.1}$$

$$x = 0, \quad \frac{dT}{dx} = 0 \tag{3.2}$$

$$x = L, \quad T = T_b \tag{3.3}$$

This problem is solved in two main cases using Differential Transformation Method (DTM). In the following subsections, the governing equations for these two cases are introduced.

A. Fin With Temperature-Dependent Internal Heat Generation and Constant Thermal Conductivity

In the first case, we assume that heat generation in the fin varies with temperature as given in Eq. (3.4) and the thermal conductivity is constant k_0.

$$\overset{*}{q} = \overset{*}{q}_\infty(1 + \varepsilon(T - T_\infty)) \tag{3.4}$$

where $\overset{*}{q}_\infty$ is the internal heat generation at temperature T_∞. With the introduction of following dimensionless quantities:

$$\theta = \frac{(T - T_\infty)}{(T_b - T_\infty)}, \quad X = \frac{x}{L}, \quad N^2 = \frac{hPL^2}{k_0 A}$$
$$G = \frac{\overset{*}{q}_\infty}{hP(T_b - T_\infty)}, \quad \varepsilon_G = \varepsilon(T_b - T_\infty) \tag{3.5}$$

Eq. (3.1)−(3.3) can be rewritten as

$$\frac{d^2\theta}{dX^2} - N^2\theta + N^2G(1 + \varepsilon_G\theta) = 0 \tag{3.6}$$

$$X = 0, \quad \frac{d\theta}{dX} = 0 \tag{3.7}$$

$$X = 1, \quad \theta = 1 \tag{3.8}$$

Now we apply DTM from Chapter 1 into Eq. (3.6) to find $\theta(x)$.

$$(k+1)(k+2)\Theta(k+2) - N^2\Theta(k) + N^2G(\delta(k) + \varepsilon_G\Theta(k)) = 0 \tag{3.9}$$

Rearranging Eq. (3.9), a simple recurrence relation is obtained as follows

$$\Theta(k+2) = \frac{N^2\Theta(k) - N^2G(\delta(k) + \varepsilon_G\Theta(k))}{(k+1)(k+2)} \tag{3.10}$$

where

$$\delta(k) = \begin{cases} 1 & \text{if } k = 0 \\ 0 & \text{if } k \neq 0 \end{cases} \tag{3.11}$$

Similarly, the transformed form of boundary conditions can be written as

$$\Theta(0) = a, \quad \Theta(1) = 0 \tag{3.12}$$

By solving Eq. (3.10) and using boundary conditions (Eq. 3.12) the DTM terms are obtained as

$$\begin{aligned}
\Theta(2) &= \frac{1}{2}N^2\Theta(0) - \frac{1}{2}N^2G(1 + \varepsilon_G\Theta(0)) \\
\Theta(3) &= \frac{1}{6}N^2\Theta(1) - \frac{1}{6}N^2G(1 + \varepsilon_G\Theta(1)) \\
\Theta(4) &= \frac{1}{12}N^2\Theta(2) - \frac{1}{12}N^2G(1 + \varepsilon_G\Theta(2)) \\
\Theta(5) &= \frac{1}{20}N^2\Theta(1) - \frac{1}{20}N^2G(1 + \varepsilon_G\Theta(1)) \\
&\vdots
\end{aligned} \tag{3.13}$$

Now by applying DTM principle equation, into Eq. (3.13), and by using Eq. (3.3) the constant parameter "a" will be obtained, so the temperature distribution equation will be estimated.

B. Fin With Temperature-Dependent Internal Heat Generation and Temperature-Dependent Thermal Conductivity

In the second case, we assume that thermal conductivity of fin as well as internal heat generation is temperature-dependent. We consider it to vary linearly with temperature, then we have

$$k = k_0[1 + \beta(T - T_\infty)] \tag{3.14}$$

The dimensionless form of Eq. (3.15) is

$$\frac{k}{k_0} = [1 + \varepsilon_c\theta] \tag{3.15}$$

where

$$\varepsilon_c = \beta(T_b - T_\infty) \tag{3.16}$$

Eq. (3.6) for this condition becomes

$$\frac{d}{dX}\left[(1 + \varepsilon_G\theta)\frac{d\theta}{dX}\right] - N^2\theta + N^2G(1 + \varepsilon_G\theta) = 0 \tag{3.17}$$

Where its boundary conditions are given by Eqs. (3.7) and (3.7). Now we must apply DTM to this governing equation.

$$(k+2)(k+1)\Theta(k+2) + \varepsilon_C\sum_{m=0}^{k}\{(k+1-m)\Theta(k+1-m)(m+1)\Theta(m+1)\} + \varepsilon_C\sum_{m=0}^{k}\{(k-m)\Theta(k-m)(m+2)\Theta(m+2) - N^2\Theta(k) + N^2G(\delta(k) + \varepsilon_G\Theta(k))\right\} = 0 \tag{3.18}$$

Rearranging Eq. (3.17), a simple relation is obtained as follows

$$\Theta(k+2) = \frac{-1}{(k+2)(k+1)}(\varepsilon_C\sum_{m=0}^{k}\{((k+1-m\Theta(k+1-m)(m+1)\Theta(m+1)\} + \varepsilon_C\sum_{m=0}^{k}\{(k-m)\Theta(k-m)(m+2)\Theta(m+2) - N^2\Theta(k) + N^2G(\delta(k) + \varepsilon_G\Theta(k))\right\} \tag{3.19}$$

Where the boundary conditions of this case is the same as that of the previous case Eq. (3.12). By solving Eq. (3.19) and using boundary conditions Eq. (3.12), the DTM terms for this case can be as

$$\Theta(2) = \frac{N^2\Theta(0) - N^2 G\varepsilon_G\Theta(0) - N^2 G}{2(1+\varepsilon_C\Theta(0))}$$

$$\Theta(3) = \varepsilon_C\Theta(2)\Theta(1) + \frac{1}{6}N^2\Theta(1) - \frac{1}{6}N^2 G\varepsilon_G\Theta(1)$$

$$\Theta(4) = -\frac{1}{12}N^2 G\varepsilon_G\Theta(2) - \frac{3}{4}\varepsilon_C\Theta(3)\Theta(1) - \frac{2}{3}\varepsilon_C\Theta^2(2) + \frac{1}{12}N^2\Theta(2)$$

.

.

.

(3.20)

Finally, by applying DTM into Eq. (3.20) and using Eq. (3.3) the constant parameter "*a*" will be obtained and the temperature distribution equation will be calculated.

A. Case1: Fin With Temperature-Dependent Internal Heat Generation and Constant Thermal Conductivity

Temperature distribution in case 1 (temperature-dependent heat generation and constant thermal conductivity) is shown in Figs. 3.2–3.5. It is common in fin design that the N parameter is considered to be 1. Fig. 3.1 shows temperature distribution for this state and $\varepsilon_G = G = 0.2$, $\varepsilon_G = G = 0.4$, and $\varepsilon_G = G = 0.6$. This choice of parameters represents a fin with moderate temperature-dependent heat generation and the thermal conductivity variation of 20% between the base and the surrounding coolant temperatures that are often used in nuclear rods. As we see in the figure by increasing ε_G and G the temperature of the fin is increased because of increasing heat generation. By comparing the results with numerical method, it was observed that DTM has a good efficiency and accuracy, error of DTM is plotted in Fig. 3.1 and it reveals this fact. As seen in Fig. 3.1 the maximum error occurs in the tip of the fin. Fig. 3.4 shows comparison results, which pertain to $N = 0.5$ (this choice is used in compact heat exchanger fin design), and this figure illustrates that fin temperature in this condition is greater than $N = 1$ state. Fig. 3.5 shows the errors for $N = 1$ and $\varepsilon_G = G = 0.2$, $\varepsilon_G = G = 0.4$, and $\varepsilon_G = G = 0.6$. As

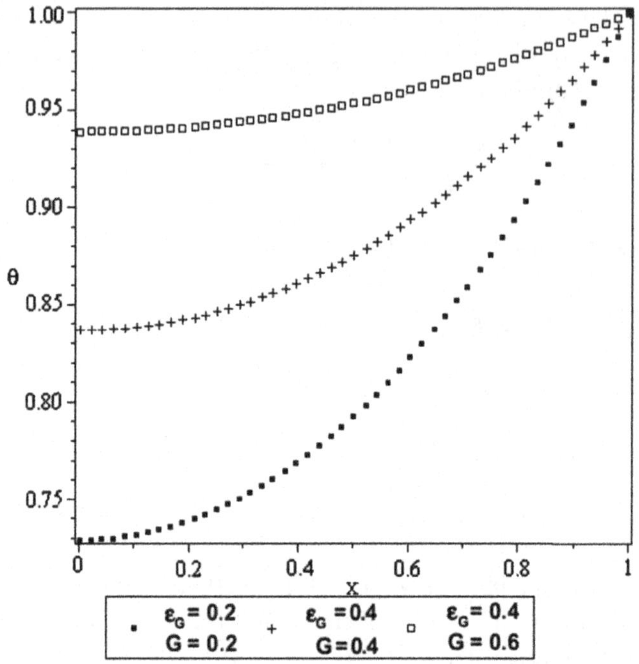

Figure 3.2 Temperature distribution in the fin with temperature-dependent internal heat generation and constant thermal conductivity for $N = 1$.

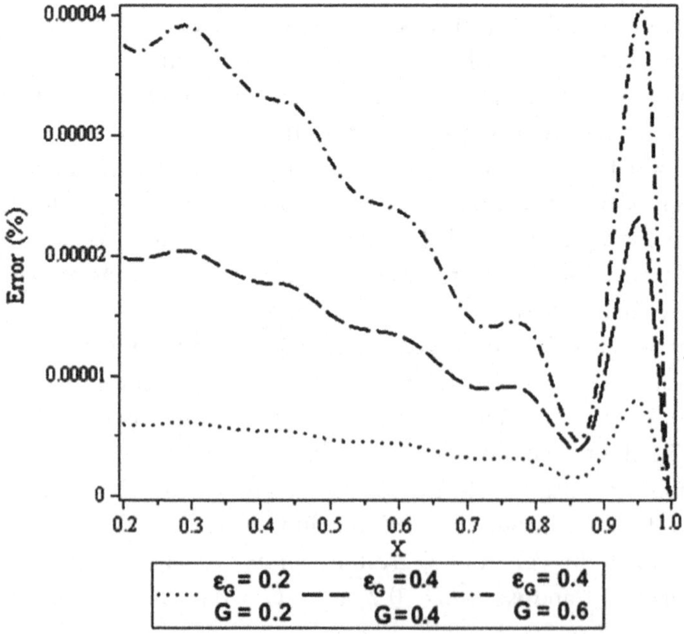

Figure 3.3 Error of differential transformation method in comparison by numerical method for case 1 and $N = 1$.

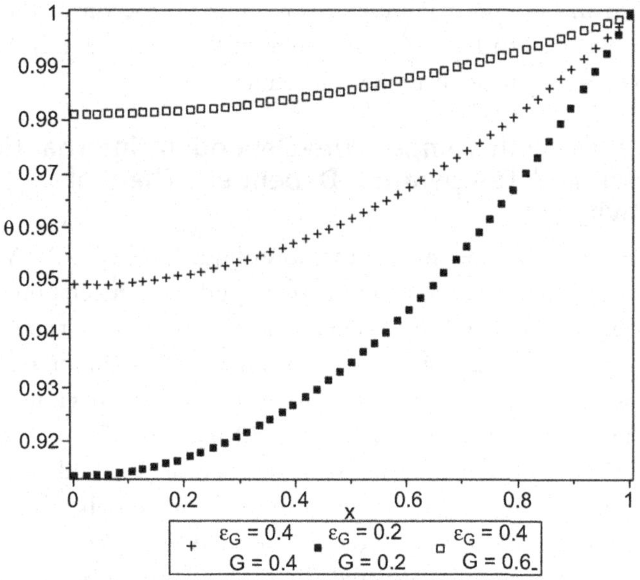

Figure 3.4 Temperature distribution in the fin with temperature-dependent internal heat generation and constant thermal conductivity for $N = 0.5$.

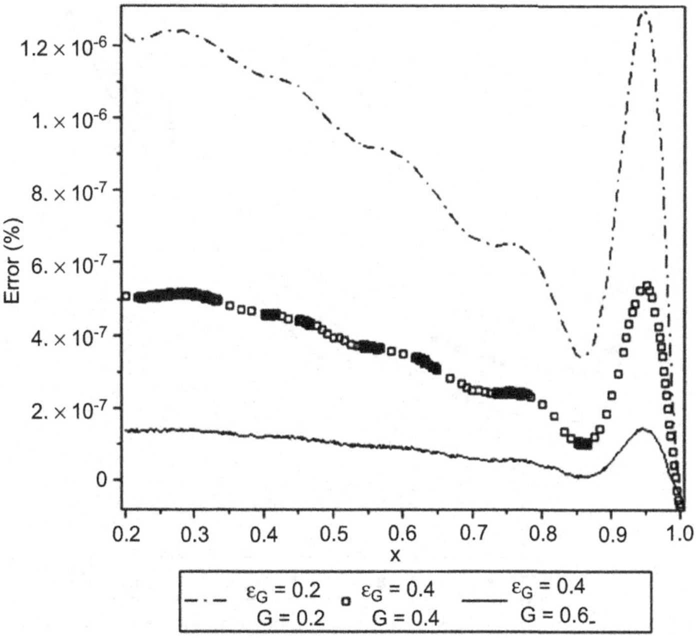

Figure 3.5 Error of differential transformation method in comparison by numerical method for case 1 and $N = 0.5$.

already seen in Fig. 3.5 maximum error occurs in the tip of the fin, this result also occurred in Fig. 3.5. The range of the errors reveals that DTM has a good agreement with numerical results.

B. Case 2: Fin with Temperature-Dependent Internal Heat Generation and Temperature-Dependent Thermal Conductivity

Figs. 3.6–3.9 show the temperature distribution in case 2. As already mentioned, in case 2, thermal conductivity and heat generation are temperature dependent. Fig. 3.6 illustrates the temperature distribution with $N = 1$, $\varepsilon_G = G = 0.4$, and ε_C increased from 0 to 0.6 with intervals 0.2. As seen in Fig. 3.6, when ε_C increases, the local fin temperature increases because the ability of the fin to conduct heat increases. Fig. 3.7 shows the error of DTM in comparison with numerical method for $N = 1$, $\varepsilon_G = G = 0.4$ and low maximum error in this figure emphasis on accuracy and efficiency of DTM. In Fig. 3.8 the N parameter is decreased to 0.5 and temperature distribution is depicted. In Fig. 3.9, the error of DTM for $N = 0.5$, $G = 0.4$, $\varepsilon_G = 0.4$ is depicted.

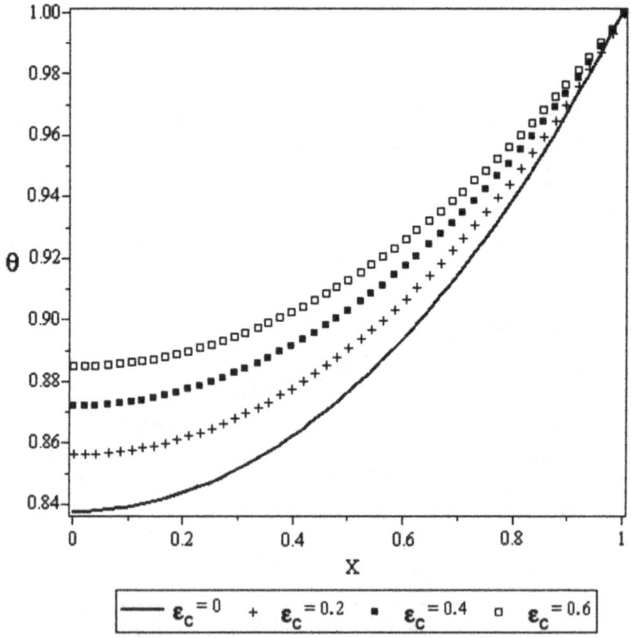

Figure 3.6 Temperature distribution in the fin with temperature-dependent internal heat generation and temperature-dependent thermal conductivity for $N = 1$, $G = 0.4$, $\varepsilon_G = 0.4$.

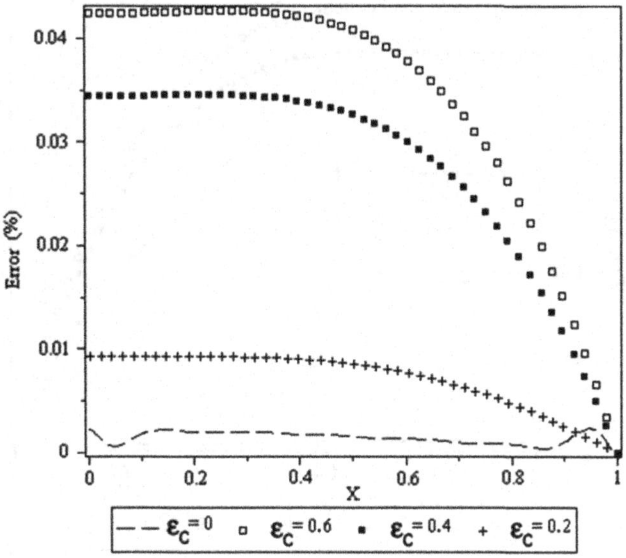

Figure 3.7 Error of differential transformation method in comparison by numerical method for case 2 and $N = 1$.

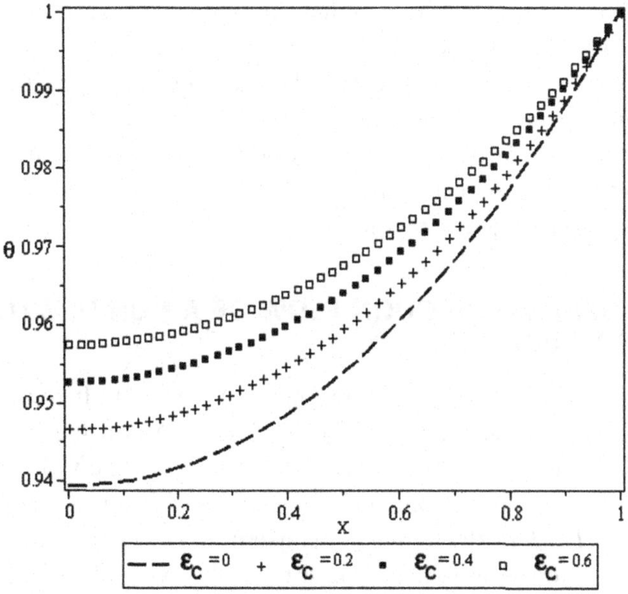

Figure 3.8 Temperature distribution in the fin with temperature-dependent internal heat generation and temperature-dependent thermal conductivity for $N = 0.5$, $G = 0.4$, $\varepsilon_G = 0.4$.

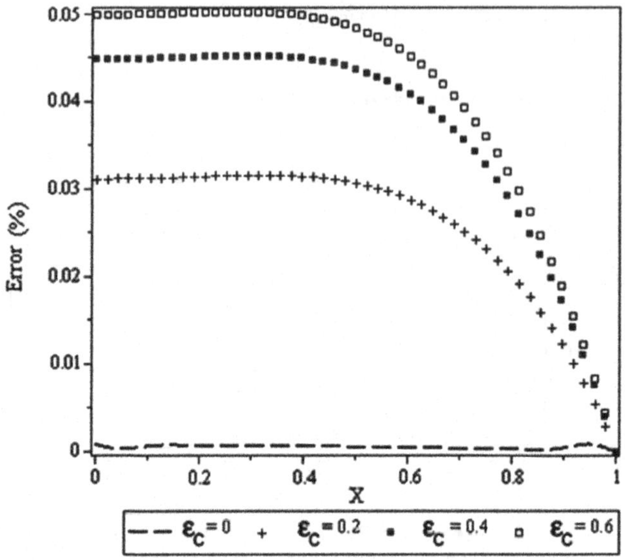

Figure 3.9 Error of differential transformation method in comparison by numerical method for case 2 and $N = 0.5$.

Finally, by a comparative assessment of figures introduced for this case and pervious case, it can be found that the local fin temperature increases as the parameters G, ε_G, and ε_C increase. The increase in parameter ε_G implies that the heat generation is increased, and hence it causes to produce a higher temperature in the fin. An increase in ε_C means the thermal conductivity of the fin is increased, and it makes more heat conducting through the fin and local temperature will increase.

3.3 NATURAL CONVECTION FLOW OF A NON-NEWTONIAN NANOFLUID

A schematic theme of the problem is shown in Fig. 3.10. It consists of two vertical flat plates separated by a distance $2b$ apart. A non-Newtonian fluid flows between the plates due to natural convection. The walls at $x = +b$ and $x = -b$ are held at constant temperatures T_2 and T_1, respectively, where $T_1 > T_2$. This difference in temperature causes the fluid near the wall at $x = -b$ to rise and the fluid near the wall at $x = +b$ to fall. The fluid is a water-based nanofluid containing Cu. It is assumed that the base fluid and the nanoparticles are in thermal equilibrium and no slip occurs between

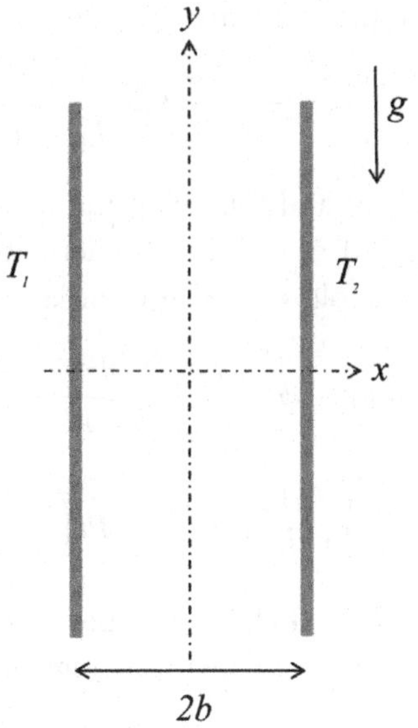

Figure 3.10 Geometry of the problem.

Table 3.1 Thermophysical Properties of Water and Nanoparticles

	ρ (kg/m^3)	C_p (j/kgk)	k (W/m·k)	$\beta \times 10^5$ (K^{-1})
Pure water	997.1	4179	0.613	21
Copper(Cu)	8933	385	401	1.67

them. The thermophysical properties of the nanofluid are given in Table 3.1 [3].

The effective density ρ_{nf}, the effective dynamic viscosity μ_{nf}, the heat capacitance $(\rho C_p)_{nf}$, and the thermal conductivity k_{nf} of the nanofluid are given as

$$\rho_{nf} = \rho_f(1-\phi) + \rho_s\phi \tag{3.21}$$

$$\mu_{nf} = \frac{\mu_f}{(1-\phi)^{2.5}} \tag{3.22}$$

$$(\rho C_p)_{nf} = (\rho C_p)_f(1-\phi) + (\rho C_p)_s\phi \tag{3.23}$$

$$\frac{k_{nf}}{k_f} = \frac{k_s + 2k_f - 2\phi(k_f - k_s)}{k_s + 2k_f + \phi(k_f - k_s)} \tag{3.24}$$

here, ϕ is the solid volume fraction.

Following, we define the similarity variables:

$$V = \frac{v}{V_0}, \eta = \frac{x}{b}, \quad \theta = \frac{T - T_m}{T_1 - T_2}, \tag{3.25}$$

Under these assumptions and following the nanofluid model proposed by Maxwell—Garnetts (MG) model, the Navier—Stokes and energy equations can be reduced to the following pair of ordinary differential equations:

$$\frac{d^2 V}{d\eta^2} + 6\delta(1 - \phi)^{2.5}\left(\frac{dV}{d\eta}\right)^2 \frac{d^2 V}{d\eta^2} + \theta = 0, \tag{3.26}$$

$$\frac{d^2\theta}{d\eta^2} + EcPr\left(\frac{(1-\phi)^{-2.5}}{A_1}\right)\left(\frac{dV}{d\eta}\right)^2 + 2\delta EcPr\left(\frac{1}{A_1}\right)\left(\frac{dV}{d\eta}\right)^2 = 0. \tag{3.27}$$

where Prandtl number (Pr), Eckert number (Ec), dimensionless non-Newtonian viscosity (δ), and A_1 have following forms:

$$Ec = \frac{\rho_f V_0^2}{(\rho C_p)_f (T_1 - T_2)}, \quad Pr = \frac{\mu_f (\rho C_p)_f}{\rho_f k_f}, \quad \delta = \frac{6\beta_3 V_0^2}{\mu_f b^2} \tag{3.28}$$

$$A_1 = \frac{k_{nf}}{k_f} = \frac{k_s + 2k_f - 2\varphi(k_f - k_s)}{k_s + 2k_f + 2\varphi(k_f - k_s)} \tag{3.29}$$

The appropriate boundary conditions are

$$\begin{aligned} \eta = -1: &\quad V = 0, \theta = 0.5 \\ \eta = +1: &\quad V = 0, \theta = -0.5 \end{aligned} \tag{3.30}$$

The average Nusselt number is defined as

$$Nu_{ave} = -\frac{k_{nf}}{k_f}\left(\frac{\theta'(1) + \theta'(-1)}{2}\right) \tag{3.31}$$

Now DTM will be applied to the governing equations. Taking the differential transform of equations with respect to χ, and considering $H = 1$ gives

$$\begin{aligned} &(k+1)(k+2)\vartheta[k+2] + \Theta[k] + \\ &6\alpha(1-\phi)^{2.5}\left(\sum_{m=0}^{k}\left(\sum_{r=0}^{m}(k-m+1)\vartheta[k-m+1](m-r+1)\right.\right. \\ &\left.\left.\times\, \vartheta[k-r+1](r+1)(r+2)\vartheta[r+2]\right)\right) \end{aligned} \tag{3.32}$$

$$\vartheta[0] = a_1, \quad \vartheta[1] = a_2 \tag{3.33}$$

$$\begin{aligned}&(k+1)(k+2)\Theta[k+2] + EcPr\left(\frac{(1-\phi)^{2.5}}{A_1}\right)\\&\times\sum_{m=0}^{k}((m+1)\vartheta[m+1](k-m+1)\vartheta[k-m+1])\\&+2\delta EcPr\left(\frac{1}{A_1}\right)\sum_{m=0}^{k}\sum_{r=0}^{m}\sum_{s=0}^{r}(k-m+1)\vartheta[k-m+1](m-r+1)\\&\times\vartheta[m-r+1](r-s+1)\vartheta[r-s+1](r+1)\vartheta[r+1])\end{aligned} \tag{3.34}$$

$$\Theta[0] = b_1, \quad \Theta[1] = b_2 \tag{3.35}$$

where $\vartheta(k)$ and $\Theta(k)$ are the differential transforms of $V(\eta)$ and $\theta(\eta)$, also a_1, a_2, b_1, and b_2 are constants which can be obtained through boundary conditions, Eqs. (3.26) and (3.27). This problem can be solved as follows:

$$\begin{aligned}&\vartheta[0] = a_1, \quad \vartheta[1] = a_2\\&\vartheta[2] = -\frac{1}{2}\frac{b_1}{1+6\delta(1-\phi)^{2.5}a_2^2}\\&\vartheta[3] = -\frac{1}{6}\frac{12\, b_1^2 a_2\, \delta(1-\phi)^{2.5} + b_2 + 12b_2\delta(1-\phi)^{2.5}a_2^2 + 36b_2\left(\delta(1-\phi)^{2.5}\right)^2 a_2^4}{\left(1+6\delta(1-\phi)^{2.5}a_2^2\right)^3}\\&,\ldots\end{aligned} \tag{3.36}$$

$$\begin{aligned}&\Theta[0] = b_1, \quad \Theta[1] = b_2\\&\Theta[2] = -\frac{1}{2}EcPr\left(\frac{(1-\phi)^{-2.5}}{A_1}\right)a_2^2 - EcPr\left(\frac{1}{A_1}\right)\delta a_2^4\\&\Theta[3] = \frac{1}{3}\frac{EcPr\left(\frac{(1-\phi)^{-2.5}}{A_1}\right)b_1\left(1+9\delta(1-\phi)^{2.5}a_2^2 + 6\left(\delta(1-\phi)^{2.5}\right)^2 a_2^4 - \delta(1-\phi)^{2.5}a_2 b_1\right)}{\left(1+6\delta(1-\phi)^{2.5}a_2^2\right)^2}\\&,\ldots\end{aligned} \tag{3.37}$$

The above process is continuous. By substituting Eqs. (3.36) and (3.37) into the main equation based on DTM, it can be obtained that the closed form of the solutions is

$$\vartheta(\eta) = a_1 + a_2\eta + \left(-\frac{1}{2}\frac{b_1}{1+6\delta(1-\phi)^{2.5}a_2^2}\right)\eta^2 + \left(-\frac{1}{6}\frac{12\,b_1^2a_2\,\delta(1-\phi)^{2.5} + b_2 + 12b_2\delta(1-\phi)^{2.5}a_2^2 + 36b_2\left(\delta(1-\phi)^{2.5}\right)^2a_2^4}{\left(1+6\delta(1-\phi)^{2.5}a_2^2\right)^3}\right)\eta^3 + \ldots \tag{3.38}$$

$$Q_n(\eta) = b_1 + b_2\eta + \left(-\frac{1}{2}EcPr\left(\frac{(1-\phi)^{-2.5}}{A_1}\right)a_2^2 - EcPr\left(\frac{1}{A_1}\right)\delta a_2^4\right)\eta^2 + \left(\frac{1}{3}\frac{EcPr\left(\frac{(1-\phi)^{-2.5}}{A_1}\right)b_1\left(1+9\delta(1-\phi)^{2.5}a_2^2 + 6\left(\delta(1-\phi)^{2.5}\right)^2a_2^4 - \delta(1-\phi)^{2.5}a_2b_1\right)}{\left(1+6\delta(1-\phi)^{2.5}a_2^2\right)^2}\right)\eta^3 + \ldots \tag{3.39}$$

By substituting the boundary condition from Eq. (3.30) into Eqs. (3.38) and (3.39) the values of a_1, a_2, b_1, and b_2 can be obtained.

$$\begin{aligned} \vartheta(-1) &= 0 \\ \vartheta(+1) &= 0 \\ \Theta(-1) &= 0.5 \\ \Theta(+1) &= -0.5 \end{aligned} \tag{3.40}$$

By solving Eq. (3.40) the values of a_1, a_2, b_1, and b_2 are given. By substituting obtained a_1, a_2, b_1, and b_2 into Eqs. (3.38)–(3.39), the expression of $\vartheta(k)$ and $\Theta(k)$ can be obtained.

For example, when $Ec = 1$, $\delta = 0.5$, $\phi = 0$, $Pr = 6.2$, a_1, a_2, b_1 and b_2 are obtained as follows:

$$a_1 = -1.719612304,\quad a_2 = 0.2493778738e-1,\quad b_1 = 2.256751261,$$

$$b_2 = -0.6092964931$$

In this example, natural convection of a non-Newtonian nanofluid between two infinite parallel vertical flat plates has been investigated. These equations are solved analytically using DTM. To verify the accuracy of the present results, we have compared these results with a numerical method (the fourth-order Runge–Kutta method). Comparison between numerical results and DTM solutions for different values of effective parameter is shown in Fig. 3.11. It shows that the results obtained by DTM are in good

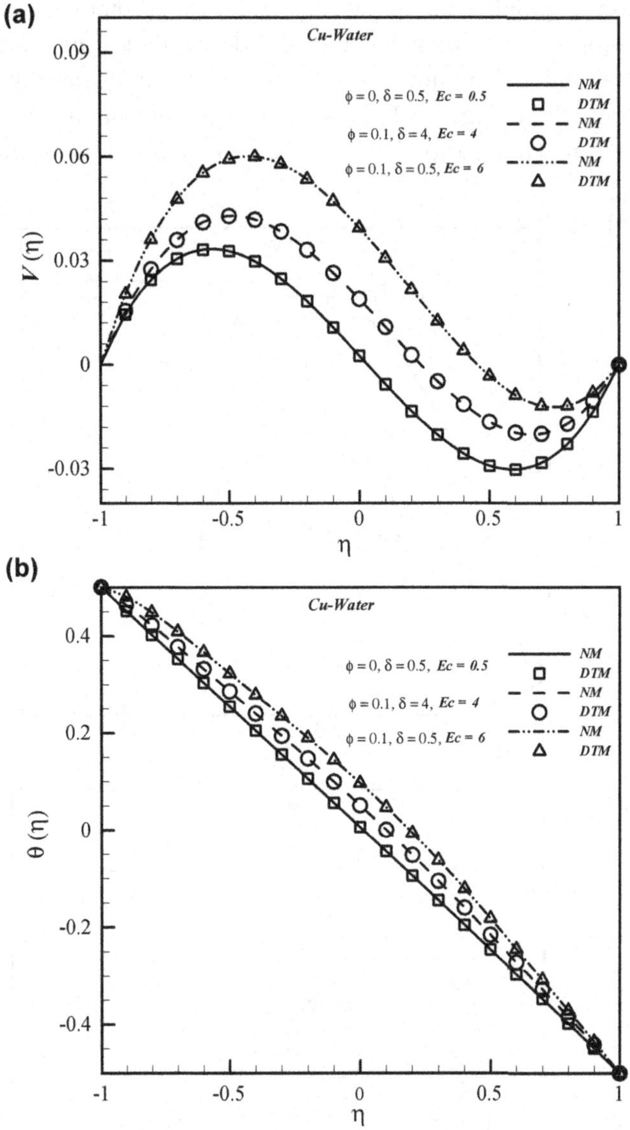

Figure 3.11 Comparison between numerical results and differential transformation method solution for different values of effective parameter.

agreement with those carried out by the numerical solution obtained using fourth-order Runge-Kutta method.

This accuracy gives high confidence to us about the validity of this method and reveals an excellent agreement of engineering accuracy. This investigation is completed by depicting the effects of important parameters such as nanoparticle volume fraction (ϕ), dimensionless non-Newtonian viscosity (δ), and Eckert number (Ec) to evaluate how these parameters influence this fluid. Fig. 3.12 shows the effect of nanoparticle volume fraction (ϕ) on: (a) velocity profiles $V(\eta)$; (b) temperature profiles $\theta(\eta)$,

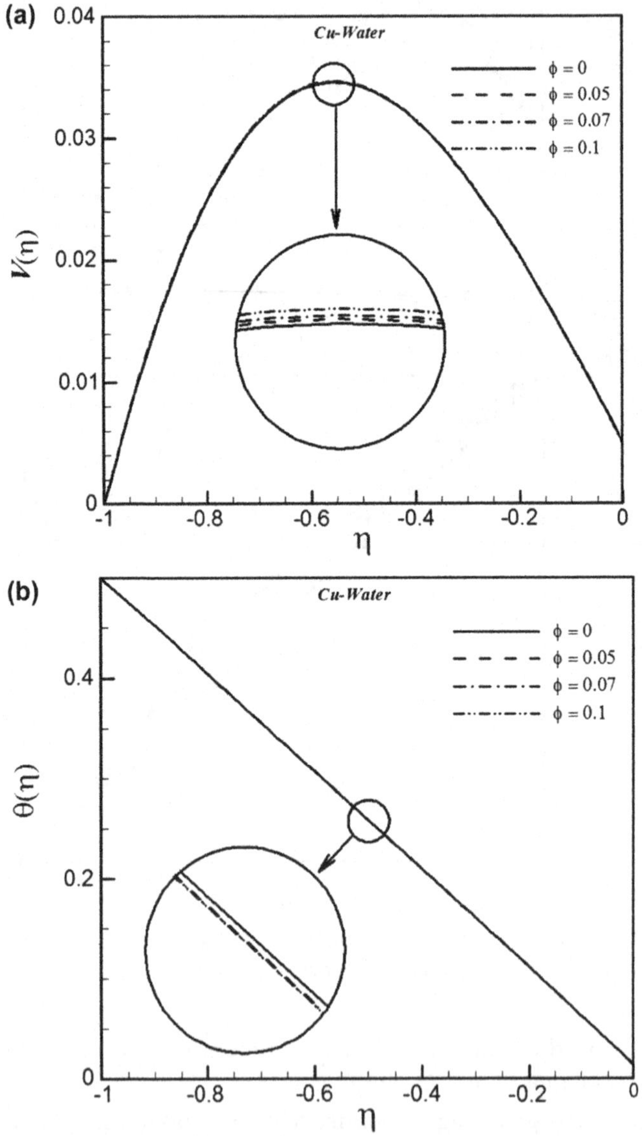

Figure 3.12 Effect of nanoparticle volume fraction (ϕ) on: (a) velocity profiles $V(\eta)$; (b) temperature profiles $\theta(\eta)$, when $\delta = 1$, $Ec = 1$, and $Pr = 6.2$.

when $\delta = 1$, $Ec = 1$, and $Pr = 6.2$. When the volume fraction of the nanoparticles increases from 0 to 0.1, the velocity also increases. Also, we can see that, with increasing volume fraction of the nanoparticles, the thermal boundary layer thickness decreases, and heat transfer rate increases at the surface (Nusselt number). Fig. 3.13 displays the effect of dimensionless non-Newtonian viscosity (δ) on: (a) velocity profiles $V(\eta)$;

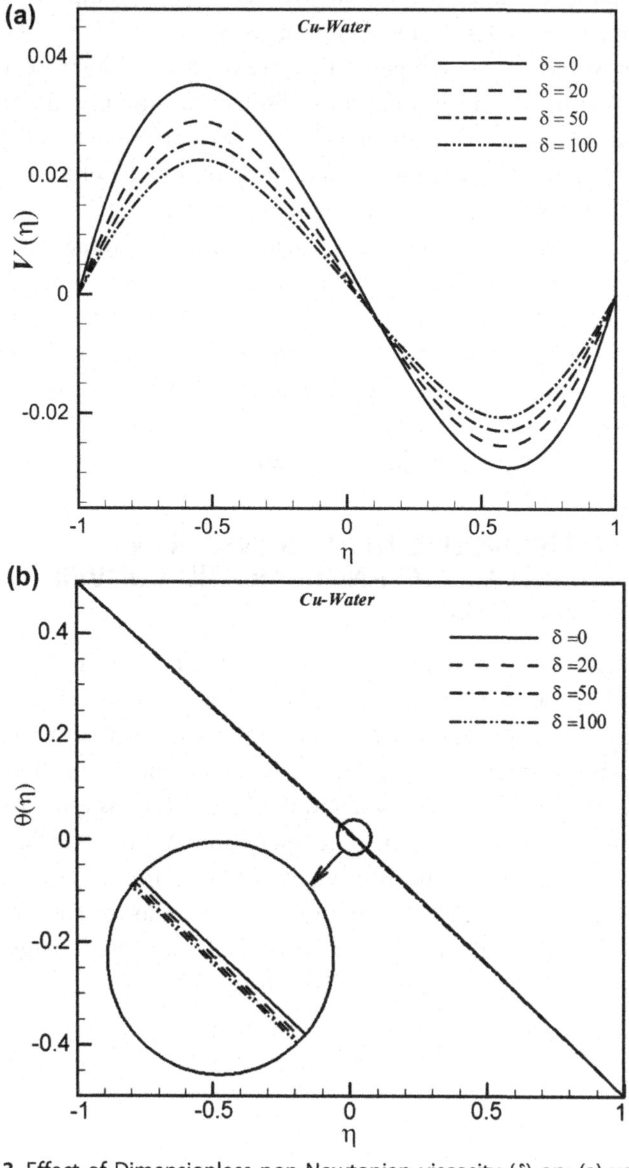

Figure 3.13 Effect of Dimensionless non-Newtonian viscosity (δ) on: (a) velocity profiles $V(\eta)$; (b) temperature profiles $\theta(\eta)$, when $\phi = 0.1$, $Ec = 1$, and $Pr = 6.2$.

(b) temperature profiles $\theta(\eta)$, when $\phi = 0.1$, $Ec = 1$, and $Pr = 6.2$. The dimensionless non-Newtonian viscosity indicates the relative significance of the inertia effect compared to the viscous effect. The magnitudes of both velocity and temperature decrease as δ increases, and in turn the magnitude of the skin friction coefficient and the Nusselt number increase. Also, it can be seen that at $|\eta| = 0.6$ the maximum value of velocity is observed. Effects of Eckert number (Ec) on: (a) velocity profiles $V(\eta)$; (b) temperature profiles $\theta(\eta)$, when $\phi = 0.1$, $\delta = 1$, and $Pr = 6.2$ are shown in Fig. 3.14. It can be found that velocity and temperature increase due to increase in Eckert number. When we neglect viscous dissipation, minimum amount of velocity and temperature are obtained. Also, it can be seen that at $|\eta| = 0.5$ maximum value of velocity and at middle point/surface of two plates, and at ($\eta = 0$) maximum value of temperature observed.

Fig. 3.15 depicts the effects of nanoparticle volume fraction (ϕ), dimensionless non-Newtonian viscosity (δ), and Eckert number (Ec) on Nusselt number when (a) $Ec = 1$; (b) $\delta = 1$ and $Pr = 6.2$. The Nusselt number has a direct relationship with dimensionless non-Newtonian viscosity parameter and volume fraction of the nanoparticles (when $\delta < 20$), but it has reverse relationship with Eckert number and volume fraction of the nanoparticles (when $\delta > 20$).

3.4 TWO-DIMENSIONAL HEAT TRANSFER IN LONGITUDINAL RECTANGULAR AND CONVEX PARABOLIC FINS

We consider a one-dimensional longitudinal fin of an arbitrary profile $F(X)$ and cross-sectional area A_c is shown in Fig. 3.16. The perimeter of the fin is denoted by P and its length by L. The fin is attached to a fixed prime surface of temperature T_b and extends to an ambient fluid of temperature T_a. The fin thickness is given by d and the base thickness is δ_b. We assume that the fin is initially at ambient temperature. At time $t = 0$, the temperature at the base of the fin is suddenly changed from T_a to T_b, and the problem is to establish the temperature distribution in the fin for all $t \geq 0$. Based on the one-dimensional heat conduction, the energy balance equation is then given by (see e.g., Ref. [4])

$$\rho c \frac{\partial T}{\partial t} = \frac{\partial}{\partial x}\left(\frac{\partial_b}{2}F(x)K(T)\frac{\partial T}{\partial X}\right) - \frac{P}{A_C}H(T)(T - T_a), 0 \leq X \leq L \tag{3.41}$$

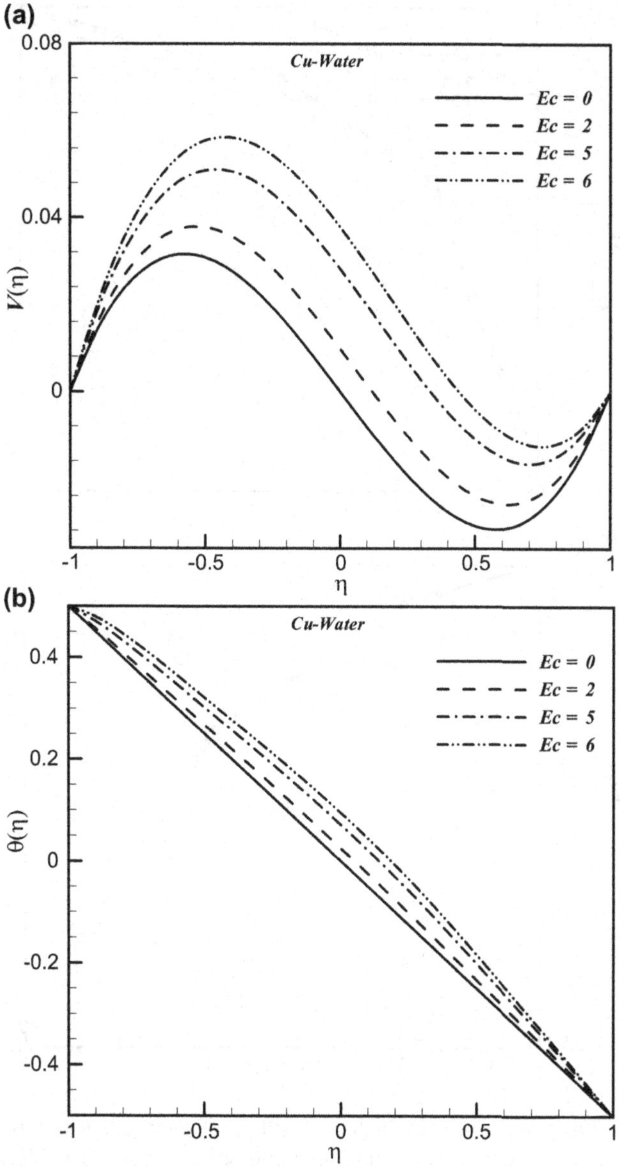

Figure 3.14 Effect of Eckert number (Ec) on: (a) velocity profiles $V(\eta)$; (b) temperature profiles $\theta(\eta)$, when $\phi = 0.1$, $\delta = 1$, and $Pr = 6.2$.

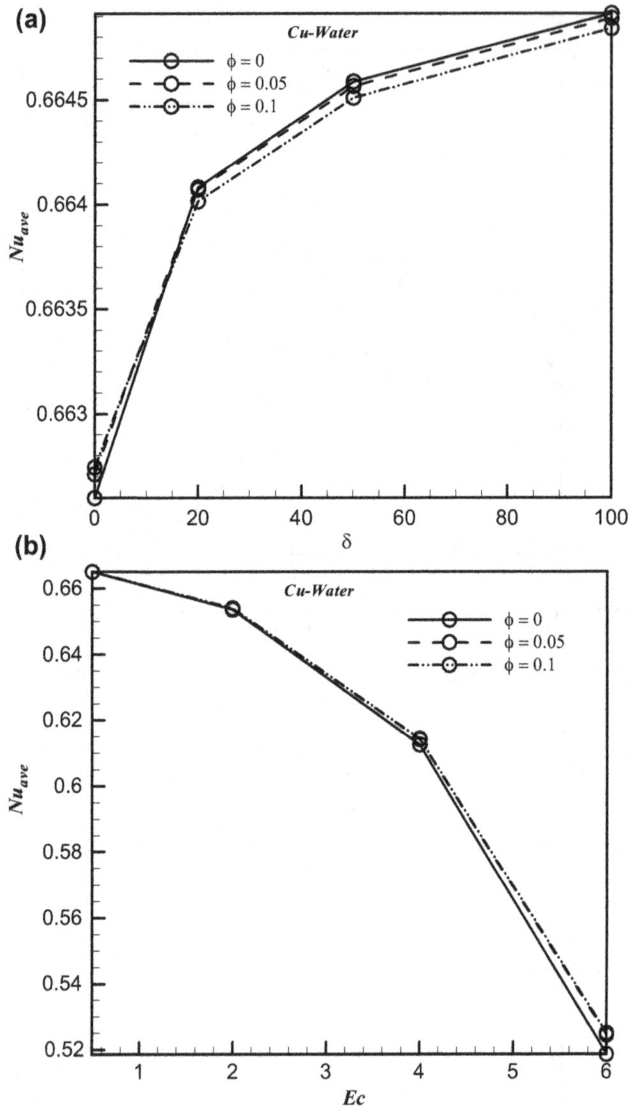

Figure 3.15 Effects of nanoparticle volume fraction (ϕ), dimensionless non-Newtonian viscosity (δ), and Eckert number (Ec) on Nusselt number, when (a) $Ec = 1$; (b) $\delta = 1$ and $Pr = 6.2$.

where K and H are the nonuniform thermal conductivity and heat transfer coefficients depending on the temperature (see e.g., Ref. [4]), q is the density, c is the specific heat capacity, T is the temperature distribution, t is the time, and X is the space variable. Assuming that the fin tip is adiabatic

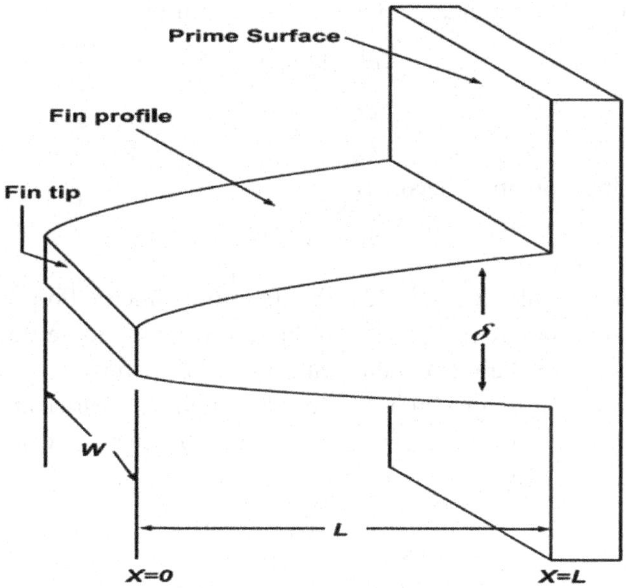

Figure 3.16 Schematic representation of a longitudinal fin of an arbitrary profile.

(insulated) and the base temperature is kept constant, then the boundary conditions are given by

$$T(t, L) = T_b \quad \text{and} \quad \left.\frac{\partial T}{\partial X}\right|_{X=0} = 0 \tag{3.42}$$

Initially the fin is kept at the ambient temperature,

$$T(0, X) = T_a \tag{3.43}$$

Introducing the following dimensionless variables:

$$x = \frac{X}{L}, \ \tau = \frac{K_a t}{\rho c_v L^2}, \ \theta = \frac{T - T_a}{T_b - T_a}, \ h = \frac{H}{k_b}, \ K = \frac{K}{K_a}, \\ M^2 = \frac{P h_b L^2}{A_C K_a}, \ f(x) = \frac{\delta_b}{2} F(X) \tag{3.44}$$

reduces Eq. (3.41) into

$$\frac{\partial \theta}{\partial \tau} = \frac{\partial}{\partial X}\left(f(x) k(\theta) \frac{\partial \theta}{\partial x} \right) - M^2 h(\theta)\theta, 0 \le x \le 1 \tag{3.45}$$

The prescribed boundary conditions are given by

$$\theta(\tau, 1) = 1 \tag{3.46}$$

$$\frac{\partial \theta}{\partial x}\Big|_{x=0} = 0 \tag{3.47}$$

and the initial condition becomes

$$\theta(0, x) = 0 \tag{3.48}$$

The dimensionless variable M is the thermo-geometric fin parameter, h is the dimensionless temperature, x is the dimensionless space variable, k is the dimensionless thermal conductivity, k_a is the thermal conductivity of the fin at ambient temperature, h_b is the heat transfer coefficient at the fin base. For most industrial application, the heat transfer coefficient maybe given as a power law [4].

$$H(T) = h_b\left(\frac{T - T_a}{T_b - T_a}\right)^n \tag{3.49}$$

where n and h_b are constants. The constant n may vary between −6.6 and 5. However, in most practical applications it lies between −3 and 3. The exponent n represents laminar film boiling or condensation when $n = 1/4$, laminar natural convection when $n = 1/4$, turbulent natural convection when $n = 1/3$, nucleate boiling when $n = 2$, radiation when $n = 3$, and $n = 0$ implies a constant heat transfer coefficient. Exact solutions may be constructed for the steady-state one-dimensional differential equation describing temperature distribution in a straight fin when the thermal conductivity is a constant and the exponent of the heat transfer coefficient is given by $n = -1$; 0; 1 or 2. Furthermore, exact solution for the steady state may be constructed when thermal conductivity is a differential consequence of the term involving heat transfer coefficient (see e.g., Ref. [4]), that is if the nonlinear ordinary differential equation is linearizable.

In dimensionless variables we have $h(\theta) = \theta^n$. Also, for many engineering applications, the thermal conductivity may depend linearly on temperature, that is

$$K(T) = K_a[1 + \gamma(T - T_a)] \tag{3.50}$$

The dimensionless thermal conductivity given by the linear function of temperature is $k(\theta) = 1 + \beta\theta$, where the thermal conductivity gradient is $\beta = \mu(T_b - T_a)$. As such the governing equation is given by

$$\frac{\partial \theta}{\partial \tau} = \frac{\partial}{\partial x}\left(f(x)(1 + \beta\theta)\frac{\partial \theta}{\partial x}\right) - M^2\theta^{n+1} \tag{3.51}$$

We will employ the two-dimensional DTM to construct the series solutions to Eq. (3.51), subject to the initial and boundary conditions Eqs. (3.46)–(3.48). A comparison on temperature distribution in longitudinal rectangular and convex parabolic fins is established.

A. Heat Transfer in Fins of Rectangular Profile

Here the governing equation becomes

$$\frac{\partial \theta}{\partial \tau} = \frac{\partial}{\partial x}\left((1+\beta\theta)\frac{\partial \theta}{\partial x}\right) - M^2\theta^{N+1} \tag{3.52}$$

Taking the two-dimensional differential transform of Eq. (3.51) for $n = 1$, we obtain the following recurrence relation,

$$\begin{aligned}(K+1)\Theta(K+1+H) &= (H+1)(H+2)\Theta(K,H+1)\Theta(K,H+2)\\ &+\beta\sum_{I=0}^{K}\sum_{J=0}^{H}\Theta(K-I,J)(H+1-J)(H+2-J)\Theta(I,H+2-J)\\ &+\beta\sum_{I=0}^{K}\sum_{J=0}^{H}(J+1)\Theta(K-I,J+1)(H+1-J)\Theta(I,H+1-J)\\ &-M^2\sum_{I=0}^{K}\sum_{J=0}^{H}\Theta(I,H-J)\Theta(K-I,J)\end{aligned} \tag{3.53}$$

where $\Theta(K,H)$ is the differential transform of $\theta(\tau,X)$. Taking the two-dimensional differential transform of the initial condition Eq. (3.48) and boundary condition Eq. (3.47) we obtain the following transformations, respectively,

$$\Theta(0,H) = 0, H = 0,1,2,.. \tag{3.54}$$

$$\Theta(K,1) = 0, K = 0,1,2,\ldots \tag{3.55}$$

We consider the other boundary condition as follows:

$$\Theta(K,1) = a,\ a\in R,\ k = 1,2,3 \tag{3.56}$$

where the constant a can be determined from the boundary Eq. (3.46) at each time step after obtaining the series solution. Substituting Eqs. (3.54)–(3.56) into (3.53) we obtain the following,

$$\Theta(1,2) = a \tag{3.57}$$

$$\Theta(2,2) = 1/2(3a - 2\beta a^2 + a^2M^2) \tag{3.58}$$

$$\Theta(3,2) = 1/2(4a - 5\beta a^2 + 2\beta^2 a^3 + 2a^2 m^2 - \beta a^3 m^2 \tag{3.59}$$

$$\Theta(1,4) = 1/12(3a - 2\beta a^2 + a^2 m^2) \tag{3.60}$$

$$\Theta(2,4) = 1/24(12a - 33\beta a^2 + 10\beta^2 a^3 - 5\beta a^3 m^2 \tag{3.61}$$

Substituting Eqs. (3.57)–(3.61) into Eq. (3.53), we obtain the following infinite series solution given by

$$\begin{aligned}\theta(\tau, x) &= a\tau + a\tau^2 + a\tau x^2 + 1/2(3a - 2\beta a^2 + a^2 m^2)\tau^2 x^2 + a\tau^3 \\ &+ 1/2\left(4a - 5\beta a^2 + 2\beta^2 a^3 + 2a^2 m^2 - \beta a^3 m^2\right)\tau^3 x^2 \\ &+ a\tau^4 + 1/12(3a - 2\beta a^2 + a^2 m^2)\tau x^4 \\ &+ 1/24\left(12a - 33\beta a^2 + 10\beta^2 a^3 + 10a^2 m^2 - 5\beta a^3 m^2\right)\tau^2 x^4\end{aligned} \tag{3.62}$$

The constant a can be determined from the boundary condition (3.46) at each time step. To obtain the value of a, we substitute the boundary condition Eq. (3.46) into Eq. (3.62) at the point $x = 1$. Thus, we have,

$$\begin{aligned}\theta(\tau, 1) &= a\tau + a\tau^2 + a\tau + 1/2(3a - 2\beta a^2 + a^2 m^2)\tau^2 + a\tau^3 \\ &+ 1/2\left(4a - 5\beta a^2 + 2\beta^2 a^3 + 2a^2 m^2 - \beta a^3 m^2\right)\tau^3 \\ &+ a\tau^4 + 1/12(3a - 2\beta a^2 + a^2 m^2)\tau \\ &+ 1/24\left(12a - 33\beta a^2 + 10\beta^2 a^3 + 10a^2 m^2 - 5\beta a^3 m^2\right)\tau^2\end{aligned} \tag{3.63}$$

We then obtain the expression for $h(x)$ upon substituting the obtained value of a into Eq. (3.62). Using the first 40 terms of the power series solution, we plot the solution Eq. (3.62) for special parameters as shown in Fig. 3.17.

B. Heat Transfer in Fins of Convex Parabolic Profile

In this case, the governing equation becomes

$$\frac{\partial \theta}{\partial \tau} = \frac{\partial}{\partial x}\left(x^{1/2}(1 + \beta\theta)\frac{\partial \theta}{\partial x}\right) - m^2\theta^{n+1} \tag{3.64}$$

The transformation $y = x^{1/2}$ reduces Eq. (3.64) into

$$4y\frac{\partial \theta}{\partial \tau} = \frac{\partial}{\partial y}\left((1 + \beta\theta)\frac{\partial \theta}{\partial y}\right) - 4ym^2\theta^{n+1} \tag{3.65}$$

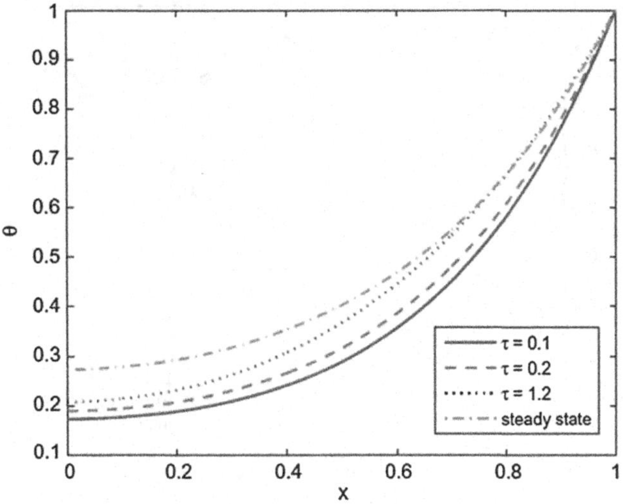

Figure 3.17 Transient temperature distribution in a longitudinal rectangular fin. Here $n = 1$; $\beta = 1$; $M = 6$.

Eq. (3.65) is more susceptible to DTM analysis. Following the techniques above we construct series solutions. When $n = 2$, the solution will be

$$\theta(\tau, x) = a\tau + a\tau^2 + \frac{4a}{3}\tau x^{3/2} - 2a/3(-3 + 2a\beta - am^2)\tau^2 x^{3/2} + a\tau^3 + 2a/3\left(4 - 5a\beta + 2a^2\beta^2 + 2am^2 - a^2\beta m^2\right)\tau^3 x^{3/2} + a\tau^4 + \ldots \quad (3.66)$$

Which is depicted in Fig. 3.18.

3.5 THERMAL BOUNDARY LAYER ON FLAT PLATE

Consider the flow of a viscous fluid over a semiinfinite flat plate, as shown in Fig. 3.19. The temperature of the wall, T_w, is uniform and constant and is greater than the free stream temperature, T_∞. It is assumed that the free stream velocity, U_∞, is also uniform and constant.

Further, assuming that the flow in the laminar boundary layer is two-dimensional, and that the temperature changes resulting from viscous dissipation are small, the continuity equation and the boundary layer equations may be expressed as [5]

$$\frac{\delta u}{\delta x} + \frac{\delta v}{\delta x} = 0, \quad (3.67)$$

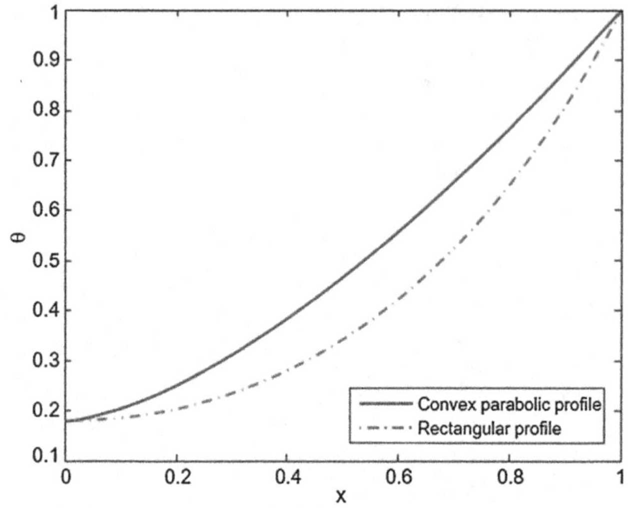

Figure 3.18 Transient temperature distribution in a rectangular and convex parabolic fin with linear thermal conductivity, $n = 2$; $M = 6$; $\beta = 1$; $\tau = 1.2$.

$$u\frac{\delta u}{\delta x} + v\frac{\delta u}{\delta y} = v\frac{\partial^2 u}{\partial y^2}, \tag{3.68}$$

$$u\frac{\delta t}{\delta x} + v\frac{\delta t}{\delta y} = \alpha\frac{\partial^2 t}{\partial y^2}, \tag{3.69}$$

where u and v are the velocity components in x- and y-direction of the fluid, a is the thermal diffusivity of the fluid, T is the temperature distribution in the vicinity of the plate, and the boundary conditions are given by

$$\text{at} \quad y = 0: \quad u = v = 0 \quad \text{and} \quad t = t_w \tag{3.70}$$

$$\text{at} \quad y \to \infty: \quad u \to u_\infty \quad \text{and} \quad t = t_\infty \tag{3.71}$$

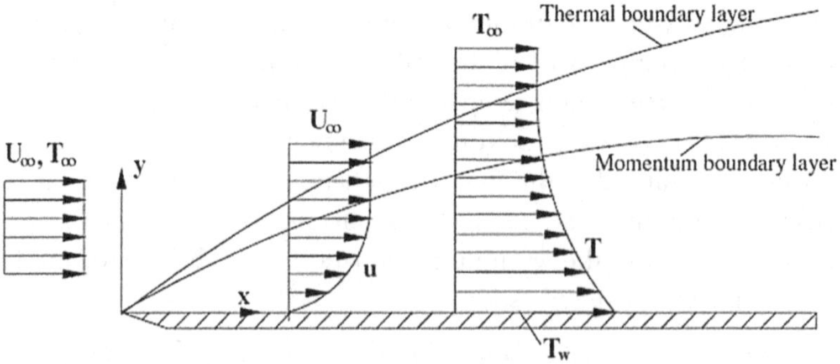

Figure 3.19 Velocity and thermal boundary layers on a flat plate.

$$\text{at} \quad x \to 0: \quad u = u_\infty \quad \text{and} \quad t = t_\infty. \tag{3.72}$$

A stream function, $\psi(x, y)$, is introduced such that

$$u = \frac{d\psi}{dy} \quad \text{and} \quad v = -\frac{d\psi}{dx}. \tag{3.73}$$

In addition to the physical considerations which require the introduction of this function, the mathematical significance of its use is that the equation of continuity, i.e., Eq. (3.67), is satisfied identically, and the momentum equation becomes

$$\frac{\delta\psi}{\delta y}\frac{\delta^2\psi}{\partial x\delta y} - \frac{\delta\psi}{\delta x}\frac{\delta^2\psi}{\delta y^2} = \nu\frac{\delta^3\psi}{\delta y^3} \tag{3.74}$$

Integrating Eq. (3.73) and introducing a similarity variable yield

$$f(\eta) = \frac{\psi}{\sqrt{u_\infty \nu x}}, \tag{3.75}$$

$$\eta = \sqrt{\frac{u_\infty}{\nu x}}. \tag{3.76}$$

Substituting Eqs. (3.68) and (3.69) into Eq. (3.67) gives

$$\frac{d^3f(\eta)}{d\eta^3} + \frac{1}{2}f(\eta)\frac{d^2f(\eta)}{d\eta^2} = 0. \tag{3.77}$$

The boundary conditions of $f(\eta)$ are given by

$$\text{at} \quad \eta = 0 \quad f(0) = \frac{df(0)}{d\eta} = 0. \tag{3.78}$$

$$\text{at} \quad \eta \to \infty: \frac{df(\infty)}{d\mu} = 1. \tag{3.79}$$

A dimensionless temperature parameter is defined as follows:

$$\theta = \frac{t - t_w}{t_\infty - t_w}. \tag{3.80}$$

If Eq. (3.73) is substituted into Eq. (3.69), the boundary layer energy equation then becomes

$$\frac{d^2\theta(\eta)}{d\eta^2}+\frac{1}{2}Prf(\eta)\frac{d\theta(\eta)}{d\eta}=0, \tag{3.81}$$

with the following boundary conditions:

$$\text{at}\quad \eta=0: \quad \theta=0, \tag{3.82}$$

$$\text{at}\quad \eta=0\rightarrow\infty: \quad \theta=1, \tag{3.83}$$

where Pr is the Prandtl number, which is equal to the ratio of the momentum diffusivity of the fluid to its thermal diffusivity (i.e., $Pr=\frac{\nu}{\alpha}$). The boundary value problems (Eqs. 3.70–3.72) can be reduced to a pair of initial value problems by means of a group of transformations [5]. The initial value problems are given by

$$\frac{d^2\theta(\eta)}{d\eta^2}+\frac{1}{2}f(\zeta)\frac{d^2f(\zeta)}{d\zeta^2}=0, \tag{3.84}$$

With initial conditions of

$$\zeta=0: \quad f(0)=\frac{df(0)}{d\zeta}=0, \quad \frac{d^2f(0)}{d^2\zeta}=1, \tag{3.85}$$

and by,

$$\frac{d^3f(\eta)}{d\eta^3}+\frac{1}{2}f(\eta)\frac{d^2f(\eta)}{d\eta^2}=0, \tag{3.86}$$

with initial conditions of

$$\eta=0: \quad f(0)=\frac{df(0)}{d\eta}=0, \quad \frac{d^2f(0)}{d\eta^2}=\left[\frac{1}{\frac{df(\infty)}{d\zeta}}\right]^{\frac{3}{2}} \tag{3.87}$$

These equations suggest a transformation of the form:

$$f(\zeta)=\lambda^{\frac{-1}{3}}f(\eta), \quad \zeta=\lambda^{\frac{1}{3}}\eta, \quad \lambda=\left[\frac{1}{\frac{df(\infty)}{d\zeta}}\right]^{\frac{3}{2}}. \tag{3.88}$$

The DTM is then used to solve the pair of initial value problems (Eqs. 3.84–3.87). The following expression is initially defined:

$$\gamma(\zeta)=\frac{df(\zeta)}{d\zeta}, \tag{3.89}$$

and

$$z(\zeta) = \frac{dy(\zeta)}{d\zeta} = \frac{d^2f(\zeta)}{d\zeta^2} = 0 \tag{3.90}$$

Thereafter, the third-order ordinary differential equation (Eq. 3.86) is reduced to a first-order ordinary differential equation with the following form:

$$\frac{dz(\zeta)}{d\zeta} + \frac{1}{2}f(\zeta)z(\zeta) = 0. \tag{3.91}$$

The initial conditions become

$$\zeta = 0: \quad f(0) = y(0) = 0, \ \ z(0) = 1. \tag{3.92}$$

Eqs. (3.89)–(3.91) undergo the differential transformation to give the following:

$$\frac{k+1}{h_i}\overline{F}_i(k+1) = Y_i(k), \tag{3.93}$$

$$\frac{k+1}{h_i}\overline{Y}_i(k+1) = Z_i(k), \tag{3.94}$$

$$\frac{k+1}{h_i}\overline{Z}_i(k+1) + \frac{1}{2}\overline{F}_i \times Z_i(k+1) + \frac{1}{2}\sum_{i=0}^{k}\overline{F}_i(k-l)Z_i(l) = 0. \tag{3.95}$$

The third-order ordinary differential equation (i.e., Eq. 3.86) becomes a first-order ordinary differential equation with the following form [5]:

$$\frac{dv(\eta)}{d\mu} + \frac{1}{2}f(\eta)v(\eta) = 0 \tag{3.96}$$

The initial conditions become

$$\eta = 0: \quad f(0) = u(0) = 0, \quad v(0) = \lambda^{\frac{-3}{2}}. \tag{3.97}$$

More details about the solution are presented in Ref. [5].

Since the solutions of Eq. (3.70) can be established from the previous calculations, $f(\eta)$ is also known and can be substituted into Eq. (3.74) to solve the boundary layer energy equation. Eq. (3.74) is a linear second-order ordinary differential equation with variable coefficients. The solution of this energy equation can be obtained by using the method of superposition. The following relationship is established.

$$\theta(\eta) = c(\eta) + sD(\eta). \tag{3.98}$$

Substituting Eq. (3.98) into Eq. (3.74) and separating the resulting equations into a group of terms give two initial value problems, i.e.,

$$\frac{D^2C(\eta)}{D\eta^2} + \frac{1}{2}Prf(\eta)\frac{dc(\eta)}{d\eta} = 0, \tag{3.99}$$

with initial conditions of

$$\eta = 0: \quad c(0) = 0, \quad \frac{dc(0)}{d\eta} = 1, \tag{3.100}$$

and

$$\frac{d^2D(\eta)}{d\eta^2} + \frac{1}{2}Prf(\eta)\frac{dD(\eta)}{d\eta} = 0, \tag{3.101}$$

with initial conditions of

$$\eta = 0: \quad D(0) = 0, \quad \frac{dD(0)}{d\eta} = -1. \tag{3.102}$$

Substituting Eqs. (3.100) and (3.102) into Eq. (3.98) gives

$$\frac{d\theta(0)}{d\eta} = 1 - s. \tag{3.103}$$

The parameter "*s*" in Eq. (3.98) can be calculated by using the boundary condition given in Eq. (3.74). This yields

$$s = \frac{1 - c(\infty)}{D(\infty)}. \tag{3.104}$$

The solutions of the pair of linear second-order ordinary differential equations (Eqs. 3.99–3.102) can be obtained from the DTM. Initially, the following relationship is defined:

$$w(\eta) = \frac{dc(\eta)}{d\eta}. \tag{3.105}$$

Substituting Eq. (3.105) into Eq. (3.99) gives

$$\frac{dw(\eta)}{d\eta} + \frac{1}{2}Prf(\eta)w(\eta) = 0. \tag{3.106}$$

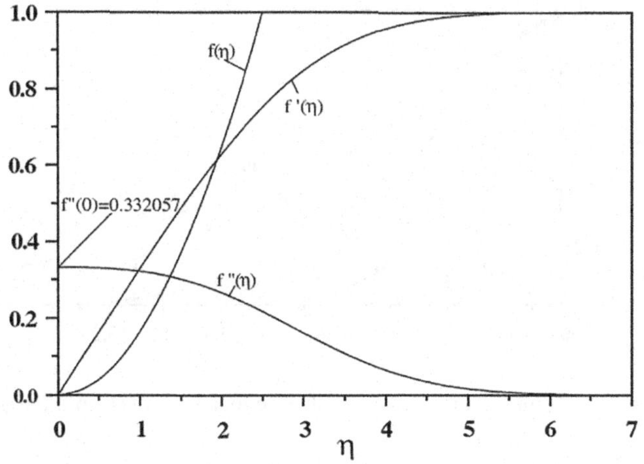

Figure 3.20 Solutions of Eq. (3.70) for function $f(\eta)$ and its derivatives.

The initial conditions become

$$\eta = 0: \quad c(0) = 0, \quad w(0) = 1. \tag{3.107}$$

By applying DTM, Eqs. (3.105) and (3.106) undergo the differential transformation to give the following:

$$\frac{k+1}{h}\overline{C_i}(k+1) = W_i(k), \tag{3.108}$$

$$\begin{aligned} &\frac{k+1}{h_i}\overline{C_i}(k+1) + \frac{1}{2}\overline{f_i}(k) \times W_i(k) \\ &= \frac{k+1}{h_i}W_i(k+1) + \frac{1}{2}\sum_{i=0}^{m}\overline{f_i}(k-l)W_i(l) = 0. \end{aligned} \tag{3.109}$$

More details of solutions and boundary conditions can be found in Ref. [5].

The variation of the values of $f(\eta)$ and its derivatives are plotted in Fig. 3.20. Since $f(\eta)$ is known, the boundary layer energy equation (Eq. 3.74) can be solved numerically for various values of Prandtl number. The temperature distributions in the thermal boundary layer over the flat plate are shown in Fig. 3.21 for a range of Prandtl numbers.

3.6 FALKNER–SKAN WEDGE FLOW

Consider the flow of an incompressible viscous fluid over a wedge, as shown in Fig. 3.22. The temperature of the wall, T_w, is uniform and

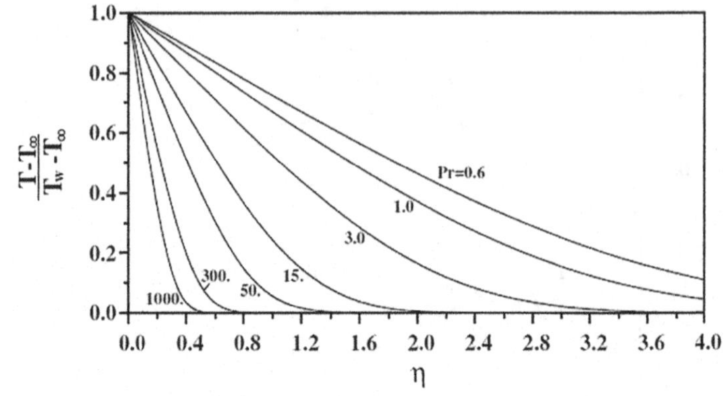

Figure 3.21 Temperature profiles in the laminar boundary on a flat plate.

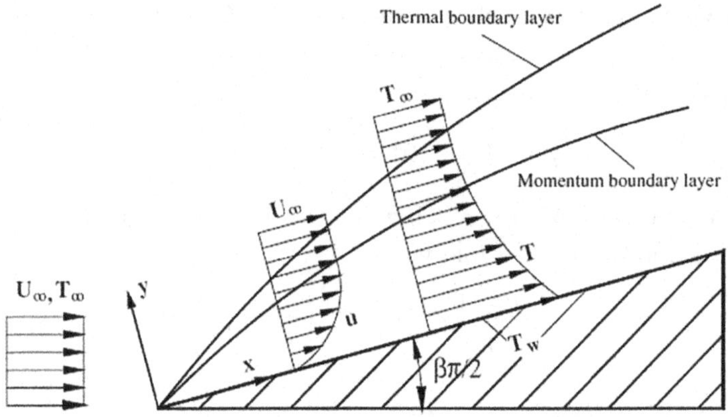

Figure 3.22 Velocity and thermal boundary layers for the Falkner–Skan wedge flow.

constant and is greater than the free stream temperature, T_1. It is assumed that the free stream velocity, U_1, is also uniform and constant. Further, assuming that the flow in the laminar boundary layer is two-dimensional, and that the temperature changes resulting from viscous dissipation are small, the continuity equation and the boundary layer equations may be expressed as [6]

$$\frac{\partial u}{\partial x}+\frac{\partial v}{\partial y}=0, \tag{3.110}$$

$$u\frac{\partial u}{\partial x}+v\frac{\partial u}{\partial y}=U\frac{dU}{dx}+\upsilon\frac{\partial^2 u}{\partial y^2}, \tag{3.111}$$

$$u\frac{\partial T}{\partial x}+\upsilon\frac{\partial T}{\partial y}=+\alpha\frac{\partial^2 T}{\partial y^2},\tag{3.112}$$

where u and v are the respective velocity components in the x- and y-direction of the fluid flow, m is the viscosity of the fluid, and U is the reference velocity at the edge of the boundary layer and is a function of x. a is the thermal diffusivity of the fluid, T is the temperature in the vicinity of the wedge, and the boundary conditions are given by

$$\text{at } y=0: u=\upsilon=0, \quad \text{and} \quad T=T_w \tag{3.113}$$

$$\text{at } \quad y\rightarrow\infty : u\rightarrow U(x)=U_\infty(x/L)^m, \quad \text{and } T=T_\infty, \tag{3.114}$$

$$\text{at } \quad x=0: u=U_\infty \quad \text{and} \quad T=T_\infty, \tag{3.115}$$

where U_∞ is the mean stream velocity, L is the length of the wedge, m is the Falkner–Skan power-law parameter, and x is measured from the tip of the wedge. A stream function, $\psi(x,y)$, is introduced such that

$$u=\frac{\partial\psi}{\partial y} \quad \text{and} \quad \upsilon=-\frac{\partial\psi}{\partial x}. \tag{3.116}$$

In addition to the physical considerations which require the introduction of this function, the mathematical significance of its use is that the equation of continuity, i.e., Eq. (3.110), is satisfied identically. The momentum equation becomes

$$\frac{\partial\psi}{\partial y}\frac{\partial^2\psi}{\partial x\partial y}-\frac{\partial\psi}{\partial x}\frac{\partial^2\psi}{\partial y^2}=U\frac{dU}{dx}+\upsilon\frac{\partial^3\psi}{\partial y^3}. \tag{3.117}$$

Integrating Eq. (3.116) and introducing a similarity variable yield

$$f(\eta)=\sqrt{\frac{1+m}{2}\frac{L^m}{\upsilon U_\infty}}\left(\psi/x^{(1+m)/2}\right), \tag{3.118}$$

$$f(\eta)=\sqrt{\frac{1+m}{2}\frac{U_\infty}{\upsilon L^m}}\left(y/x^{(1-m)/2}\right) \tag{3.119}$$

Substituting Eqs. (3.10) and (3.119) into Eq. (3.117) gives

$$\frac{d^3f(\eta)}{d\eta^3}+f(\eta)\frac{d^2f(\eta)}{d\eta^2}+\beta\left[1-\left(\frac{df(\eta)}{d\eta}\right)^2\right]=0, \tag{3.120}$$

which is known as the Falkner–Skan boundary layer equation [6]. The boundary conditions of $f(\eta)$ are given by

$$\text{at} \quad \eta = 0: f(0) = \frac{df(0)}{d\eta} = 0, \tag{3.121}$$

$$\text{at} \quad \eta \to \infty \quad \frac{df(\infty)}{d\eta} = 1. \tag{3.122}$$

Note that in the equations above, parameters β and m are related through the expression $\beta = 2\,m/(1+m)$. A dimensionless temperature is defined as follows:

$$\theta = \frac{T - T_\infty}{T_\infty - T_w} \tag{3.123}$$

If Eq. (3.123) is substituted into Eq. (3.112), the boundary layer energy equation then becomes

$$\frac{d^2\theta(\eta)}{d\eta^2} + Prf(\eta, \beta)\frac{d\theta(\eta)}{d\eta} = 0 \tag{3.124}$$

with the following boundary conditions:

$$\text{at} \quad \eta = 0: \theta = 0 \tag{3.125}$$

$$\text{at} \quad \eta \to \infty : \theta = 1 \tag{3.126}$$

where Pr is the Prandtl number, which is equal to the ratio of the momentum diffusivity of the fluid to its thermal diffusivity (i.e., $Pr = \frac{\upsilon}{\alpha}$). Eqs. (3.120) and (3.124) present a system of ordinary differential equations for the Falkner–Skan boundary layer problem. Simultaneous solution of these two equations yields the velocity and temperature profiles for the flow of a viscous fluid passing a wedge. To solve the Falkner–Skan boundary layer equation for a family of values of β, it is first necessary to define a dependent variable, $g(\eta)$, i.e.,

$$g(\eta) = \frac{\partial f(\eta)}{\partial \beta}. \tag{3.127}$$

Differentiating Eqs. (3.120)–(3.122) with respect to β gives

$$\begin{aligned} &\frac{d^3g(\eta)}{d\eta^3} + f(\eta)\frac{d^2g(\eta)}{d\eta^2} + g(\eta)\frac{d^2g(\eta)}{d\eta^2} + \left[1 - \left(\frac{df(\eta)}{d\eta}\right)^2\right] \\ &- 2\beta\frac{df(\eta)}{d\eta}\frac{dg(\eta)}{d\eta} = 0, \end{aligned} \tag{3.128}$$

The boundary conditions are given by

$$g(0) = \frac{dg(0)}{d\eta} = 0, \frac{dg(\infty)}{d\eta} = 0 \tag{3.129}$$

The method of superposition is used together with a group of transformations to solve the boundary layer equation given in Eq. (3.128).

$$g(\eta) = P(\eta) + C_1 Q(\eta), \tag{3.130}$$

where C_1 is a constant to be determined. Substituting Eq. (3.130) into Eq. (3.128) gives the following pair of initial value problems:

$$\frac{d^3P(\eta)}{d\eta^3} + f(\eta)\frac{d^2P(\eta)}{d\eta^2} + P(\eta)\frac{d^2g(\eta)}{d\eta^2} + -2\beta\frac{df(\eta)}{d\eta}\frac{dP(\eta)}{d\eta} = \left(\frac{df(\eta)}{d\eta}\right)^2 - 1, \tag{3.131}$$

with initial conditions of

$$P(0) = \frac{dP(0)}{d\eta} = \frac{d^2P(\eta)}{d\eta^2} = 0, \tag{3.132}$$

and

$$\frac{d^3Q(\eta)}{d\eta^3} + f(\eta)\frac{d^2Q(\eta)}{d\eta^2} + Q(\eta)\frac{d^2f(\eta)}{d\eta^2} + -2\beta\frac{df(\eta)}{d\eta}\frac{dQ(\eta)}{d\eta} = 0, \tag{3.133}$$

with initial conditions of

$$Q(0) = \frac{dQ(\infty)}{d\eta} = 0, \quad \frac{d^2Q(\eta)}{d\eta^2} = 1. \tag{3.134}$$

Details of this method can be found in Ref. [6]. The solution of this energy equation can be obtained by using the method of superposition. The following relationship is established:

$$\theta(\eta) = C(\eta) + C_2 D(\eta). \tag{3.135}$$

Substituting Eq. (3.135) into Eq. (3.124), and separating the resulting equations into a group of terms gives two initial value problems, i.e.,

$$\frac{d^2C(\eta)}{d\eta^2} + Prf(\eta\beta)\frac{dC(\eta)}{d\eta} = 0, \tag{3.136}$$

with initial conditions of

$$\eta = 0: \quad C(0) = 0, \quad \frac{dC(\eta)}{d\eta} = 1 \tag{3.137}$$

and

$$\frac{d^2D(\eta)}{d\eta^2} + Prf(\eta\beta)\frac{dD(\eta)}{d\eta} = 0, \tag{3.138}$$

with initial conditions of

$$\eta = 0: \quad D(0) = 0, \quad \frac{dD(0)}{d\eta} = -1. \tag{3.139}$$

Substituting Eqs. (3.137) and (3.139) into Eq. (3.135) gives

$$\frac{d\theta(0)}{d\eta} = 1 - C_2. \tag{3.140}$$

The parameter "C_2" in Eq. (3.135) can be calculated by using the boundary condition given in Eq. (3.126). This yields

$$C_2 = \frac{1 - C(\infty)}{D(\infty)}. \tag{3.141}$$

By solving Eqs. (3.136)–(3.139) then gives $C(\eta)$, $D(\eta)$, and their derivatives. The value of C_2, the values of $\theta(\eta)$ are derived from Eq. (3.135). Hence, we have been determined the solutions of the Falkner–Skan wedge flow.

The DTM is then used to solve the pair of initial value problems (when $\beta = 0$). Initially, the following expressions are defined [6]:

$$y(\zeta) = \frac{dF(\zeta)}{d\zeta}, \tag{3.142}$$

and

$$z(\zeta) = \frac{dy(\zeta)}{d\zeta} = \frac{d^2F(\zeta)}{d\zeta^2}. \tag{3.143}$$

Thereafter, the third-order ordinary differential equation is reduced to a first-order ordinary differential equation with the following form:

$$\frac{dz(\zeta)}{d\zeta} + F(\zeta)z(\zeta) = 0. \tag{3.144}$$

The initial conditions become

$$\zeta = 0: \quad F(0) = y(0) = 0, \quad z(0) = 1. \tag{3.145}$$

By a process of inverse differential transformation, performing differential transformation of Eqs. (3.142)−(3.144) gives the following:

$$\frac{k+1}{H_i}\overline{F}_i(k+1) = Y_i(k), \tag{3.146}$$

$$\frac{k+1}{H_i}Y_i(k+1) = Z_i(k), \tag{3.147}$$

$$\begin{aligned} &\frac{k+1}{H_i}Z_i(k+1) + \overline{F}_i(k) \times Z_i(k) \\ &= \frac{k+1}{H_i}Z_i(k+1) + \sum_{l=1}^{k}(k-1)Z_i(L) = 0 \end{aligned} \tag{3.148}$$

By changing the third-order ordinary differential equation into a first-order ordinary differential equation the solution can be found [6]. By the current method, solution of other governing equation (energy) is obtained and presented [6]. Fig. 3.23 plots the variation in the values of $f(\eta)$ and its derivatives for various values of β. Fig. 3.24 plots the dimensionless temperature distributions of the Falkner–Skan boundary layer problem for the Prandtl number range of 0.001–10,000.

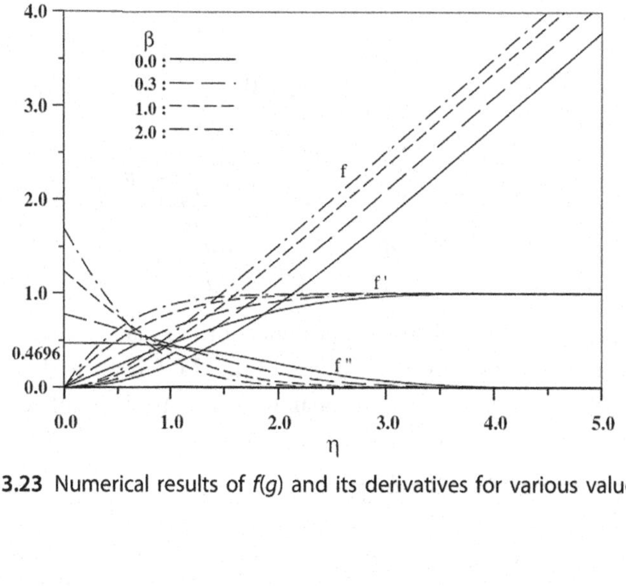

Figure 3.23 Numerical results of *f*(*g*) and its derivatives for various values of *b*.

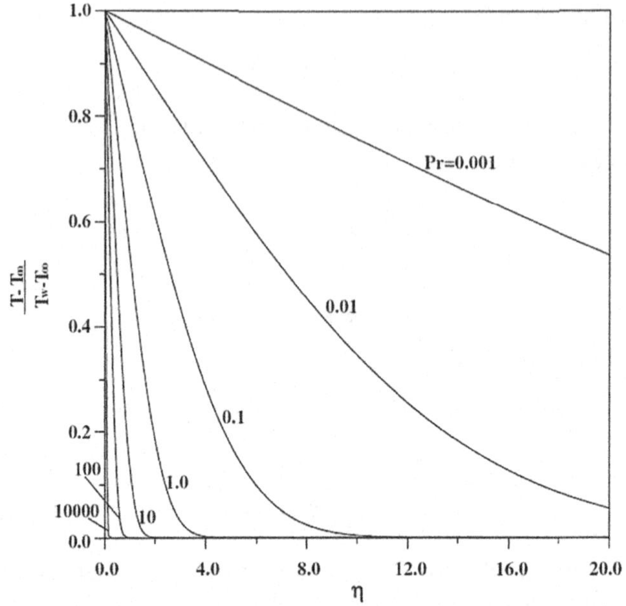

Figure 3.24 Dimensionless temperature profiles for $\beta = 0$ and various Prandtl number.

3.7 FREE CONVECTION PROBLEM

Consider the flow of an incompressible viscous fluid over a vertical plate. Assuming that the flow in the laminar boundary layer is two-dimensional, the continuity equation and the boundary layer equations may be expressed as [7],

$$\frac{\partial u}{\partial x} + \frac{\partial v}{\partial y} = 0 \tag{3.149}$$

$$u\frac{\partial u}{\partial x} + v\frac{\partial u}{\partial y} = \nu\frac{\partial^2 u}{\partial y^2} + g\frac{T_w - T_\infty}{T_\infty}\theta \tag{3.150}$$

$$u\frac{\partial \theta}{\partial x} + v\frac{\partial \theta}{\partial y} = \alpha\frac{\partial^2 \theta}{\partial y^2} \tag{3.151}$$

where u and v are the velocity components in the x- and y-direction of the fluid flow, respectively, ν is the viscosity of the fluid, α is the thermal diffusivity of the fluid, g is the gravitational acceleration of the plate, T_w is the temperature of the wall and is assumed to be uniform and constant, and T_∞ is the free stream temperature (Fig. 3.25).

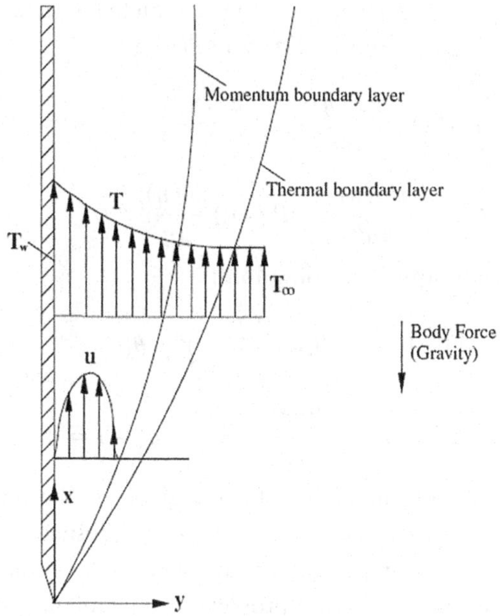

Figure 3.25 Velocity and temperature profiles in free convection flow over a vertical plate.

A dimensionless temperature can be defined as follows:

$$\theta = \frac{T - T_w}{T_\infty - T_w} \tag{3.152}$$

where T is the temperature distribution in the vicinity of the plate, and the boundary conditions are given by

$$\text{at} \quad y = 0: u = v = 0, \quad \text{and} \quad \theta = 1 \tag{3.153}$$

$$\text{at} \quad y \to \infty: u = 0, \quad \text{and} \quad \theta = 0 \tag{3.154}$$

A stream function, $\Psi(x,y)$, is introduced such that

$$u = \frac{\partial \psi}{\partial y} \quad \text{and} \quad v = -\frac{\partial \psi}{\partial x} \tag{3.155}$$

Introducing the similarity variables yields

$$\eta = Cy/x^{1/4}, \psi = 4\nu C x^{3/4} F(\eta), \quad C = \left[g(T_w - T_\infty)/4\nu^2 T_\infty\right]^{1/4} \tag{3.156}$$

Eqs. (3.149)–(3.151) then become a coupled system of differential equations expressed in terms of $F(\eta)$ and $\theta(\eta)$, i.e.,

$$\frac{d^3F(\eta)}{d\eta^3} + 3F(\eta)\frac{d^2F(\eta)}{d\eta^2} + 2\left(\frac{dF(\eta)}{d\eta}\right)^2 + \theta(\eta) = 0 \tag{3.157}$$

$$\frac{d^2\theta(\eta)}{d\eta^2} + 3PrF(\eta)\frac{d\theta(\eta)}{d\eta} = 0 \tag{3.158}$$

with the following boundary conditions:

$$\text{at} \quad \eta = 0: F(0) = 0, \frac{dF(0)}{d\eta} = 0, \quad \theta(0) = 1, \tag{3.159}$$

$$\text{at} \quad \eta \rightarrow \infty: \frac{dF(\infty)}{d\eta} = 0, \quad \theta(\infty) = 0, \tag{3.160}$$

where Pr is the Prandtl number, which is given by the ratio of the momentum diffusivity of the fluid to its thermal diffusivity (i.e., $Pr = \nu/\alpha$). Eqs. (3.157) and (3.158) represent a coupled system of ordinary differential equations for the free convection problem. Simultaneous solution of these two equations yields the velocity and temperature profiles for the flow of a viscous fluid passing the vertical plate. To solve the free convection equations for a family of values of Pr, it is first necessary to define the dependent variables, $g(\eta)$ and $S(\eta)$, i.e.,

$$g(\eta) = \frac{\partial F(\eta)}{\partial Pr}, \quad S(\eta) = \frac{\partial \theta(\eta)}{\partial Pr}. \tag{3.161}$$

Differentiating Eqs. (3.157) and (3.158) with respect to Pr gives

$$\frac{d^3g(\eta)}{d\eta^3} + 3F(\eta)\frac{d^2g(\eta)}{d\eta^2} - 4\frac{dF(\eta)}{d\eta}\frac{dg(\eta)}{d\eta} + 3\frac{d^2F(\eta)}{d\eta^2}g(\eta) + S(\eta) = 0 \tag{3.162}$$

$$\frac{d^2S(\eta)}{d\eta^2} + 3F(\eta)\frac{d\theta(\eta)}{d\eta} + 3Prg(\eta)\frac{d\theta(\eta)}{d\eta} + 3PrF(\eta)\frac{dS(\eta)}{d\eta} = 0 \tag{3.163}$$

The boundary conditions are given by

$$\text{at} \quad \eta = 0: g(0) = 0, \quad \frac{dg(0)}{d\eta} = 0, \quad S(0) = 0, \tag{3.164}$$

$$\text{at} \quad \eta \rightarrow \infty: \frac{dg(\infty)}{d\eta} = 0, \quad S(\infty) = 0. \tag{3.165}$$

The boundary layer equations given in Eqs. (3.162) and (3.163) are solved by means of the superposition method and a group of transformations. Initially, the following expressions are defined in terms of the two parameters λ and μ:

$$g(\eta) = g_1(\eta) + \lambda g_2(\eta) + \mu g_3(\eta), \tag{3.166}$$

$$S(\eta) = S_1(\eta) + \lambda g_2(\eta) + \mu S_3(\eta). \tag{3.167}$$

Substituting these expressions into Eqs. (3.162)–(3.165) and then separating the resulting equations give the three sets of initial value problems presented below

$$\frac{d^3 g_1(\eta)}{d\eta^3} + 3F(\eta)\frac{d^2 g_1(\eta)}{d\eta^2} - 4\frac{dF(\eta)}{d\eta}\frac{dg_1(\eta)}{d\eta} + 3\frac{d^2F(\eta)}{d\eta^2}g_1(\eta) + S_1(\eta) = 0 \tag{3.168}$$

$$\frac{d^2 S_1(\eta)}{d\eta^2} + 3F(\eta)\frac{d\theta(\eta)}{d\eta} + 3Prg_1(\eta)\frac{d\theta(\eta)}{d\eta} + 3PrF(\eta)\frac{dS_1(\eta)}{d\eta} = 0 \tag{3.169}$$

with initial conditions of

$$g_1(0) = \frac{dg_1(0)}{d\eta} = \frac{d^2 g_1(0)}{d\eta^2} = 0 \tag{3.170}$$

$$S_1(0) = \frac{dS_1(0)}{d\eta} = 0 \tag{3.171}$$

and

$$\frac{d^3 g_2(\eta)}{d\eta^3} + 3F(\eta)\frac{d^2 g_2(\eta)}{d\eta^2} - 4\frac{dF(\eta)}{d\eta}\frac{dg_2(\eta)}{d\eta} + 3\frac{d^2F(\eta)}{d\eta^2}g_2(\eta) + S_2(\eta) = 0 \tag{3.172}$$

$$\frac{d^2 S_2(\eta)}{d\eta^2} + 3Prg_2(\eta)\frac{d\theta(\eta)}{d\eta} + 3PrF(\eta)\frac{dS_2(\eta)}{d\eta} = 0 \tag{3.173}$$

with initial conditions of

$$g_2(0) = \frac{dg_2(0)}{d\eta} = \frac{d^2 g_2(0)}{d\eta^2} = 0, \tag{3.174}$$

$$S_2(0) = 0, \frac{dS_2(0)}{d\eta} = 1 \tag{3.175}$$

and

$$\frac{d^3g_3(\eta)}{d\eta^3} + 3F(\eta)\frac{d^2g_3(\eta)}{d\eta^2} - 4\frac{dF(\eta)}{d\eta}\frac{dg_3(\eta)}{d\eta} + 3\frac{d^2F(\eta)}{d\eta^2}g_3(\eta) + S_3(\eta) = 0 \tag{3.176}$$

$$\frac{d^2S_3(\eta)}{d\eta^2} + 3Prg_3(\eta)\frac{d\theta(\eta)}{d\eta} + 3PrF(\eta)\frac{dS_3(\eta)}{d\eta} = 0 \tag{3.177}$$

with initial conditions of

$$g_3(0) = \frac{dg_3(0)}{d\eta} = 0, \quad \frac{d^2g_3(0)}{d\eta^2} = 1, \tag{3.178}$$

$$S_3(0) = \frac{dS_3(0)}{d\eta} = 0, \tag{3.179}$$

Substituting the boundary conditions at infinity from Eq. (3.165) into Eqs. (3.166) and (3.167) gives the values of parameters λ and μ as,

$$\lambda = \frac{S_1(\infty)dg_3(\infty)/d\eta - S_3(\infty)dg_1(\infty)/d\eta}{S_3(\infty)dg_2(\infty)/d\eta - S_2(\infty)dg_3(\infty)/d\eta}, \tag{3.180}$$

$$\mu = \frac{S_2(\infty)dg_1(\infty)/d\eta - S_1(\infty)dg_2(\infty)/d\eta}{S_3(\infty)dg_2(\infty)/d\eta - S_2(\infty)dg_3(\infty)/d\eta}. \tag{3.181}$$

More details about the solution process can be found in Ref. [7]. To solve Eqs. (3.157) and (3.158) using the DTM, it is first necessary to solve these equations for $Pr = 1$. The boundary conditions used are

$$\begin{aligned} &\text{at} \quad \eta = 0{:}\ F(0) = 0, \quad \frac{dF(0)}{d\eta} = 0, \quad \frac{d^2F(0)}{d\eta^2} = 0.6421, \\ &\theta(0) = 1, \quad \frac{d\theta(0)}{d\eta} = -0.5671 \end{aligned} \tag{3.182}$$

The DTM is then used to solve the system of initial value problems for $Pr = 1$. Initially, the following expressions are defined:

$$y(\eta) = \frac{dF(\eta)}{d\eta}, \quad z(\eta) = \frac{dy(\eta)}{d\eta} = \frac{d^2F(\eta)}{d\eta^2} \tag{3.183}$$

and

$$s(\eta) = \frac{d\theta(\eta)}{d\eta} \tag{3.184}$$

Thereafter, the third-order and second-order ordinary differential equations (Eqs. 3.157 and 3.158) are reduced to the following first-order ordinary differential equations:

$$\frac{dz(\eta)}{d\eta} + 3F(\eta)z(\eta) - 2y^2(\eta) + \theta(\eta) = 0, \tag{3.185}$$

$$\frac{ds(\eta)}{d\eta} + 3F(\eta)s(\eta) = 0, \tag{3.186}$$

The initial conditions become

$$\begin{aligned} \eta = 0: \quad & F(0) = y(0) = 0, \quad z(0) = 0.6421 \\ & \theta(0) = 1, \quad s(0) = -0.5671 \end{aligned} \tag{3.187}$$

Performing differential transformation of Eqs. (3.185)–(3.187) yields the following:

$$\frac{k+1}{H_i}\overline{F}(k+1) = Y_i(k), \quad \frac{k+1}{H_i}Y_i(k+1) = Z_i(k), \tag{3.188}$$

$$\frac{k+1}{H_i}\Theta_i(k+1) = S_i(k), \tag{3.189}$$

$$\frac{k+1}{H_i}Z_i(k+1) + 3\sum_{\iota=0}^{k}\overline{F}(k-\iota)Z_i(\iota) - 2\sum_{\iota=0}^{k}Y(k-\iota)Y(\iota) + \Theta(k) = 0 \tag{3.190}$$

$$\frac{k+1}{H_i}S_i(k+1) + 3\sum_{\iota=0}^{k}\overline{F}(k-\iota)S_i(\iota) = 0. \tag{3.191}$$

various values of $F_i(k)$, $Y_i(k)$, $Z_i(k)$, $\Theta_i(k)$, and $S_i(k)$ are obtained by using Eqs. (3.188)–(3.191) and the transformed initial conditions.

To solve Eqs. (3.168)–(3.171) using the DTM, the following expressions are first defined:

$$A(\eta) = \frac{dg_1(\eta)}{d\eta}, \quad B(\eta) = \frac{dA(\eta)}{d\eta} = \frac{d^2g_1(\eta)}{d\eta^2} \tag{3.192}$$

and

$$C(\eta) = \frac{dS_1(\eta)}{d\eta}. \tag{3.193}$$

Hence, the third-order and second-order ordinary differential equations (Eqs. 3.168 and 3.169) become two first-order ordinary differential equations with the following forms:

$$\frac{dB(\eta)}{d\eta}+3F(\eta)B(\eta)-4\frac{dF(\eta)}{d\eta}A(\eta)+3\frac{d^2F(\eta)}{d\eta^2}g_1(\eta)+S_1(\eta)=0 \tag{3.194}$$

$$\frac{dC(\eta)}{d\eta}+3F(\eta)\frac{d\theta(\eta)}{d\eta}+3Prg_1(\eta)\frac{d\theta(\eta)}{d\eta}+3PrF(\eta)+C(\eta)=0 \tag{3.195}$$

with initial conditions of

$$g_1(0)=A(0)=B(0)=0 \tag{3.196}$$

$$S_1(0)=C(0)=0 \tag{3.197}$$

As in the previous procedure, a process of inverse differential transformation is used to yield the following solutions:

$$g_{1_i}(\eta)=\sum_{k=0}^{m}\left(\frac{\eta}{H_i}\right)^k G_{1_i}(k),\quad A_i(\eta)=\sum_{k=0}^{m}\left(\frac{\eta}{H_i}\right)^k \overline{A}_i(k),$$
$$B_i(\eta)=\sum_{k=0}^{m}\left(\frac{\eta}{H_i}\right)^k \overline{B}_i(k), \tag{3.198}$$

$$S_{1_i}(\eta)=\sum_{k=0}^{m}\left(\frac{\eta}{H_i}\right)^k \overline{S}_{1_i}(k),\quad C_i(\eta)=\sum_{k=0}^{m}\left(\frac{\eta}{H_i}\right)^k \overline{C}_i(k), \tag{3.199}$$

where $0\le\eta\le H_i$,

where, as before, $i=0,1,2,\ldots,n$ indicates the ith subdomain, $k=0,1,2,\ldots,m$ represents the number of terms of the power series. From the initial conditions (Eqs. 3.195 and 3.196) and the solution equations (Eqs. 3.198 and 3.199), it can be shown that

$$G_1(0)=\overline{A}(0)=\overline{B}(0)=0 \tag{3.200}$$

$$\overline{S}_1(0)=\overline{C}(0)=0 \tag{3.201}$$

Eqs. (3.192)−(3.195) undergo a process of differential transformation to give the following:

$$\frac{k+1}{H_i}G_{1_i}(k+1)=\overline{A}_i(k),\quad \frac{k+1}{H_i}\overline{A}_i(k+1)=\overline{B}_i(k) \tag{3.202}$$

$$\frac{k+1}{H_i}\overline{S}_{1_i}(k+1) = \overline{C}_i(k), \tag{3.203}$$

$$\begin{aligned}&\frac{k+1}{H_i}\overline{B}_i(k+1) + 3\sum_{\iota=0}^{k}\overline{B}_i(k-\iota)\overline{F}_i(\iota) - 4\sum_{\iota=0}^{k}\overline{A}_i(k-\iota)Y(\iota)\\ &\quad + 3\sum_{\iota=0}^{k} G_{1_i}(k-\iota)Z(\iota)\overline{S}_{1_i}(k) = 0\end{aligned} \tag{3.204}$$

$$\begin{aligned}&\frac{k+1}{H_i}\overline{C}_i(k+1) + 3\sum_{\iota=0}^{k} S_i(k-\iota)\overline{F}_i(\iota) + 3Pr\sum_{\iota=0}^{k} S_i(k-\iota)G_{1_i}(\iota)\\ &\quad + 3Pr\sum_{\iota=0}^{k}\overline{C}_i(k-\iota)\overline{F}_i(k) = 0.\end{aligned} \tag{3.205}$$

As in the previous procedures, Eqs. (3.172)–(3.175) can be solved using the DTM. The following expressions are defined:

$$D(\eta) = \frac{dg_2(\eta)}{d\eta}, \quad E(\eta) = \frac{dD(\eta)}{d\eta} = \frac{d^2g_2(\eta)}{d\eta^2} \tag{3.206}$$

and

$$H(\eta) = \frac{dS_2(\eta)}{d\eta} \tag{3.207}$$

Hence, the third-order and second-order ordinary differential equations (Eqs. 3.172 and 3.173) become two first-order ordinary differential equations with the following forms:

$$\frac{dE(\eta)}{d\eta} + 3F(\eta)E(\eta) - 4\frac{dF(\eta)}{d\eta}D(\eta) + 3\frac{d^2F(\eta)}{d\eta^2}g_2(\eta) + S_2(\eta) = 0 \tag{3.208}$$

$$\frac{dH(\eta)}{d\eta} + 3Prg_2(\eta)\frac{d\theta(\eta)}{d\eta} + 3PrF(\eta)H(\eta) = 0 \tag{3.209}$$

with initial conditions of

$$g_2(0) = D(0) = E(0) = 0 \tag{3.210}$$

$$S_2(0) = 0, H(0) = 1 \tag{3.211}$$

As in the previous procedure, a process of inverse differential transformation is used to yield the following solutions:

$$g_{2_i}(\eta)=\sum_{k=0}^{m}\left(\frac{\eta}{H_i}\right)^k G_{2_i}(k),\quad D_i(\eta)=\sum_{k=0}^{m}\left(\frac{\eta}{H_i}\right)^k \overline{D}_i(k),$$
$$E_i(\eta)=\sum_{k=0}^{m}\left(\frac{\eta}{H_i}\right)^k \overline{E}_i(k) \tag{3.212}$$

$$S_{2_i}(\eta)=\sum_{k=0}^{m}\left(\frac{\eta}{H_i}\right)^k \overline{S}_{2_i}(k),\quad H_i(\eta)=\sum_{k=0}^{m}\left(\frac{\eta}{H_i}\right)^k \overline{H}_i(k), \tag{3.213}$$

where $0\le\eta\le H_i$

From the initial conditions (Eqs. 3.210 and 3.211) and the solution equations (Eqs. 3.212 and 3.213), it can be shown that

$$G_2(0)=\overline{D}(0)=\overline{E}(0)=0, \tag{3.214}$$

$$\overline{S}_2(0)=0,\quad \overline{H}(0)=\delta(k),\quad \text{where}\quad \delta(k)=\begin{cases}1 & k=0\\ 0 & k\neq 0\end{cases} \tag{3.215}$$

Eqs. (3.206)–(3.209) undergo a process of differential transformation to give the following:

$$\frac{k+1}{H_i}G_{2i}(k+1)=\overline{D}_i(k),\quad \frac{k+1}{H_i}\overline{D}_i(k+1)=\overline{E}_i(k) \tag{3.216}$$

$$\frac{k+1}{H_i}\overline{S}_{2i}(k+1)=\overline{H}_i(k), \tag{3.217}$$

$$\frac{k+1}{H_i}\overline{E}_i(k+1)+3\sum_{\iota=0}^{k}\overline{E}_i(k-\iota)\overline{F}(\iota)-4\sum_{\iota=0}^{k}\overline{D}_i(k-\iota)Y(\iota)$$
$$+3\sum_{\iota=0}^{k}G_{2i}(k-\iota)Z(\iota)+\overline{S}_{2i}(k)=0 \tag{3.218}$$

$$\frac{k+1}{H_i}\overline{H}_i(k+1)+3Pr\sum_{\iota=0}^{k}\overline{H}_i(k-\iota)\overline{F}(\iota)+3Pr\sum_{t=0}^{k}S_i(k-\iota)G_{2i}(\iota)=0 \tag{3.219}$$

As in the previous procedures, Eqs. (3.176)–(3.179) can be solved using the DTM. The following expressions are defined:

$$I(\eta)=\frac{dg_3(\eta)}{d\eta},\quad J(\eta)=\frac{dI(\eta)}{d\eta}=\frac{d^2g_3(\eta)}{d\eta^2} \tag{3.220}$$

and

$$K(\eta) = \frac{dS_3(\eta)}{d\eta} \tag{3.221}$$

Hence, the third-order and second-order ordinary differential equations (Eqs. 3.176 and 3.177) become two first-order ordinary differential equations with the following forms:

$$\frac{dJ(\eta)}{d\eta} + 3F(\eta)J(\eta) - 4\frac{dF(\eta)}{d\eta}I(\eta) + 3\frac{d^2F(\eta)}{d\eta^2}g_3(\eta) + s_3(\eta) = 0 \tag{3.222}$$

$$\frac{dK(\eta)}{d\eta} + 3Prg_3(\eta)\frac{d\theta(\eta)}{d\eta} + 3PrF(\eta)K(\eta) = 0 \tag{3.223}$$

with initial conditions of

$$g_3(0) = I(0) = 0, \quad J(0) = 1 \tag{3.224}$$

$$S_3(0) = 0, K(0) = 0. \tag{3.225}$$

As in the previous procedure, inverse differential transformation is used to yield the following solutions:

$$g_{3i}(\eta) = \sum_{k=0}^{m}\left(\frac{\eta}{H_i}\right)^k G_{3i}(k), \quad I_i(\eta) = \sum_{k=0}^{m}\left(\frac{\eta}{H_i}\right)^k \bar{I}_i(k),$$
$$J(\eta) = \sum_{k=0}^{m}\left(\frac{\eta}{H_i}\right)^k \bar{J}_i(k), \tag{3.226}$$

$$S_{3i}(\eta) = \sum_{k=0}^{m}\left(\frac{\eta}{H_i}\right)^k \bar{S}_{3i}(k), \quad K_i(\eta) = \sum_{k=0}^{m}\left(\frac{\eta}{H_i}\right)^k \bar{K}_i(k), \tag{3.227}$$

where $0 \le \eta \le H_i$

From the initial conditions (Eqs. 3.224 and 3.225) and the solution equations (Eqs. 3.226 and 3.227), it can be shown that

$$G_3(0) = \bar{I}(0) = 0, \quad \bar{J}(0) = \delta(k), \quad \text{where } \delta(k) = \begin{cases} 1 & k = 0 \\ 0 & k \neq 0 \end{cases} \tag{3.228}$$

$$\bar{S}_3(0) = \bar{K}(0) = 0. \tag{3.229}$$

Eqs. (3.222)–(3.225) undergo a process of differential transformation to give the following:

$$\frac{k+1}{H_i}G_{3i}(k+1) = \bar{I}_i(k), \quad \frac{k+1}{H_i}\bar{I}_i(k+1) = \bar{J}_i(k), \tag{3.230}$$

$$\frac{k+1}{H_i}\overline{S}_{3i}(k+1) = \overline{K}_i(k), \tag{3.231}$$

$$\frac{k+1}{H_i}\overline{J}_i(k+1) + 3\sum_{\iota=0}^{k}\overline{J}_i(k-1)\overline{F}_i(\iota) - 4\sum_{\iota=0}^{k}\overline{I}_i(k-\iota)Y(\iota) + 3\sum_{\iota=0}^{k} G_{3i}(k-\iota)Z(\iota) + \overline{S}_{3i}(k) = 0 \tag{3.232}$$

$$\frac{k+1}{H_i}\overline{K}_i(k+1) + 3Pr\sum_{\iota=0}^{k}\overline{K}_i(k-\iota)\overline{F}_i(\iota) + 3Pr\sum_{\iota=0}^{k} S_i(k-\iota)G_{3i}(\iota) = 0 \tag{3.233}$$

Fig. 3.26 plots the velocity distributions obtained by the present method for free convection over a vertical plate for various values of Prandtl number. It is observed that the maximum values of the velocity profiles occur at larger values of g as the value of Pr decreases, and the velocity values decrease with increasing Prandtl number. Fig. 3.27 plots the temperature distributions obtained by the present method for free convection boundary layer flow over a vertical plate for various values of Prandtl number.

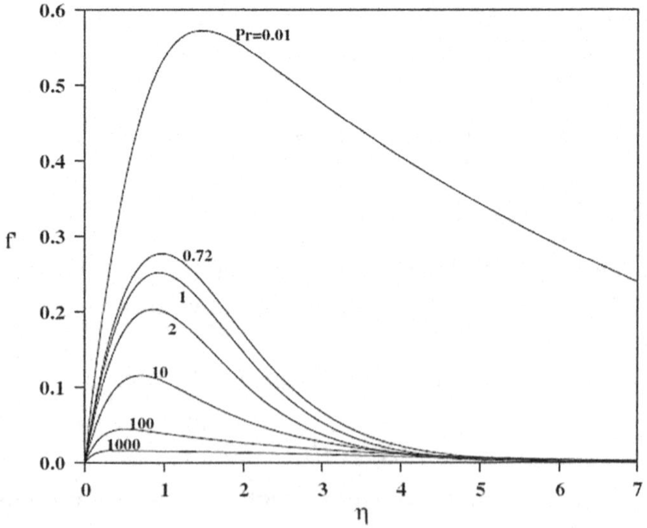

Figure 3.26 Dimensionless velocity distributions for free convection flow over a vertical plate for various values of Prandtl number.

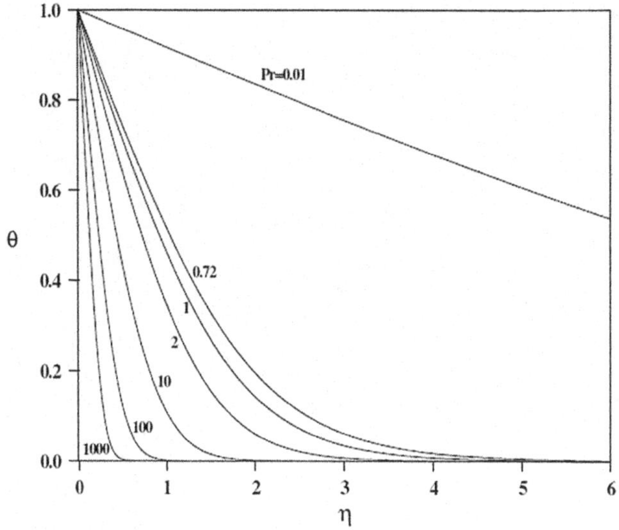

Figure 3.27 Dimensionless temprature distributions for free convection flow over a vertical plate for various values of Prandtl number.

REFERENCES

[1] Kraus AD, Aziz A, Welty JR. Extended surface heat transfer. New York: John Wiley; 2002.

[2] Ghasemi SE, Hatami M, Ganji DD. Thermal analysis of convective fin with temperature-dependent thermal conductivity and heat generation. Case Studies in Thermal Engineering 2014;4:1–8.

[3] Domairry D, Sheikholeslami M, Ashorynejad HR, Gorla RSR, Khani M. Natural convection flow of a non-Newtonian nanofluid between two vertical flat plates. Proceedings of the Institution of Mechanical Engineers, Part N: Journal of Nano-engineering and Nanosystems 2011;225(3):115–22.

[4] Ndlovu PL, Moitsheki RJ. Application of the two-dimensional differential transform method to heat conduction problem for heat transfer in longitudinal rectangular and convex parabolic fins. Communications in Nonlinear Science and Numerical Simulation 2013;18(10):2689–98.

[5] Kuo B-L. Thermal boundary-layer problems in a semi-infinite flat plate by the differential transformation method. Applied Mathematics and Computation 2004;150(2):303–20.

[6] Kuo B-L. Heat transfer analysis for the Falkner–Skan wedge flow by the differential transformation method. International Journal of Heat and Mass Transfer 2005;48(23):5036–46.

[7] Kuo B-L. Application of the differential transformation method to the solutions of the free convection problem. Applied Mathematics and Computation 2005;165(1):63–79.

CHAPTER 4

DTM for Fluids Flow Analysis

4.1 INTRODUCTION

Studies of fluid transport in biological organisms often concern the flow of a particular fluid inside an expanding or contracting vessel with permeable walls. For a valve vessel exhibiting deformable boundaries, alternating wall contractions produce the effect of a physiological pump. The flow behavior inside the lymphatic exhibits a similar character. In such models, circulation is induced by successive contractions of two thin sheets that cause the downstream convection of the sandwiched fluid. Seepage across permeable walls is clearly important to the mass transfer between blood, air, and tissue [1]. Therefore, a substantial amount of research work has been invested in the study of the flow in different geometries in both Newtonian and non-Newtonian form. This chapter introduces Differential Transformation Method (DTM) to solve these problems which contains the following sections:

4.1 Introduction
4.2 Two-Dimensional Viscous Flow
4.3 Magnetohydrodynamic Boundary Layer
4.4 Nanofluid Flow Over a Flat Plate
4.5 Non-Newtonian Fluid Flow Analysis

4.2 TWO-DIMENSIONAL VISCOUS FLOW

Consider the laminar, isothermal, and incompressible flow in a rectangular domain bounded by two permeable surfaces that enable the fluid to enter or exit during successive expansions or contractions. A schematic diagram of the problem is shown in Fig. 4.1. The walls expand or contract uniformly at a time-dependent rate $\dot{a}$. At the wall, it is assumed that the fluid inflow velocity V_w is independent of position. The equations of continuity and motion for the unsteady flow are given as follows [2]:

$$\frac{\partial u^*}{\partial x^*} + \frac{\partial v^*}{\partial y^*} = 0, \tag{4.1}$$

Differential Transformation Method for Mechanical Engineering Problems
ISBN 978-0-12-805190-0
http://dx.doi.org/10.1016/B978-0-12-805190-0.00004-8

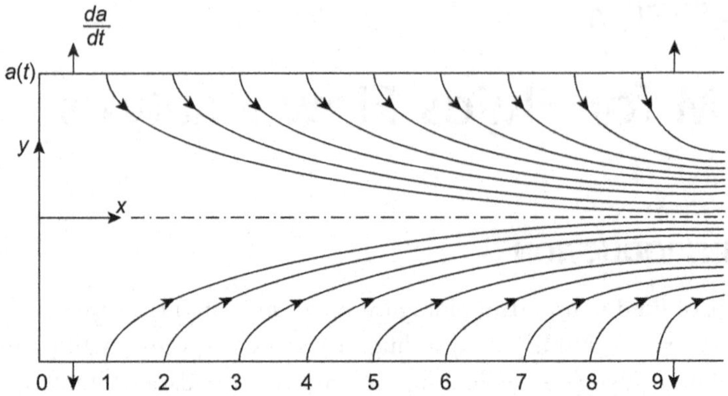

Figure 4.1 Two-dimensional domain with expanding or contracting porous walls.

$$\frac{\partial u^*}{\partial t} + u^* \frac{\partial u^*}{\partial x^*} + v^* \frac{\partial u^*}{\partial y^*} = -\frac{1}{\rho}\frac{\partial p^*}{\partial x^*} + \upsilon\left[\frac{\partial^2 u^*}{\partial x^{*2}} + \frac{\partial^2 u^*}{\partial y^{*2}}\right], \tag{4.2}$$

$$\frac{\partial v^*}{\partial t} + u^* \frac{\partial v^*}{\partial x^*} + v^* \frac{\partial v^*}{\partial y^*} = -\frac{1}{\rho}\frac{\partial p^*}{\partial y^*} + \upsilon\left[\frac{\partial^2 v^*}{\partial x^{*2}} + \frac{\partial^2 v^*}{\partial y^{*2}}\right]. \tag{4.3}$$

In the above equations u^* and v^* indicate the velocity components in x and y directions; p^* denotes the dimensional pressure; ρ, υ, and t are the density, kinematic viscosity, and time, respectively. The boundary conditions will be:

$$\begin{gathered} y^* = a(t): \quad u^* = 0,\ v^* = -V_w = -\frac{\dot{a}}{c}, \\ y^* = 0: \quad \frac{\partial u^*}{\partial y^*} = 0,\ v^* = 0, \\ x^* = 0: \quad u^* = 0\cdot \end{gathered} \tag{4.4}$$

where $c = \frac{\dot{a}}{V_w}$ is the wall presence or injection/suction coefficient, which is a measure of wall permeability. The stream function and mean flow vorticity can be introduced by putting,

$$\begin{gathered} u^* = \frac{\partial \psi^*}{\partial y^*},\ v^* = \frac{\partial \psi^*}{\partial x^*},\ \xi^* = \frac{\partial v^*}{\partial x^*} - \frac{\partial u^*}{\partial y^*} \\ \frac{\partial \xi^*}{\partial t} + u^* \frac{\partial \xi^*}{\partial x^*} + v^* \frac{\partial \xi^*}{\partial y^*} = \upsilon\left[\frac{\partial^2 \xi^*}{\partial x^{*2}} + \frac{\partial^2 \xi^*}{\partial y^{*2}}\right]. \end{gathered} \tag{4.5}$$

Due to mass conservation, a similar solution can be developed with respect to x^*. Starting with:

$$\psi^* = \frac{\nu x^* f^*(y,t)}{a},\ u^* = \frac{\nu x^* f_y^*}{a^2},\ v^* = \frac{-\nu f^*(y,t)}{a},$$
$$y = \frac{y^*}{a},\ f_y^* \equiv \frac{\partial f^*}{\partial y}. \tag{4.6}$$

Substitution of Eq. (4.6) into Eq. (4.5) yields,

$$u^*_{y^*t} + u^* u^*_{y^*x^*} + v^* u^*_{y^*y^*} = \nu u^*_{y^*y^*y^*} \tag{4.7}$$

To solve Eq. (4.7), one uses the chain rule to obtain,

$$f^*_{yyyy} + \alpha\left(y f^*_{yyy} + 3 f^*_{yy}\right) + f^* f^*_{yyy} - f^*_y f^*_{yy} - a^2 \upsilon^{-1} f^*_{yyt} = 0, \tag{4.8}$$

With the following boundary conditions:

$$\text{at } y = 0:\ f^* = 0,\ f^*_{yy} = 0,$$
$$\text{at } y = 1:\ f^* = Re,\ f^*_y = 0, \tag{4.9}$$

where $\alpha(t) \equiv \frac{a\dot{a}}{\upsilon}$ is the nondimensional wall dilation rate defined positive for expansion and negative for contraction. Furthermore, $Re = \frac{aV_w}{\upsilon}$ is the permeation Reynolds number defined positive for injection and negative for suction through the walls. Eqs. (4.6), (4.8), and (4.9) can be normalized by putting,

$$\psi = \frac{\psi^*}{a\dot{a}},\ u = \frac{u^*}{\dot{a}},\ v = \frac{v^*}{a},\ f = \frac{f^*}{Re}, \tag{4.10}$$

And so,

$$\psi = \frac{xf}{c},\ u = \frac{xf'}{c},\ v = \frac{-f}{c},\ c = \frac{\alpha}{Re}, \tag{4.11}$$

$$f^{IV} + \alpha\left(y f''' + 3f''\right) + Re\, f f''' - Re\, f' f'' = 0 \tag{4.12}$$

The boundary conditions (4.9) will be:

$$y = 0:\ f = 0,\ f'' = 0$$
$$y = 1:\ f = 1,\ f' = 0 \tag{4.13}$$

The resulting Eq. (4.12) is the classic Berman's formula [2], with $\alpha = 0$ (channel with stationary walls).

After the flow field is found, the normal pressure gradient can be obtained by substituting the velocity components into Eqs. (4.1)–(4.3). Hence it is,

$$p_y = -[Re^{-1}f'' + ff' + \alpha Re^{-1}(f + yf')],$$
$$p = \frac{p^*}{\rho V_w^2}. \tag{4.14}$$

We can determine the normal pressure distribution, if we integrate Eq. (4.14). Let p_c be the centerline pressure, hence,

$$\int_{p_c}^{p(y)} dp = \int_0^y -[Re^{-1}f'' + ff' + \alpha\, Re^{-1}(f + yf')], \tag{4.15}$$

Then using $ff' = (f^2)'/2$ and $(f + yf') = (yf)'$, the resulting normal pressure drop will be,

$$\Delta p_n = Re^{-1}f'(0) - \left[Re^{-1}f' + \frac{f^2}{2} + \alpha\, Re^{-1}\, yf\right]. \tag{4.16}$$

Another important quantity is the shear stress. The shear stress can be determined from Newton's law for viscosity:

$$\tau^* = \mu\left(v^*_{x^*} + u^*_{y^*}\right) = \frac{\rho v^2 x^* f^{*''}}{a^3}. \tag{4.17}$$

Introducing the nondimensional shear stress $\tau = \frac{\tau^*}{\rho V_w^2}\, w$, we have,

$$\tau = \frac{xf''}{Re}. \tag{4.18}$$

Now DTM into governing equations has been applied. Taking the differential transform of Eqs. (4.12) and (4.13) with respect to χ and considering $H = 1$ gives,

$$(k+1)(k+2)(k+3)(k+4)F[k+4]$$
$$+\alpha\sum_{m=0}^{k}(\delta[m](k-m+1)(k-m+2)(k-m+3)F[k-m+3])$$
$$+3\alpha(k+1)(k+2)F[k+2]$$
$$+Re\sum_{m=0}^{k}(F[k-m](m+1)(m+2)(m+3)F[m+3])$$
$$-Re\sum_{m=0}^{k}((k-m+1)F[k-m+1](m+1)(m+2)F[m+2])=0,$$
$$\delta[m]=\begin{cases}1 & m=1\\ 0 & m\neq 1\end{cases} \tag{4.19}$$

$$F[0]=0,\ F[1]=a_0,\ F[2]=0,\ F[3]=a_1 \tag{4.20}$$

where $F(k)$ are the differential transforms of $f(\eta)$ and a_0, a_1 are constants, which can be obtained through boundary conditions from Eq. (4.13). This problem can be solved as follows:

$$F[0]=0,\ F[1]=a_0,\ F[2]=0,\ F[3]=a_1,\ F[4]=F[6]=F[8]=0$$
$$F[5]=-\frac{3}{20}\alpha\,a_0$$
$$F[7]=\frac{3}{280}\alpha^2 a_1+\frac{1}{70}Re\,a_1^2+\frac{1}{140}Re\,a_0a_1\,\alpha$$
$$F[9]=-\frac{1}{2240}\alpha^3 a_1-\frac{1}{560}\alpha\,Re\,a_1^2-\frac{1}{1120}Re\,a_0\,a_1\,\alpha^2-\frac{1}{1260}a_0\,a_1\,\alpha^2$$
$$-\frac{1}{1260}a_0\,Re\,a_1^2-\frac{1}{2520}Re\,a_0^2\,\alpha\,a_1$$
$$\vdots \tag{4.21}$$

The above process is continuous. By substituting Eq. (4.21) into the main equation based on DTM, it can be obtained that the closed form of the solutions is,

$$F(\eta) = a_0\eta + a_1\eta^3 + \left(-\frac{3}{20}\right)\eta^5 + \left(\frac{3}{280}\alpha^2 a_1 + \frac{1}{70}Re\ a_1^2 + \frac{1}{140}Re\ a_0 a_1\ \alpha\right)\eta^7 + \left(-\frac{1}{2240}\alpha^3 a_1\left(-\frac{1}{560}\alpha\ Re\ a_1^2 - \frac{1}{1120}Re\ a_0\ a_1\ \alpha^2 - \frac{1}{1260}a_0\ a_1\ \alpha^2 - \frac{1}{1260}a_0\ Re\ a_1^2 - \frac{1}{2520}Re\ a_0^2\ \alpha\ a_1\right)\eta^9 + \ldots \tag{4.22}$$

By substituting the boundary conditions from Eq. (4.13) into Eq. (4.22) in point $\eta = 1$ the values of a_0, a_1 can be obtained.

$$F(1) = a_0 + a_1 + \left(-\frac{3}{20}\right) + \left(\frac{3}{280}\alpha^2 a_1 + \frac{1}{70}Re\ a_1^2 + \frac{1}{140}Re\ a_0 a_1\ \alpha\right) + \left(-\frac{1}{2240}\alpha^3 a_1 - \frac{1}{560}\alpha\ Re\ a_1^2 - \frac{1}{1120}Re\ a_0\ a_1\ \alpha^2 - \frac{1}{1260}a_0\ a_1\ \alpha^2 - \frac{1}{1260}a_0\ Re\ a_1^2 - \frac{1}{2520}Re\ a_0^2\ \alpha\ a_1\right) + \ldots = 1 \tag{4.23}$$

$$F'(1) = a_0\eta + 3a_1 + 5\left(-\frac{3}{20}\right) + 7\left(\frac{3}{280}\alpha^2 a_1 + \frac{1}{70}Re\ a_1^2 + \frac{1}{140}Re\ a_0 a_1\ \alpha\right) + 9\left(-\frac{1}{2240}\alpha^3 a_1 - \frac{1}{560}\alpha\ Re\ a_1^2 - \frac{1}{1120}Re\ a_0\ a_1\ \alpha^2 - \frac{1}{1260}a_0\ a_1\ \alpha^2 - \frac{1}{1260}a_0\ Re\ a_1^2 - \frac{1}{2520}Re\ a_0^2\ \alpha\ a_1\right) + \ldots = 0 \tag{4.24}$$

Solving Eqs. (4.23) and (4.24) gives the values of a_0, a_1. By substituting obtained a_0, a_1 into Eq. (4.22), it can be obtained the expression of $F(\eta)$.

The objective of the present example was to apply DTM compared to homotopy perturbation method (HPM) and optimal homotopy asymptotic method (OHAM) to obtain an explicit analytic solution of laminar, isothermal, incompressible viscous flow in a rectangular domain bounded by two moving porous walls that enable the fluid to enter or exit during successive expansions or contractions (Fig. 4.1). Also, error percentage is introduced as follows:

$$\%\text{Error} = \left| \frac{f(\eta)_{\text{NM}} - f(\eta)_a}{f(\eta)_{\text{NM}}} \right| \times 100 \tag{4.25}$$

where $f(\eta)_a$ is a value obtained using different analytical methods. Fig. 2.2 shows the comparison between numerical method and other analytical solutions results for $f(y)$ when $\alpha = Re = 1$. It verifies that, there is an acceptable agreement between the numerical solution obtained by fourth-order Runge–kutta method and these methods. Tables 4.1 and 4.2 confirm the last conclusion. Comparison between obtained results showed that HPM and DTM are more accurate and acceptable than two other methods, as can be seen in Fig. 4.3 and Tables 4.3 and 4.4. After this validiation, results are given for the velocity profile, normal pressure distribution, and wall shear stress for various values of permeation Reynolds number and nondimensional wall dilation rate.

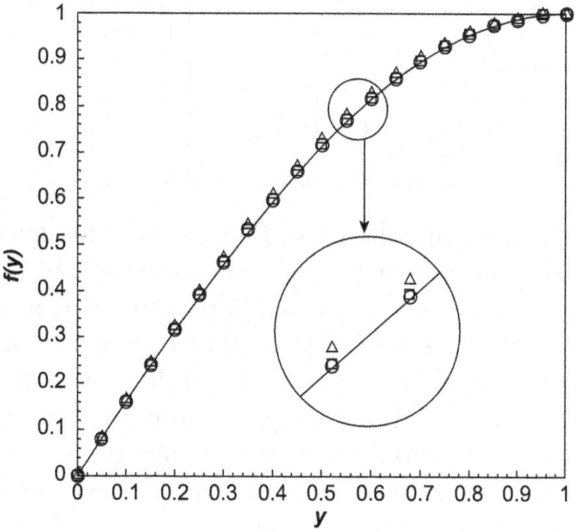

Figure 4.2 Comparison between numerical method and other analytical solutions for $f(y)$, when $\alpha = Re = 1$.

Table 4.1 Constant Values With Different Nondimensional Parameters

Re	α	a_0	a_1
1	1	1.586849	−0.68113
1	2	1.696949	−0.93264
2	1	1.585146	−0.67823

Table 4.2 The HPM, Differential Transformation Method (DTM), OHAM and Numerical Solution Results for *f(y)*, When $\alpha = 2$, $Re = 1$

	f(y)			
y	NM	HPM	DTM	OHAM
0	0	0	0	0
0.05	0.086267	0.086267	0.084731	0.097431
0.1	0.17176	0.17176	0.168765	0.193764
0.15	0.255721	0.255721	0.251416	0.28792
0.2	0.337421	0.337421	0.332018	0.378868
0.25	0.416176	0.416176	0.409935	0.465637
0.3	0.491359	0.491359	0.484572	0.547345
0.35	0.562408	0.562408	0.555383	0.623212
0.4	0.628838	0.628838	0.621875	0.692581
0.45	0.690244	0.690244	0.683621	0.754934
0.5	0.746307	0.746307	0.740259	0.809906
0.55	0.796792	0.796792	0.791502	0.8573
0.6	0.841553	0.841553	0.837137	0.897093
0.65	0.880522	0.880522	0.87703	0.929448
0.7	0.91371	0.91371	0.911122	0.954712
0.75	0.9412	0.9412	0.939433	0.973425
0.8	0.963137	0.963137	0.962059	0.986309
0.85	0.979721	0.979721	0.979164	0.994267
0.9	0.991199	0.991199	0.990982	0.998368
0.95	0.997854	0.997854	0.997811	0.99983
1	1	1	1	1

Fig. 4.4 illustrates the behavior of $f'(y)$ (or uc/x) for different permeation Reynolds number, over a range of nondimensional wall dilation rate. For every level of injection or suction, in the case of an expanding wall, increasing α leads to higher axial velocity near the center and the lower axial velocity near the wall. The reason is that the flow toward the center becomes greater to make up for the space caused by the expansion of the wall and as a result, the axial velocity also becomes greater near the center.

The pressure distribution in the normal direction for various permeation Reynolds numbers over a range of nondimensional wall dilation rates is plotted in Fig. 4.5. Fig. 4.5 shows that for every level of injection or suction,

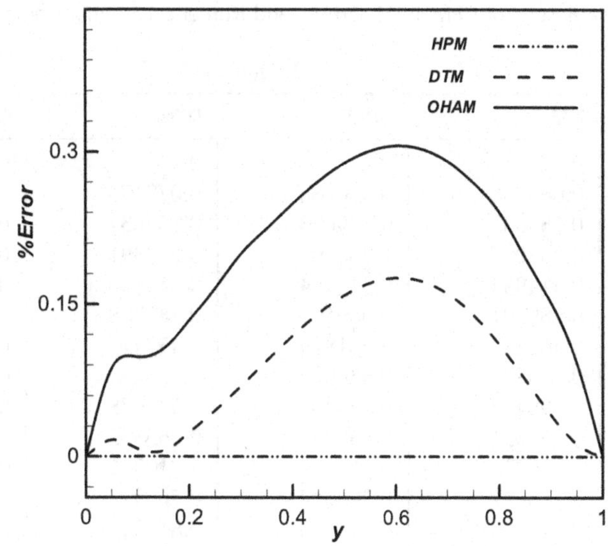

Figure 4.3 Comparison between error percentages of HPM, DTM, and OHAM for *f*(*y*), when $\alpha = Re = 1$.

the absolute pressure change in the normal direction is lowest near the central portion. Furthermore, by increasing nondimensional wall dilation rates the absolute value of pressure distribution in the normal direction increases. The wall shear stress $\left(\tau = \frac{xf''(1)}{Re}\right)$ for permeation Reynolds number $Re = -1$ and $Re = 1$ over a range of nondimensional wall dilation rates is plotted in Fig. 4.6. We can observe from Fig. 4.6 that the absolute shear stress along the wall surface increases in proportion to x. Furthermore, by increasing nondimensional wall dilation rates the absolute value of shear stress increases.

4.3 MAGNETOHYDRODYNAMIC BOUNDARY LAYER

Let us consider the Magnetohydrodynamic (MHD) flow of an incompressible viscous fluid over a stretching sheet at $y = D$. The fluid is electrically conducting under the influence of an applied magnetic field $B(x)$ normal to the stretching sheet. The induced magnetic field is neglected. The boundary layer equations are as follows [3]:

$$u_x + v_y = 0 \tag{4.26}$$

$$uu_x + vu_y = \upsilon u_{yy} - \sigma B^2(x)u/\rho \tag{4.27}$$

where u and v are the velocity components in the x and y directions, respectively, υ is the kinematic viscosity, ρ is the fluid density, and σ is

Table 4.3 The Results of HPM, DTM, OHAM, and Numerical Solution for *f*(*y*), When $\alpha = 1$, $Re = 2$

		f(*y*)		
y	**NM**	**HPM**	**DTM**	**OHAM**
0	0	0	0	0
0.05	0.080088	0.080088	0.079173	0.08608
0.1	0.159633	0.159633	0.157837	0.17144
0.15	0.238097	0.238097	0.235491	0.255368
0.2	0.314954	0.314954	0.311636	0.337174
0.25	0.389691	0.389691	0.385788	0.416193
0.3	0.461819	0.461819	0.457477	0.491799
0.35	0.530872	0.530872	0.52625	0.563412
0.4	0.596412	0.596412	0.591678	0.630506
0.45	0.658037	0.658037	0.653355	0.692621
0.5	0.715375	0.715375	0.710901	0.749366
0.55	0.768097	0.768097	0.76397	0.800432
0.6	0.815908	0.815908	0.812245	0.845595
0.65	0.858559	0.858559	0.855444	0.884723
0.7	0.895837	0.895837	0.893323	0.917782
0.75	0.927572	0.927572	0.925674	0.944843
0.8	0.953638	0.953638	0.952328	0.966079
0.85	0.973944	0.973944	0.973157	0.981772
0.9	0.988441	0.988441	0.988071	0.992311
0.95	0.997119	0.997119	0.997021	0.998189
1	1	1	1	1

the electrical conductivity of the fluid. In Eq. (4.27), the external electric field and the polarization effects are negligible [3].

$$B(x) = B_0 x^{(n-1)/2} \tag{4.28}$$

The boundary conditions corresponding to the nonlinear stretching of a sheet are:

$$\begin{aligned} u(x,0) &= cx^n \quad v(x,0) = 0 \\ u(x,y) &\to 0 \quad \text{as } y \to 0 \end{aligned} \tag{4.29}$$

Using the similarity variables,

$$\begin{aligned} t &= \sqrt{\frac{c(n+1)}{2v}} x^{\frac{n-1}{2}} y, \quad u = cx^n f'(x) \\ v &= \sqrt{\frac{cv(n+1)}{2v}} x^{\frac{n-1}{2}} \left[f(t) + \frac{n-1}{n+1} t f'(t) \right] \end{aligned} \tag{4.30}$$

Table 4.4 The Comparison Among Error Percentages of HPM, DTM, and OHAM for *f*(*y*), when $\alpha = Re = 1$

		f(*y*)	
y	**HPM**	**DTM**	**OHAM**
0	0	0	0
0.05	3.35E-06	0.016579	0.086579
0.1	1.19E-06	0.008088	0.098088
0.15	1.13E-06	0.00559	0.10559
0.2	1.26E-06	0.02375	0.13375
0.25	1.12E-06	0.045427	0.165427
0.3	9.41E-07	0.069424	0.199424
0.35	8.64E-07	0.094337	0.224337
0.4	7.35E-07	0.118603	0.248603
0.45	6.08E-07	0.140557	0.270557
0.5	5.19E-07	0.158494	0.288494
0.55	4.14E-07	0.170769	0.300769
0.6	3.66E-07	0.175905	0.305905
0.65	4.3E-07	0.172741	0.302741
0.7	2.51E-07	0.160611	0.290611
0.75	2.52E-07	0.139584	0.269584
0.8	2.62E-07	0.110756	0.240756
0.85	1E-07	0.076631	0.196631
0.9	1.42E-07	0.041614	0.151614
0.95	4.18E-07	0.012641	0.092641
1	2E-07	2E-08	0

Eqs. (4.26)−(4.29) are transformed into,

$$f'''(t) + f(t)f''(t) - \beta f'(t)^2 - Mf'(t) = 0 \tag{4.31}$$

$$f(0) = 0, \quad f'(0) = 1, \quad f'(+\infty) = 0 \tag{4.32}$$

where the primes denote differentiation with respect to t and

$$\beta = \frac{2n}{1+n}, \qquad M = \frac{2\sigma B_0^2}{\rho c(1+n)} \tag{4.33}$$

We shall solve this nonlinear differential equation using the DTM, the DTM−Pade and the numerical methods (by using a fourth-order Runge−Kutta and shooting method).

Taking differential transform of Eq. (4.31), one can obtain

$$
\begin{aligned}
&(k+1)(k+2)(k+3)F(k+3)\\
&\quad+\sum_{r=0}^{k}[-\beta(r+1)(k-r+1)F(r+1)F(k-r+1)\\
&\qquad\qquad+(k-r+1)(k-r+2)F(r)F(k-r+2)]\\
&\quad-M(k+1)F(k+1)=0
\end{aligned}
\tag{4.34}
$$

by using the DTM, the boundary conditions (Eq. 4.32) are transformed into a recurrence equation that finally leads to the solution of a system of algebraic equations. After finding the DTM solutions, the Pade approximant must be applied. We can consider the boundary conditions (Eq. 4.32) as follows:

$$
f(0)=0,\quad f'(0)=1,\quad f''(0)=\alpha \tag{4.35}
$$

The differential transform of the boundary conditions are as follows:

$$
F(0)=0,\quad F(1)=1,\quad F(2)=\alpha \tag{4.36}
$$

Moreover, substituting Eq. (4.36) into Eq. (4.34) and by recursive method we can calculate other values of $F(k)$. Hence, substituting all $F(k)$ into Eq. (4.33), we have series solutions. After finding the series solutions, the Pade approximation must be applied, by using asymptotic boundary condition $(f'(+\infty)=0)$ we can obtain α. For an analytical solution, the convergence analysis was performed and the i value is selected equal to 20. The order of Pade approximation $[L, M]$, [10, 10] has sufficient accuracy; on the other hand if the order of Padé approximation increases, the accuracy of the solution increases. For $\beta=1.5$ the analytical solutions are as follows:

$$
\begin{aligned}
f(t)_{[10,10],M=0} = {} & (t+0.952076t^2+0.469526t^3+0.141262t^4\\
&+0.0289081t^5+0.00399352t^6+0.000382091t^7\\
&+0.0000221858t^8+8.00793\times10^{-7}t^9-1.77343\times10^{-12}t^{10})/\\
&(1+1.52943t+1.10254t^2+0.491682t^3+6.67233\times10^{-8}t^{10})
\end{aligned}
\tag{4.37}
$$

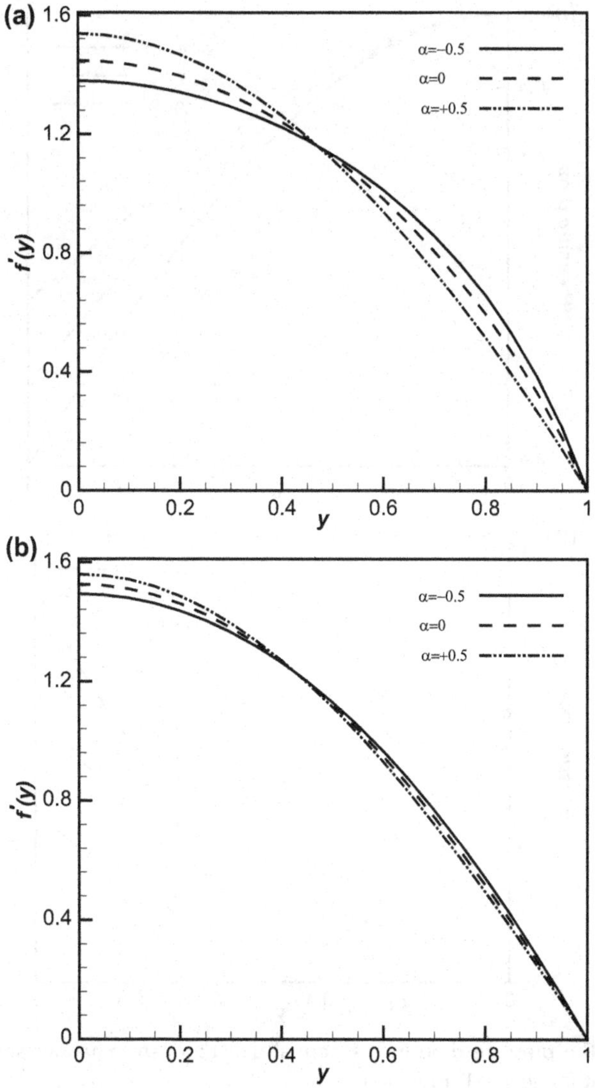

Figure 4.4 $f'(y)$ changes shown over a range of α at (a) $Re = -5$ (b) $Re = 5$.

$$\begin{aligned} f(t)_{[10,10],M=1} = {} & (t + 0.682855t^2 + 0.271323t^3 + 0.054809t^4 + 0.0086483t^5 \\ & + 0.00011291t^6 - 0.000121408t^7 + 0.0000478684t^8 \\ & - 2.34566 \times 10^{-7}t^9 - 1.12636 \times 10^{-7}t^{10})/ \\ & (1 + 1.44545t + 0.95696t^2 + 0.372964t^3 + 0.089476t^4 \\ & + 0.0119558t^5 + 0.000181381t^6 - 0.000251727t^7 \\ & - 0.0000467392t^8 - 3.73478 \times 10^{-6}t^9 + 3.19695 \times 10^{-8}t^{10}) \end{aligned} \quad (4.38)$$

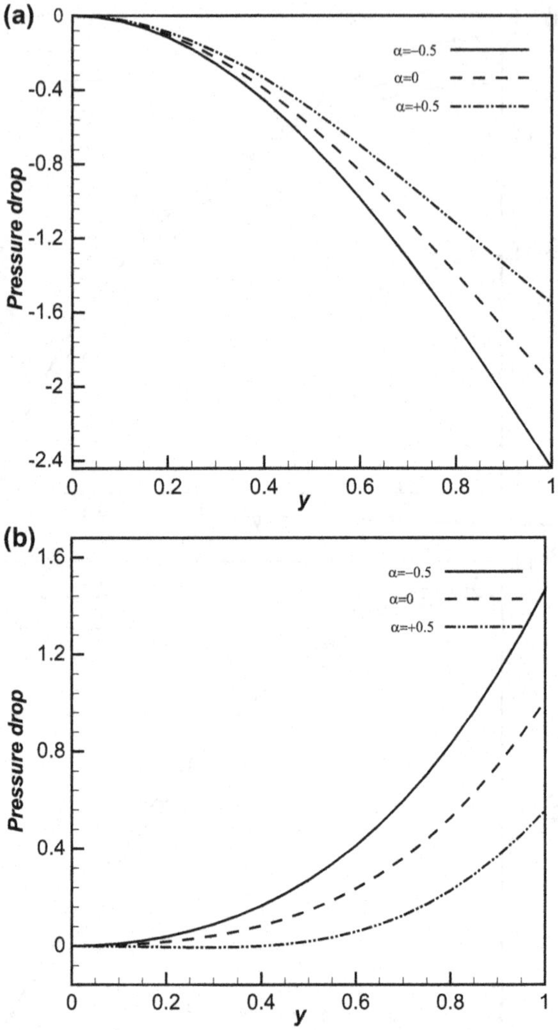

Figure 4.5 The pressure drop in the normal direction (Δp_n) changes shown over a range of α at (a) $Re = -1$ (b) $Re = 1$.

$$
\begin{aligned}
f(t)_{[10,10],M=5} = {} & (t + 0.398393t^2 + 0.134866t^3 - 0.0046991t^4 \\
& - 0.00243265t^5 - 0.00941575t^6 + 0.000501479t^7 \\
& + 0.000144631t^8 + 0.0000729533t^9 + 0.0000112928t^{10})/ \\
& (1 + 1.65644t + 1.13542t^2 + 0.3631t^3 + 0.00941431t^4 \\
& 0.0349225t^5 - 0.0121299t^6 - 0.000911216t^7 + \\
& 0.000530204t^8 + 0.00018197t^9 + 0.0000273471t^{10})
\end{aligned}
\tag{4.39}
$$

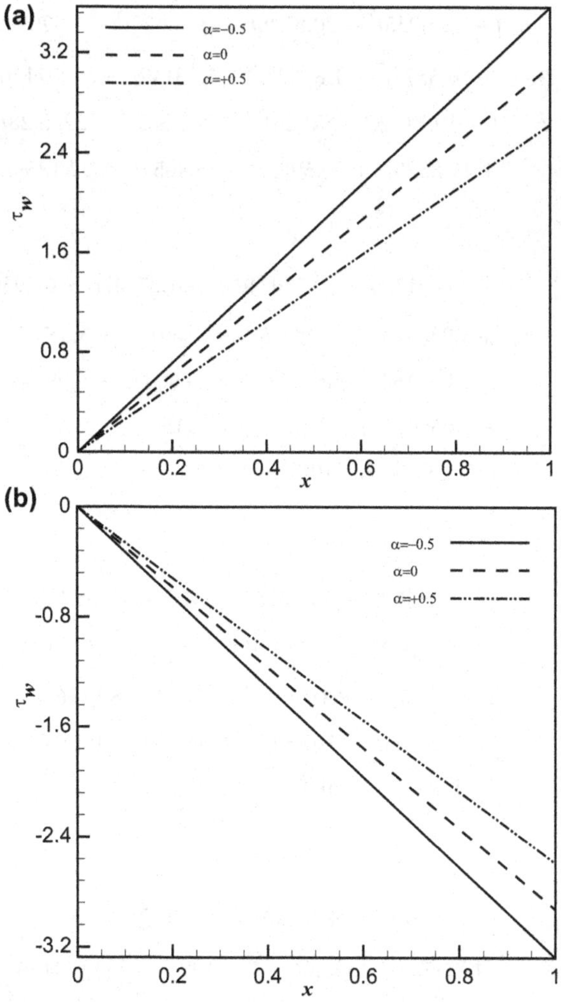

Figure 4.6 Shear stress changes shown over a range of α at (a) $Re = -1$ (b) $Re = 1$.

$$
\begin{aligned}
f(t)_{[10,10],M=10} = {} & (t + 0.0799135t^2 + 0.440267t^3 + 0.157782t^4 \\
& + 0.034105t^5 - 0.0206912t^6 - 0.00395085t^7 \\
& - 0.000579711t^8 - 0.000251482t^9 + 0.000045754t^{10})/ \\
& (1 + 2.48228t + 2.70166t^2 + 1.63052t^3 + 0.535353t^4 \\
& + 0.046855t^5 - 0.0374833t^6 - 0.0176568t^7 \\
& - 0.00331496t^8 - 0.000172074t^9 + 0.0000417687t^{10})
\end{aligned}
\tag{4.40}
$$

$$
\begin{aligned}
f(t)_{[10,10],M=50} = {} & (t - 22.6935t^2 - 20.8798t^3 - 31.0628t^4 - 19.2098t^5 - 0.841719t^6 \\
& + 16.6574t^7 + 1.82323t^8 + 5.78142t^9 - 0.00715648t^{10})/ \\
& (1 - 19.1112t - 97.9261t^2 - 202.307t^3 - 222.828t^4 - 119.697t^5 \\
& + 11.8529t^6 + 70.5985t^7 + 55.4051t^8 + 22.1859t^9 + 4.17935t^{10})
\end{aligned} \tag{4.41}
$$

$$
\begin{aligned}
f(t)_{[10,10],M=100} = {} & (t - 0.441265t^2 + 0.3802484t^3 + 1.27347t^4 - 6.29308t^5 \\
& + 4.93044t^6 - 0.418915t^7 - 4.56918t^8 + 3.63737t^9 - 3.7257t^{10})/ \\
& (1 + 4.59754t + 6.62968t^2 - 0.266082t^3 - 9.99615t^4 \\
& - 7.66542t^5 + 4.24617t^6 + 8.29156t^7 \\
& 1.73129t^8 - 3.29106t^9 - 2.16751t^{10})
\end{aligned} \tag{4.42}
$$

$$
\begin{aligned}
f(t)_{[10,10],M=500} = {} & (t - 2.5452t^2 + 14.9318t^3 - 34.3945t^4 + 21.1412t^5 - 3.17934t^6 \\
& + - 106.258t^7 + 312.014t^8 - 172.341t^9 + 190.748t^{10})/ \\
& (1 + 8.65t + 28.1869t^2 + 26.501t^3 - 89.1766t^4 \\
& -315.503t^5 - 307.616t^6 + 390.192t^7 + 1505.88t^8 \\
& +1883.72t^9 + 1023.08t^{10})
\end{aligned} \tag{4.43}
$$

$$
\begin{aligned}
f(t)_{[10,10],M=1000} = {} & (t - 3.67488t^2 + 31.4012t^3 - 103.767t^4 + 135.989t^5 - \\
& 170.603t^6 - 530.6t^7 + 2608.13t^8 - 2330.81t^9 + 3216.62t^{10})/ \\
& (1 + 12.147t + 56.6735t^2 + 86.5044t^3 - 278.191t^4 - 1561.55t^5 \\
& 2423.89t^6 + 2469.64t^7 + 16765.5t^8 + 29253t^9 + 20837.3t^{10})
\end{aligned} \tag{4.44}
$$

The system of Eq. (4.34) with transformed boundary conditions was solved analytically using the DTM—Padé and numerically using the fourth-order Runge—Kutta and shooting method. It was shown in Fig. 4.7 the analytical and the exact solution of $f(t)$ for different values of a magnetic parameter. It is clear that as the magnetic parameter increases, the thickness

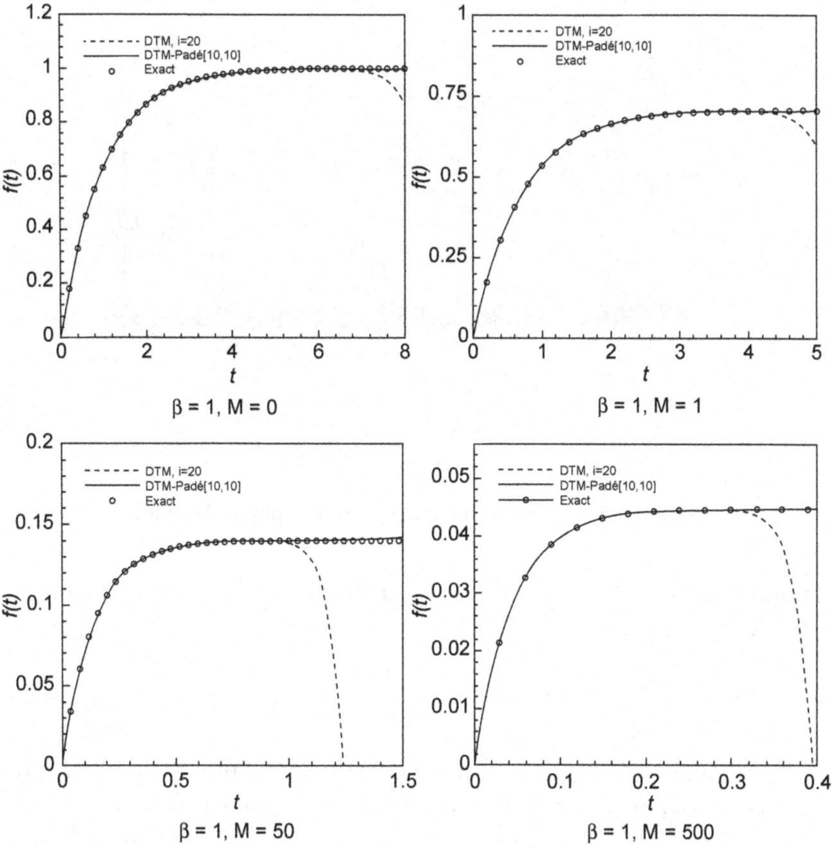

Figure 4.7 The analytical solution of $f(t)$ obtained by the differential transformation method (DTM), the DTM–Pade, and the exact solution.

of boundary layer decreases, and so does the accuracy of the DTM for large values of t. It seems that the DTM solutions are only valid for small values of independent variable (t).

4.4 NANOFLUID FLOW OVER A FLAT PLATE

Consider a nanoliquid film flow and heat transfer in the neighborhood of a thin elastic sheet. To understand more, the related physical model has been depicted in Fig. 4.8. The Cartesian coordinate (x,y) is chosen in a manner that the x-axis is measured in the direction of wall stretching and the y-axis

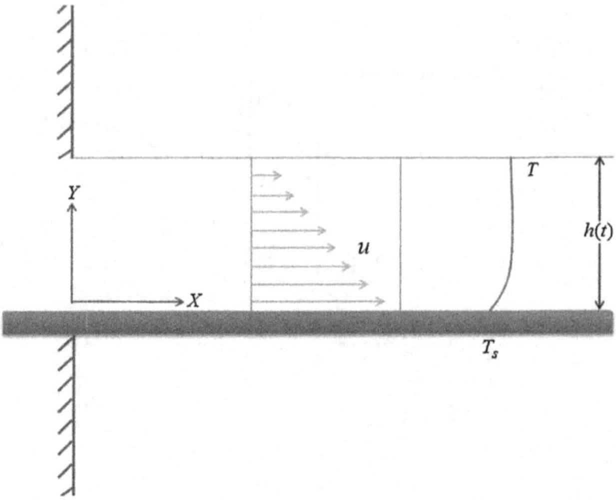

Figure 4.8 The schematic diagram of the physical model.

is normal to the wall. The continuous surface at $y = 0$ is stretched with the velocity defined as:

$$U_w = \frac{bx}{1 - \alpha t} \tag{4.45}$$

In the above equation b and α are constants with dimensions t^{-1}. After that the temperature distribution on the sheet is given as follows:

$$T_s = T_0 - T_r \left(\frac{bx^2}{2v_f} \right) (1 - \alpha t)^{-3/2} \tag{4.46}$$

In Eq. (4.46), T_0 is the temperature at the slit, T_r can be considered either as a constant reference temperature or a constant temperature difference, and finally v_f is the kinematic viscosity of the base fluid. Suppose that the film is uniform and stable, and the gravity and end effects are negligible. Therefore, the governing differential equations for this problem are expressed as follows:

$$\frac{\partial u}{\partial x} + \frac{\partial v}{\partial y} = 0, \quad \frac{\partial u}{\partial t} + u\frac{\partial u}{\partial x} + v\frac{\partial u}{\partial y} = \frac{\mu_{nf}}{\rho_{nf}} \frac{\partial^2 u}{\partial y^2}, \quad \frac{\partial T}{\partial t} + u\frac{\partial T}{\partial x} + v\frac{\partial T}{\partial y} = \alpha_{nf}\frac{\partial^2 T}{\partial y^2} \tag{4.47}$$

In Eq. (4.47), u and v are the velocity components along x and y directions, T is defined as the temperature, μ_{nf}, ρ_{nf} and then α_{nf} are the viscosity, the density, and the thermal diffusivity of nanofluid which are expressed as follows:

$$\alpha_{nf}=\frac{k_{nf}}{(\rho c_p)_{nf}},\quad \mu_{nf}=\frac{\mu_f}{(1-\phi)^{5/2}},\quad \rho_{nf}=(1-\phi)\rho_f+\phi\rho_s \tag{4.48}$$

Afterward:

$$(\rho\, c_p)_{nf}=(1-\phi)(\rho\, c_p)_f+\phi(\rho\, c_p)_s,\quad \frac{k_{nf}}{k_f}=\frac{(k_s+2k_f)-2\phi(k_f-k_s)}{(k_s+2k_f)+\phi(k_f-k_s)} \tag{4.49}$$

In the aforementioned equations ϕ is the solid volume fraction of the nanofluid, k_{nf} and $(\rho c_p)_{nf}$ are the thermal conductivity and the heat capacitance of the nanofluid, respectively. Then, the related boundary conditions for the differential equation governing on the mentioned system are defined as:

$$\text{when}\quad y=0\rightarrow u=U_w,\ v=0,\ T=T_s \tag{4.50}$$

and then,

$$\text{at}\quad y=h(t)\rightarrow\frac{\partial u}{\partial y}=0,\ \frac{\partial T}{\partial y}=0,\ v=\frac{dh}{dt} \tag{4.51}$$

In the above equation, $h(t)$ is the thickness of the film and x is assumed to be a nonnegative quantity. The boundary layer thickness $\delta(x)$ is proportional to $(x\,\nu_f/U_w)^{0.5}$, so that the similarity variable η is defined as follows:

$$\eta=y\sqrt{\frac{U_w}{x\,\nu_f}}=y\sqrt{\frac{b}{(1-\alpha t)\nu_f}} \tag{4.52}$$

In accordance with this explanation and substituting $U(x)=U_w$, it is possible to define ξ , η and $\psi(x,y)$. As regards the above explanations, the following new variables are introduced as [4]:

$$\psi=\beta\left[\frac{\nu_f b}{1-\alpha t}\right]^{1/2}x f(\eta) \tag{4.53}$$

$$T=T_0-T_r\left(\frac{bx^2}{2\nu_f}\right)(1-\alpha t)^{3/2}\theta(\eta) \tag{4.54}$$

$$\eta = \frac{1}{\beta}\left[\frac{b}{v_f(1-\alpha t)}\right]^{1/2} y \tag{4.55}$$

where $\psi(x, y)$ is the stream function explained in the usual way, take for example $u = \frac{\partial \psi}{\partial y}$, $v = -\frac{\partial \psi}{\partial x}$, and $\beta > 0$ is the dimensionless film thickness defined by $\beta = (hb/v_f)(1-\alpha t)^{-1/2}$. Note that for the limiting cases $\beta = 0$ and $\beta = \infty$ this transformation is no longer effective and particular approaches are needed to give solutions. As a result, the velocity components u and v can be explicitly defined as:

$$u = \frac{\partial \psi}{\partial y} = \left(\frac{bx}{1-\alpha t}\right) f'(\eta), \quad v = -\frac{\partial \psi}{\partial x} = -\beta\left(\frac{v_f b}{1-\alpha t}\right)^{1/2} f(\eta) \tag{4.56}$$

After substituting the obtained similarity variables from Eqs. (4.53)–(4.55) into Eq. (4.47), the continuity equation is automatically satisfied and the momentum and energy equations are reduced to,

$$\varepsilon_1 f''' + \beta^2\left[f f'' - (f')^2 - S\left(f' + \frac{\eta}{2} f''\right)\right] = 0 \tag{4.57}$$

$$\frac{\varepsilon_2}{\mathrm{Pr}}\theta'' - \beta^2\left[\frac{S}{2}(3\theta + \eta\theta') + 2\theta f' - f\theta'\right] = 0 \tag{4.58}$$

And then, the relevant boundary conditions for this problem can be expressed as follows:

$$\begin{aligned} &\text{at } \eta = 0 \rightarrow \quad f(0) = 0,\ f'(0) = 1,\ \theta(0) = 1 \\ &\text{at } \eta = 1 \rightarrow \quad f''(1) = 0,\ \theta'(1) = 0,\ f(1) = \frac{S}{2} \end{aligned} \tag{4.59}$$

In the aforementioned equations, $Pr = (v_f/\alpha_f)$ is the Prandtl number, $S = (\alpha/b)$ is the unsteadiness parameter, and finally ε_1 and ε_2 are two constants explained in the following form:

$$\varepsilon_1 = \frac{1}{(1-\phi)^{2.5}[(1-\phi) + \phi\, \rho_s/\rho_f]}, \quad \varepsilon_2 = \frac{(k_{nf}/k_f)}{[1 - \phi + \phi(\rho c_p)_s/(\rho c_p)_f]} \tag{4.60}$$

To understand more, it is better to indicate that the physical quantities for this problem are the skin friction coefficient C_f and the Nusselt number Nu that are as follows:

$$C_f = \frac{\tau_w}{\rho_f (U_w)^2}, \quad Nu = \frac{q_w x}{k_f (T_s - T_0)} \tag{4.61}$$

In the above equation, the skin friction at the surface and the heat flux from the surface are defined in the following form:

$$\tau_w = \mu_{nf}\left(\frac{\partial \psi}{\partial y}\right)_{y=0}, \quad q_w = -k_{nf}\left(\frac{\partial T}{\partial y}\right)_{y=0} \tag{4.62}$$

Therefore, after substituting Eq. (4.62) into Eq. (4.61), we will have the final form of the physical quantities as follows:

$$C_{fx}Re_x^{-1/2} = \frac{1}{\beta(1-\phi)^{5/2}}f''(0), \quad Nu_x Re_x^{1/2} = -\frac{k_{nf}}{k_f}\frac{1}{\beta}\theta'(0) \tag{4.63}$$

It is notable that in Eq. (4.63) $Re_x = U_w x/\nu_f$ is the local Reynolds number. For the current nanoliquid film flow, it is found that the similarity solutions are available in the same range of S. Beyond this region, no solutions can be found. On the other hand, it is citable that the film thickness β decreases monotonically as S increases for both the Newtonian fluids ($\phi = 0$) and the nanofluids ($\phi \neq 0$) as shown in Table 4.5.

This issue refers that the constants α and b as well as the wall stretching velocity U_w have significant effects on β. For a fixed value of b, the larger α, the smaller value of b. Moreover, it is concluded from Table 4.5 that the decaying rate for β between any two prescribed values of S remains almost the same for all of the considered nanofluids, take, for example, the decaying rate for β between $S = 0.6$ and $S = 0.8$ is 31.2838% and this rate between $S = 1.0$ and $S = 1.8$ is 76.2191%. On the basis of the above explanations, a linear formula for evaluating the variation of β in terms of

Table 4.5 The Dimensionless Film Thickness (β) for Various Values of S and ϕ, When $Pr = 1$

Types of Fluid	S	$\phi = 0.0$	$\phi = 0.1$	$\phi = 0.2$
Cu-water	0.6	3.31710	2.66586	2.57109
	0.8	2.15199	1.83187	1.76676
	1.0	1.54361	1.31399	1.26729
	1.2	1.12778	0.96002	0.92589
	1.4	0.82103	0.69890	0.67405
	1.6	0.57617	0.49046	0.47303
	1.8	0.35638	0.30337	0.29259

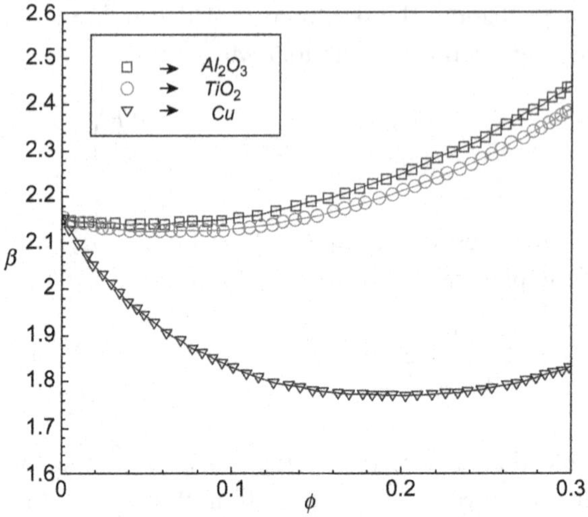

Figure 4.9 The result of varying β in terms of ϕ for three different kinds of nanoparticles, when $S = 1.2$.

nanoparticle volume fraction (ϕ) for three different kinds of nanofluids is used from Ref. [4] and depicted in Fig. 4.9:

In accordance with the Chapter 1 explanations, DTM has been applied to solve the presented problem as follows:

$$\begin{aligned}
\mathrm{DTM}_1 = {} & \varepsilon_1(k+1)(k+2)(k+3)(k+4)F(k+4) \\
& + \beta^2\left[\left(\sum_{l=0}^{k}(l+1)F(l+1)(k+1-l)(k+2-l)F(k+2-l)\right)\right. \\
& + \left(\sum_{l=0}^{k}F(l)(k+1-l)(k+2-l)(k+3-l)F(k+3-l)\right) \\
& - 2\left(\sum_{l=0}^{k}(l+1)F(l+1)(k+1-l)(k+2-l)F(k+2-l)\right) \\
& - S\left(\left(\frac{3}{2}\right)(k+1)(k+2)F(k+2)\right. \\
& \left.\left. + \frac{1}{2}\left(\sum_{l=0}^{k}(\delta(l-1)(k+1-l)(k+2-l)(k+3-l)F(k+3-l))\right)\right)\right] = 0
\end{aligned} \tag{4.64}$$

$$\mathrm{DTM}_2 = \frac{\varepsilon_2}{Pr}(k+1)(k+2)\Theta(k+2) - \beta^2\left(\frac{S}{2}\left(3\,\Theta(k) + \left(\sum_{l=0}^{k}\delta(l-1)(k+1-l)\Theta(k+1-l)\right)\right) + 2\left(\sum_{l=0}^{k}\Theta(l)(k+1-l)F(k+1-l)\right) - \sum_{l=0}^{k}F(l)(k+1-l)\Theta(k+1-l)\right) = 0 \tag{4.65}$$

where F and Θ represent the DTM transformed form of f and θ, respectively. After that the given boundary conditions can be transformed as follows:

$$F(0) = 0,\ F(1) = 1,\ F(2) = a,\ F(3) = b, \Theta(0) = 1,\ \Theta(1) = c \tag{4.66}$$

where a, b and c are constant coefficients that can be determined after specifying $F(\eta)$ and $\Theta(\eta)$ and applying the three remained boundary conditions from Eq. (4.59) on the obtained solutions. When our base fluid is pure water and the nanoparticles are copper and the constant coefficients of the governing equations are assumed to be $\beta = 0.5$, $\varepsilon_1 = -0.332$, $\varepsilon_2 = 8.35$, $Pr = 1$ and finally $S = 1$, we will have,

$$a = -0.6481199115,\ b = 0.06070499715,\ c = -0.07341467053 \tag{4.67}$$

In this step, to avoid the repeated mathematical operation for its ease of understanding, we omit the detailed DTM procedure and only mention the final solution in the following form:

$$f(\eta) = \eta - 0.6481199115\eta^2 + 0.06070499715\eta^3 + 0.1016754379\eta^4 - 0.01397202970\eta^5 - 0.0002884940148\eta^6 \tag{4.68}$$

and then

$$\theta(\eta) = 1 - 0.07341467053\eta + 0.05239520958\eta^2 - 0.01385237686\eta^3 + 0.001460996324\eta^4 + 0.001072271513\eta^5 - 0.0001706601508\eta^6 \tag{4.69}$$

To be sure about the precision of the achieved solutions, a comparison between numerical method (Runge–Kutta) and DTM has been presented in Table 4.6.

To start, the obtained solution of the given set of differential equations consisted of Eq. (4.68) and Eq. (4.69) should be substituted into the main set of differential equations , $g(\eta)$ and $h\ (\eta)$, and then it is necessary to depict the yielded equations in Cartesian coordinates. Afterward, the yielded errors of the given set of differential equations can be observed from the obtained charts. To understand more, reading the following lines is recommended. Consider a set of differential equations in the following form:

$$\begin{cases} g(\eta) = g(f(\eta), f'(\eta), \ldots) \\ h(\eta) = h(\theta(\eta),\ \theta'(\eta), \ldots) \end{cases} \tag{4.70}$$

And the answer of the aforementioned set of equations is assumed to be a function of η in the form of:

$$f = h_1(\eta),\ \theta = h_2(\eta) \tag{4.71}$$

Thus by substituting Eq. (4.71) into Eq. (4.70), the computational error of the obtained solution by each analytical or semianalytical method can be achieved as follows:

$$\begin{aligned} g(\eta) &= g(f(h_1(\eta)), f'(h_2(\eta)), \ldots) \\ h(\eta) &= h(\theta(h_1(\eta)),\ \theta'(h_2(\eta)), \ldots) \end{aligned} \tag{4.72}$$

Eventually, with regard to the given physical values and by substituting the yielded solutions using DTM, which are Eqs. (4.68)–(4.69) into Eqs. (4.57)–(4.58), the computational errors of DTM are depicted in the forms of Figs. 4.10–4.11.

The aforementioned figures (Figs. 4.10 and 4.11) show that the yielded solutions by DTM on the basis of the given physical values such as $Pr = 1$ in the specified domain are appropriate approximations for solving the presented problem. In this case study, Cu-water nanofluid is chosen as a convenient example for illustration and attempts have been made to discuss the effects of some physical parameters such as Pr, S and ϕ on the velocity and temperature distribution as follows:

It is clear from Figs. 4.12 and 4.13 that by increasing the amount of unsteadiness parameter S, the values of $f(\eta)$ increases but vice versa, the temperature profile diminishes smoothly with η in the specified domain. In accordance with Fig. 4.14, the velocity profile $f'(\eta)$ decreases uniformly

Table 4.6 A Comparison Between the Obtained Values by Differential Transformation Method and Numerical Solution and the Related Errors for $f(\eta)$ and $\theta(\eta)$ in the Specified Domain, When $\beta = 0.5$, $\varepsilon_1 = -0.332$, $\varepsilon_2 = 8.35$, $Pr = 1$ and Finally $S = 1$

	$f(\eta)$			$\theta(\eta)$		
η	NUM	DTM	The Error of DTM (%)	NUM	DTM	The Error of DTM (%)
0	0.0	0.0	0	0.999999999999	1	0
0.1	0.09345612174	0.0935895334	0.1427	0.993171519127	0.9931687892	0.00027
0.2	0.17425663412	0.1747190347	0.2653	0.987307832315	0.9873047251	0.00031
0.3	0.24321770321	0.2440976516	0.3617	0.982331096286	0.9823314688	0.000037
0.4	0.30136110613	0.3026445699	0.4258	0.978170169099	0.9781784958	0.0008512
0.5	0.34989036992	0.3514717280	0.4519	0.974761540350	0.9747820742	0.0021
0.6	0.39016599675	0.3918663230	0.4357	0.972050131446	0.9720861224	0.0037
0.7	0.42368002460	0.4252731100	0.3760	0.969989968411	0.9700429419	0.0054
0.8	0.45203012617	0.4532764928	0.2757	0.968544729388	0.9686138292	0.0071
0.9	0.47689340438	0.4775824081	0.1444	0.967688168589	0.9677695631	0.0084
1	0.49999999999	0.4999999998	0.00000004	0.967404418106	0.9674907698	0.0089

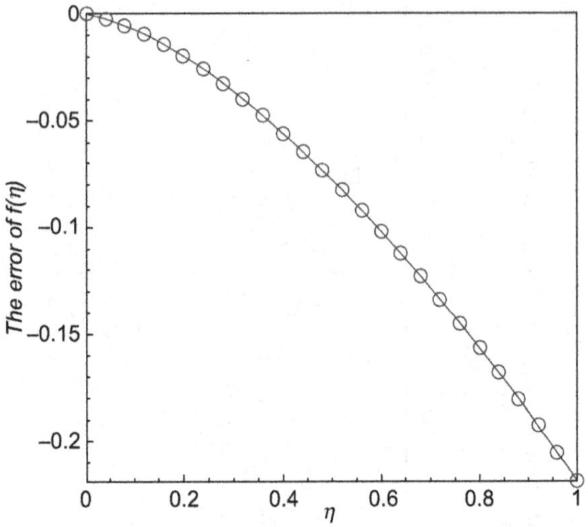

Figure 4.10 The achieved computational error for $f(\eta)$ by differential transformation method.

with η for different amounts of unsteadiness parameter and it is notable that the larger value of S, the higher velocity profile is. The mentioned variation will be clearer while increasing the amount of η. Because the unsteadiness parameter is dependent to the α and b ($S = \alpha/b$), it is observed that the wall

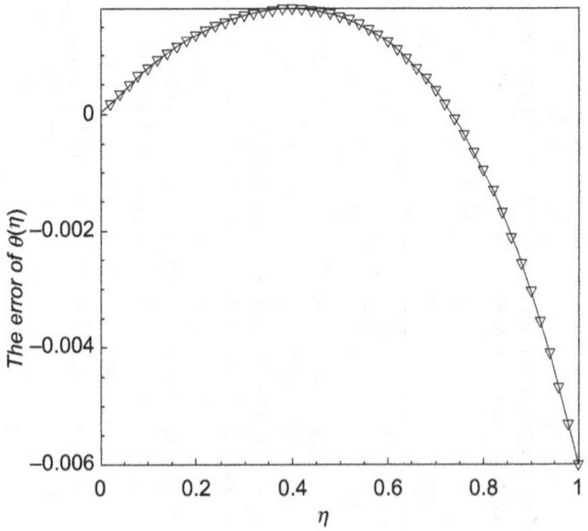

Figure 4.11 The resulted computational error for $\theta(\eta)$ by differential transformation method.

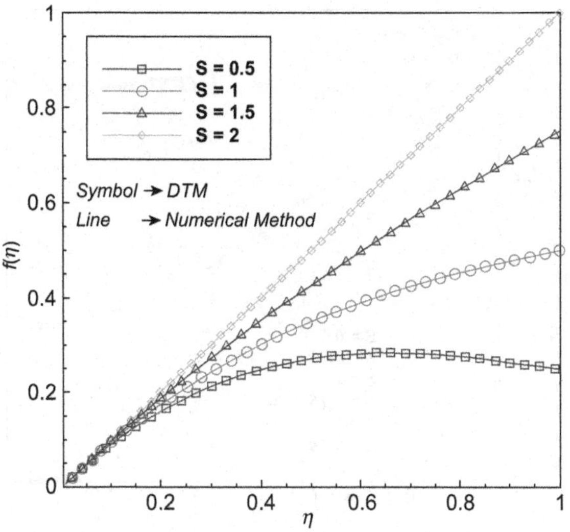

Figure 4.12 A comparison between the obtained results by differential transformation method and numerical solution in terms of varying $f(\eta)$ for different values of unsteadiness parameter(S).

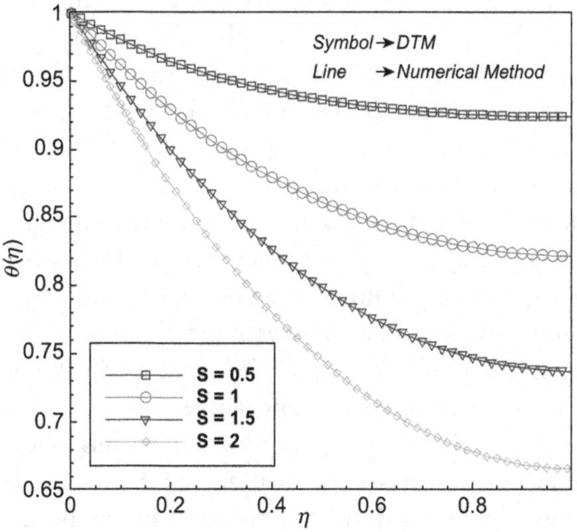

Figure 4.13 Comparing the achieved results by differential transformation method and numerical solution in terms of varying $\theta(\eta)$ for different amounts of unsteadiness parameter(S).

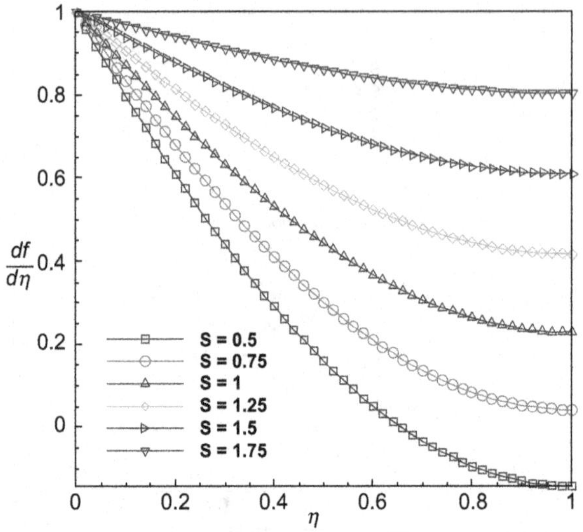

Figure 4.14 The variation of velocity profile for some values of unsteadiness parameter.

stretching velocity is an important factor for determining the velocity profile. As regards Fig. 4.15 which describes the variation of temperature distribution for any considered value of unsteadiness parameter, $\theta(\eta)$ enlarges along with increasing ϕ. This issue refers to the fact that nanofluid plays a vital role on the heat transfer properties. Afterward, the effects of Prandtl number have been investigated on $f(\eta)$ and $\theta(\eta)$ completely and the results are shown in Figs. 4.16 and 4.17. Regarding to the given figures, Pr does not have any effect on $f(\eta)$ and it enhances uniformly by increasing η, but the temperature distribution decreases significantly by growing the values of Prandtl number and it vanishes at the free surface ($\eta = 1$) for large amount of Prandtl number. Therefore, it is citable that the temperature at the free surface is equal to the ambient temperature.

Moreover, the variation of $f(\eta)$ in terms of various amounts of nanoparticle volume fraction has been analyzed in Fig. 4.18. Based on the graphical results, it is revealed that $f(\eta)$ enhances in the specified domain of η and for the effect of ϕ, we can indicate that $f(\eta)$ decreases monotonically by increasing ϕ.

In this step, the effects of solid volume fraction of the nanofluid on the velocity distribution have been presented in Fig. 4.19. In accordance with

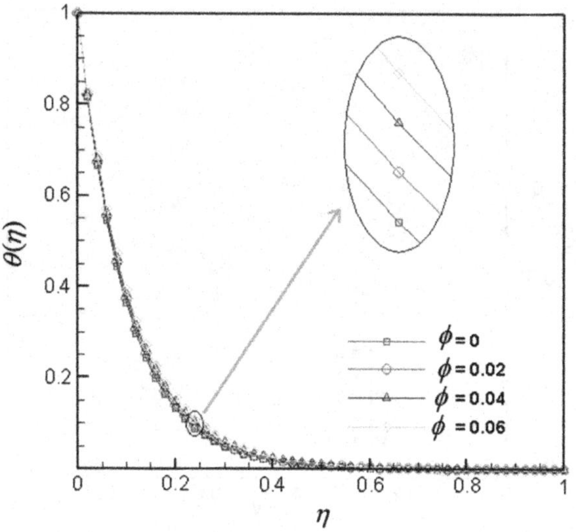

Figure 4.15 The result of varying temperature distribution for four different values of solid volume fraction of the nanofluid (ϕ).

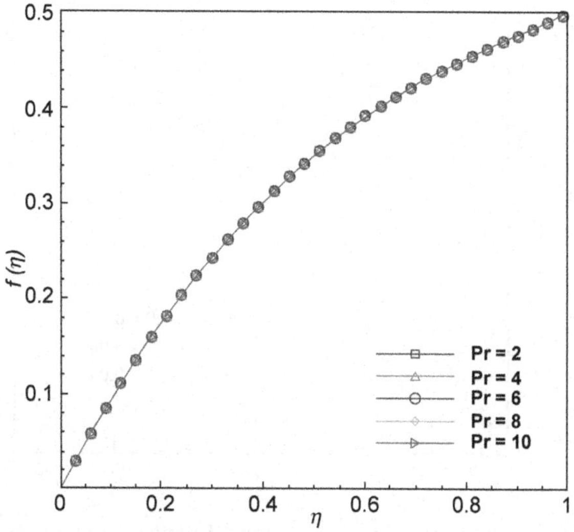

Figure 4.16 Comparing the variation of $f(\eta)$ in terms of different values of Prandtl Number.

the given information by increasing the values of $\phi, f'(\eta)$ decreases in terms of $\eta \in \{0, 0.5\}$ but this tendency inverses for $\eta \in \{0.5, 1\}$,which means the velocity profile enhances by increasing the values of ϕ. Eventually, the variations of local skin friction coefficient and also the local Nusselt number

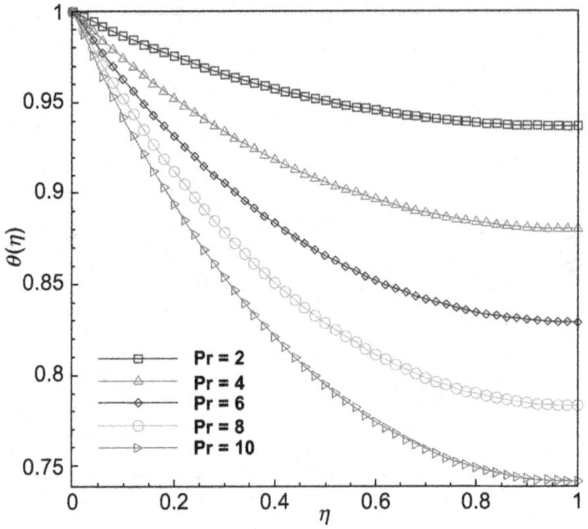

Figure 4.17 The result of varying $\theta(\eta)$ in terms of various amounts of Prandtl Number.

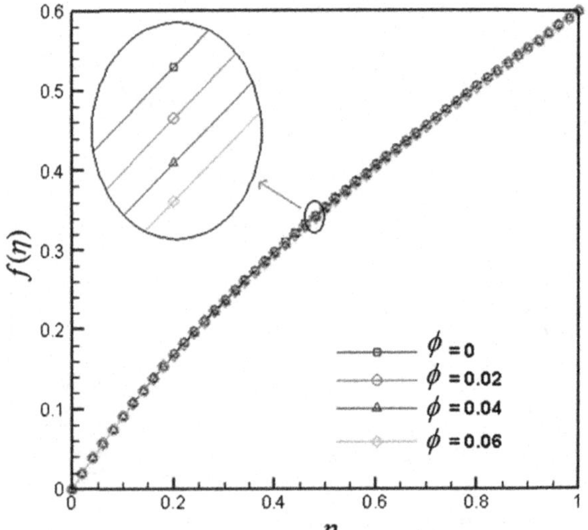

Figure 4.18 The result of varying $f(\eta)$ in terms of various amounts of solid volume fraction of the nanofluid (ϕ).

presented as the physical quantities of the mentioned problem in terms of ϕ are graphically represented in Figs. 4.20 and 4.21.

It is observed that by enlarging the unsteadiness parameter the amount of local skin friction coefficient enhances significantly, and the local skin

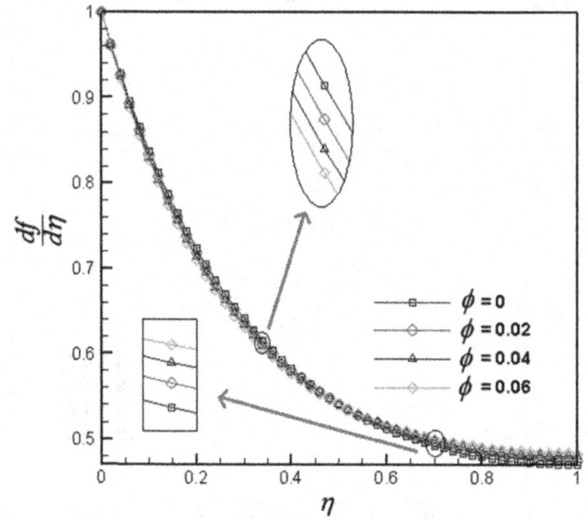

Figure 4.19 Comparing the variation of velocity profile in terms of four different values of (ϕ).

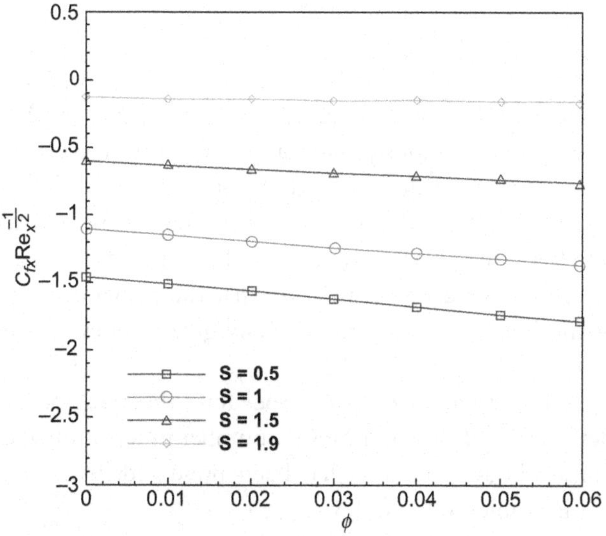

Figure 4.20 The obtained graphical results for the variation of local skin friction coefficient in terms of four different amounts of S.

friction coefficient reduces uniformly with increasing ϕ in the specified domain. This issue proves that nanofluids are sufficiently useful for reducing the drag force of fluid flow. Then, the variation of local Nusselt number with solid volume fraction of the nanofluid for different amounts

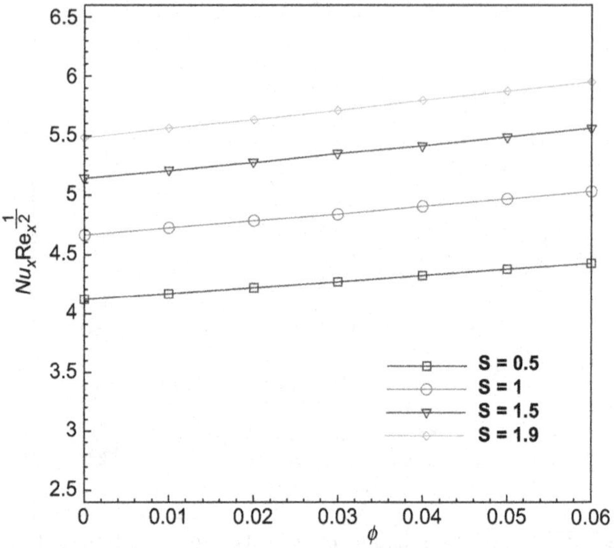

Figure 4.21 Investigating the effect of (ϕ) on the local Nusselt number in the specified domain.

of S is presented in Fig. 4.21. It can easily be seen that the local Nusselt number increases continuously with ϕ defined from 0 to 0.6. This item indicates that more the value of ϕ, the more convective heat transfer of the nanofluids. Furthermore, it is necessary to mention that the local Nusselt number enlarges by increasing the unsteadiness parameter S. Therefore, the obtained results indicate that the suspended nanoparticles mainly enhance the heat transfer rate at any given values of S and Prandtl number.

To deeply understand the above procedure, the variation of local skin friction coefficient and the local Nusselt number in terms of ϕ for different values of unsteadiness parameter has been presented in the form of numerical data in Tables 4.7 and 4.8.

4.5 NON-NEWTONIAN FLUID FLOW ANALYSIS

Consider an unsteady, incompressible, non-Newtonian fluid (such as blood) as a third-grade fluid in an artery. Table 4.9 shows the properties of the blood as considered non-Newtonian fluid. The flow is considered to take place axially through the circular tube of radius R under periodic body

Table 4.7 The Variation of $C_{fx}Re_x^{\frac{-1}{2}}$ in Terms of ϕ for Different Values of Parameter S

φ	$S = 0.5$	$S = 1$	$S = 1.5$	$S = 1.9$
0	−1.4566	−1.1022	−0.6073	−0.1295
0.01	−1.5094	−1.1459	−0.6325	−0.1350
0.02	−1.5631	−1.1902	−0.6580	−0.1405
0.03	−1.6177	−1.2349	−0.6837	−0.1461
0.04	−1.6733	−1.2802	−0.7096	−0.1518
0.05	−1.7299	−1.3262	−0.7359	−0.1575
0.06	−1.7875	−1.3729	−0.7625	−0.1633

Table 4.8 The Variation of $Nu_xRe_x^{\frac{1}{2}}$ in Terms of ϕ for Different Values of Parameter S

φ	$S = 0.5$	$S = 1$	$S = 1.5$	$S = 1.9$
0	4.1166	4.6547	5.1325	5.4835
0.01	4.1617	4.7158	5.2027	5.5604
0.02	4.2178	4.7772	5.2733	5.6375
0.03	4.2186	4.8388	5.3441	5.7150
0.04	4.3196	4.9007	5.4153	5.7928
0.05	4.3709	4.9630	5.4868	5.8710
0.06	4.4224	5.0256	5.5588	5.9496

acceleration and a pulsatile pressure gradient. Cauchy stress in a third-grade fluid is given by:

$$\tau = -pI + \mu A_1 + \alpha_1 A_2 + \alpha_2 A_1^2 + \beta_1 A_3 + \beta_2(A_1A_2 + A_2A_1) + \beta_3\left(TrA_1^2\right)A_1 \tag{4.73}$$

where $-pI$ shows the spherical stress due to the restraint of incompressibility and α_1, α_2, β_1, β_2, and β_3 are the material modules and are considered to be functions of temperature generally. In Eq. (4.73), the kinematical tensors A_1, A_2, and A_3 can be defined by following equations [5]:

$$A_1 = (\nabla V) + (\nabla V)^t \tag{4.74}$$

$$A_n = \frac{dA_{n-1}}{dt} + A_{n-1}(\nabla V) + (\nabla V)^t A_{n-1}, \quad n = 2, 3, \tag{4.75}$$

where $V = [0,0,v(r)]$ denotes the velocity field, superscript t stands for matrix transposition and $\frac{D}{Dt}$ is the material time derivative, which is defined by

$$\frac{D(.)}{Dt} = \frac{\partial(.)}{\partial t} + [grad(.)]V \tag{4.76}$$

The momentum equation for an incompressible, unsteady, axisymmetric (with z-axis as the axis of symmetry) and fully developed flow in a cylindrical polar coordinate (r, θ, z) is,

$$\rho \frac{\partial \overline{u}}{\partial t} = \frac{\partial p}{\partial z} + \rho G + \frac{1}{\overline{r}} \frac{\partial}{\partial \overline{r}} [\overline{r} \tau_{rz}] \tag{4.77}$$

where $\rho, \overline{u}, p, \tau_{rz}, t$, and G denote density, axial velocity, pressure, shear stress, time, and body acceleration in axial direction, respectively. The shear stress τ_{rz} for a third-grade fluid in an axisymmetric and thermodynamically compatible flow situation can be written as,

$$\tau_{rz} = \left[\mu + \beta \left(\frac{\partial \overline{u}}{\partial \overline{r}} \right)^2 \right] \frac{\partial \overline{u}}{\partial \overline{r}} \tag{4.78}$$

In human beings the pressure gradient $(\frac{\partial p}{\partial z})$ produced by the pumping action of the heart takes the approximate form [5].

$$-\frac{\partial p}{\partial z} = A_0 + A_1 \cos \omega_p t \tag{4.79}$$

where A_0, A_1, $\omega_p = 2\pi f_p$ and f_p are respectively the constant component of the pressure gradient, the amplitude of the fluctuating component (giving rise to the systolic and diastolic pressures), the circular frequency, and the pulse frequency. The body acceleration G is assumed to be:

$$G = A_g \cos(\omega_b t + \phi) \tag{4.80}$$

where A_g, is the amplitude $\omega_b = 2\pi f_b$, f_b is the frequency, and ϕ is the lead angle of G with respect to the heart action. By using Eqs. (4.78)–(4.80), equation of motion (4.77) can be written as,

$$\begin{aligned} \rho \frac{\partial \overline{u}}{\partial t} &= A_0 + A_1 \cos \omega_p t + \rho A_g \cos(\omega_b t + \phi) \\ &+ \frac{1}{\overline{r}} \frac{\partial}{\partial \overline{r}} \left\{ \overline{r} \left[\mu + \beta \left(\frac{\partial \overline{u}}{\partial \overline{r}} \right)^2 \right] \frac{\partial \overline{u}}{\partial \overline{r}} \right\} \end{aligned} \tag{4.81}$$

With the corresponding to initial and boundary conditions:

$$\begin{aligned} &\overline{r} = R \quad \overline{u} = 0 \\ &\overline{r} = 0 \quad \frac{\partial \overline{u}}{\partial \overline{r}} = 0 \\ &t = 0 \quad \overline{u} = 0 \end{aligned} \tag{4.82}$$

The initial condition is essential for the numerical scheme adapted to estimate the time at which the pulsatile steady state sets in. The nondimensional form of Eqs. (4.81)–(4.82) are respectively

$$\frac{\alpha^2}{2\pi}\frac{\partial u}{\partial T} = B_1(1 + e\cos 2\pi T) + B_2\cos(2\pi\omega_r T + \phi) + \frac{1}{r}\frac{\partial}{\partial r}\left\{ r\frac{\partial u}{\partial r}\left[1 + B\left(\frac{\partial u}{\partial r}\right)^2\right]\right\} \tag{4.83}$$

And boundary conditions:

$$\begin{aligned} r &= 1 \quad u = 0 \\ r &= 0 \quad \frac{\partial u}{\partial r} = 0 \\ T &= 0 \quad u = 0 \end{aligned} \tag{4.84}$$

where

$$\begin{aligned} &\alpha^2 = \frac{\rho\omega_p R^2}{\mu} \quad B_1 = \frac{A_0 R^2}{\mu u_0} \quad B_2 = \frac{\rho A_g R^2}{\mu u_0} \\ &B = \frac{\beta u_0^2}{\mu R^2} \quad \omega_r = \frac{\omega_b}{\omega_p} \quad u = \frac{\overline{u}}{u_0} \quad r = \frac{\overline{r}}{R} \\ &T = \frac{\omega_p t}{2\pi} \quad e = \frac{A_1}{A_0} \quad u_0 = \frac{A_0 R^2}{8\mu} \end{aligned} \tag{4.85}$$

Here u_0 is the cross-sectional average velocity of flow under steady state pressure gradient A_0. The second-order nonlinear partial differential Eq. (4.83) with the boundary conditions (4.85) should be solved by efficient analytical or numerical methods.

In this example analytical hybrid-DTM is used for the nondimensional time (T) and the r directions for obtaining the solution of Eq. (4.83) with boundary condition (4.85). Hybrid-DTM which is the combination of finite difference method (FDM) and multi-step differential transformation method (Ms-DTM) can solve the PDEs problem easily. To proceed with its computations, FDM is applied to uniform points in the T (nondimensional time) and Ms-DTM is based on r (nondimensional radius) directions, and solutions are performed. Obtained results by hybrid-DTM are compared with Crank–Nicolson method (CNM) [5] in Fig. 4.22, which shows an excellent agreement between the analytical and numerical methods. In this

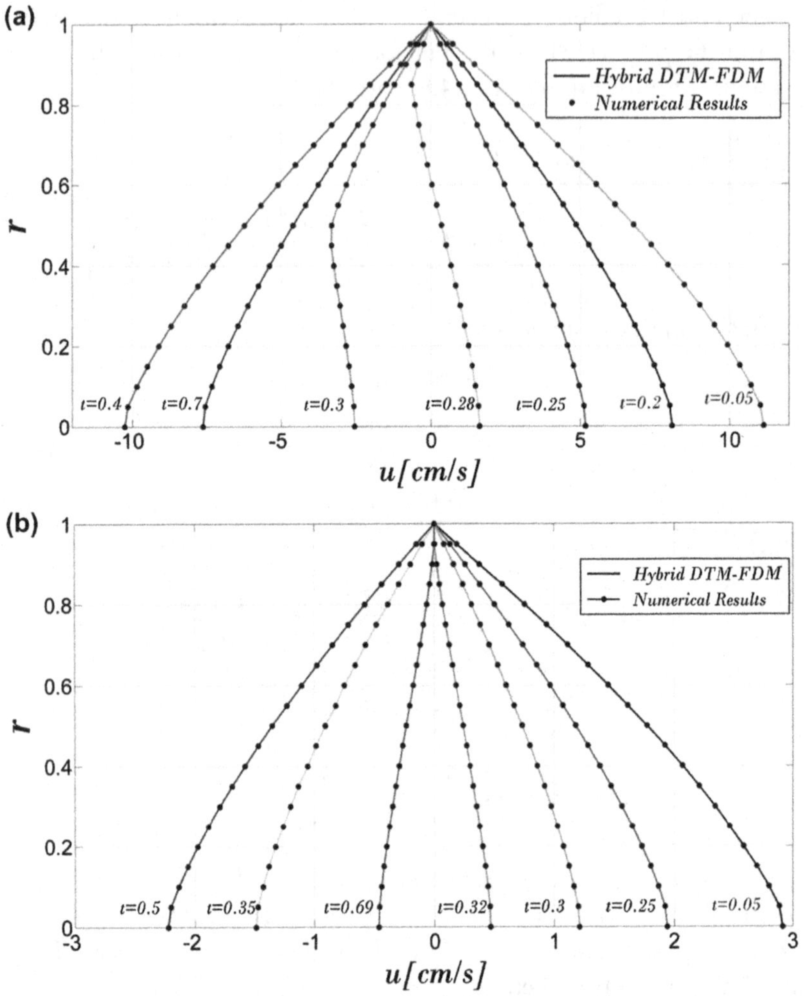

Figure 4.22 Comparison between hybrid differential transformation method (present study) with Crank–Nicolson method, when $\phi = 0$, $e = \omega_r = 1$, $A_g = g$ and $f_b = 1.2$ Hz for (a) femoral and (b) coronary arteries.

example, two different arteries are considered, femoral and coronary arteries. Femoral artery and coronary arteries have diameters 1 and 0.3 cm respectively. Also, the constant component of the pressure gradient, A_0, for coronary and femoral arteries are taken as 698.65 and 32 dyne/cm, respectively as presented in the literature. In whole solution, physical properties of blood are considered constant as shown in Table 4.9 and g is the acceleration due to gravity which considered being 10. Comparison between hybrid-DTM and CNM [5] is depicted in Fig. 4.22 for both

Table 4.9 Some Properties of Non-Newtonian Blood [5]

Specification	ρ (kg/m^3)	C_p (J/kg·k)	k (W/m·k)	μ	β
Blood	1060	3617	0.52	0.003	0.001

femoral and coronary arteries, when $\phi = 0$, $e = \omega_r = 1$, $A_g = g$ and $f_b = 1.2$ Hz in various times. As seen, hybrid-DTM has excellent agreement with previous method. This is necessary to inform that due to the fluctuating flow, increasing the time makes the velocity take negative values and increasing it further will increase the velocity and it reaches positive values again. This process occurs continuously. For a better perception, Fig. 4.23 is

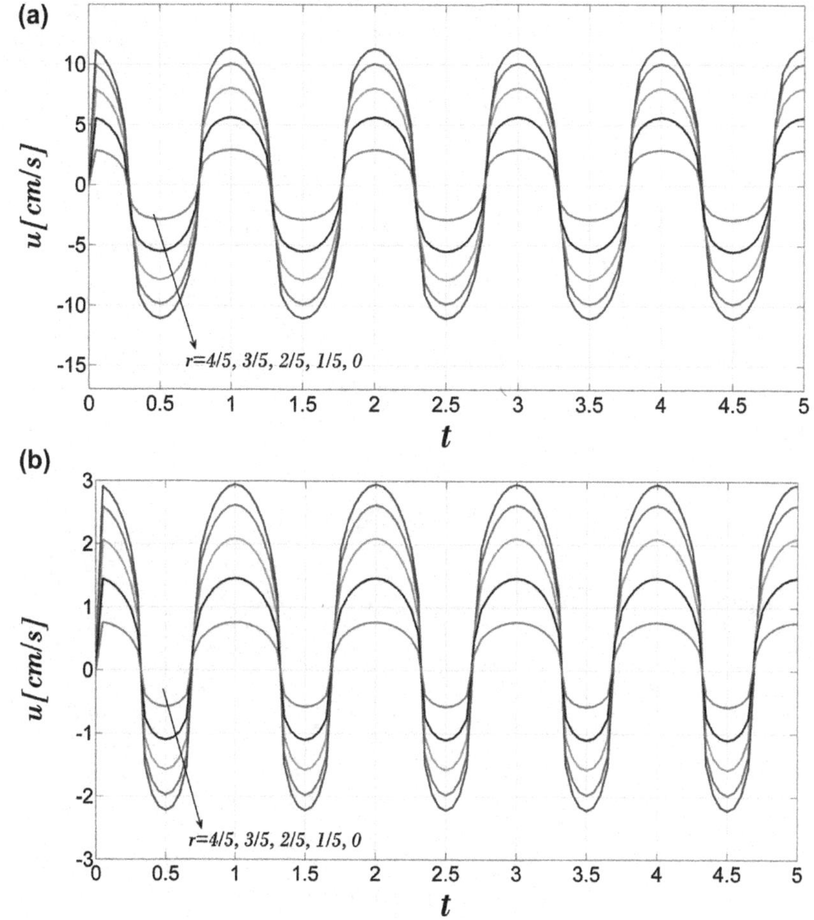

Figure 4.23 Velocity profiles versus time in different radii for (a) femoral and (b) coronary arteries, when $\phi = 0$, $e = \omega_r = 1$, $A_g = g$, $f_b = 1.2$.

depicted for coronary and femoral arteries. This figure shows the velocity versus time in different radii, as seen, the maximum velocities occur in the center of tube ($r = 0$), and in the whole domain, it has a fluctuating nature. Also, it's obvious that when the velocity becomes negative, the maximum value is seen in the region beyond the center and near the tube wall, and maximum values take place in lower time duration compared to positive values. Fig. 4.24 which is depicted for femoral and coronary arteries,

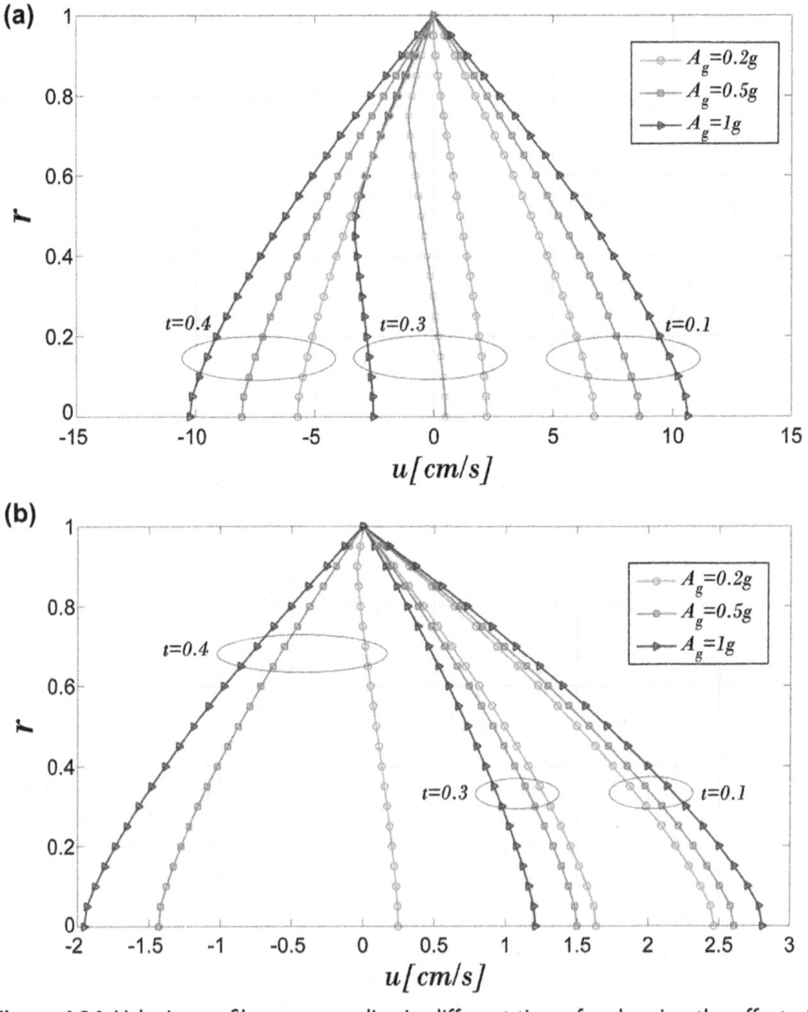

Figure 4.24 Velocity profiles versus radius in different times for showing the effect of A_g when $\phi = 0$, $e = \omega_r = 1$, and $f_b = 1.2$ Hz for (a) femoral and (b) coronary arteries.

demonstrate the A_g effects on velocity profiles. It can be concluded that increasing the amplitude, A_g makes higher velocity profiles in both negative and positive values. These effects on velocity and wall shear stress during the time viewpoint are presented in Fig. 4.25 for femoral artery when $\phi = \pi/6$, $e = 20$, $\omega_r = 1$, and $f_b = 1.2$ Hz. As A_g increases, the maximum magnitude of the velocity along the axis increases and assumes both positive and negative values during a cycle. Also, the maximum wall shear stress increases when A_g increases. These effects are also observed in coronary artery as

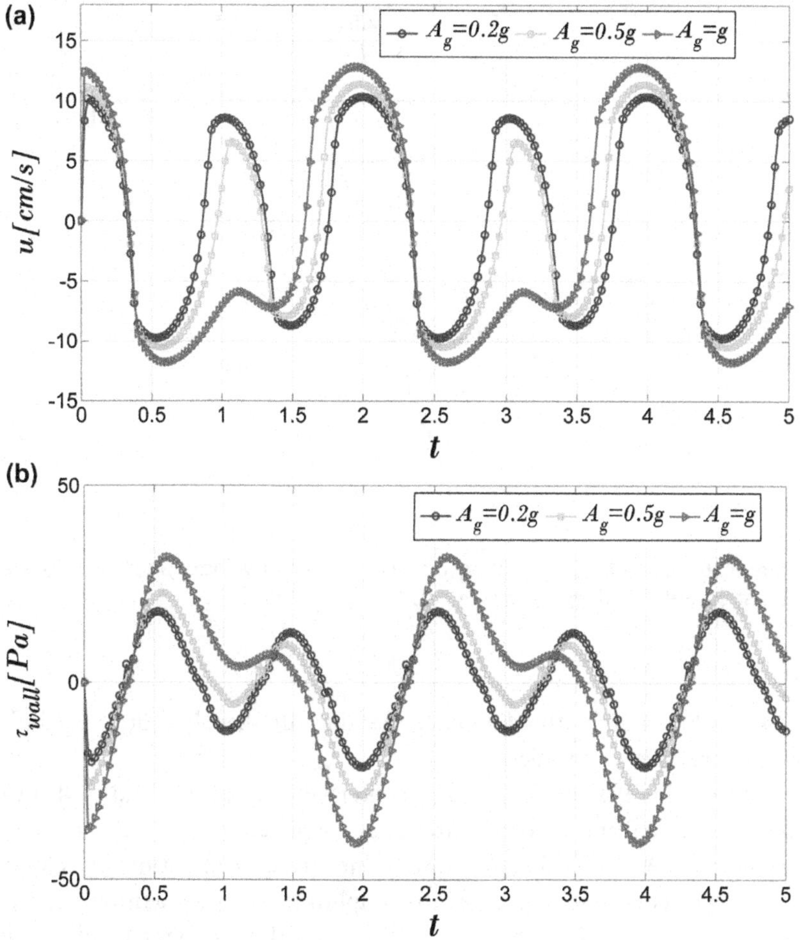

Figure 4.25 Effect of A_g for femoral artery on (a) center velocity profile and (b) wall shear stress values, when $\phi = \pi/6$, $e = 20$, and $f_b = 1.2$ Hz.

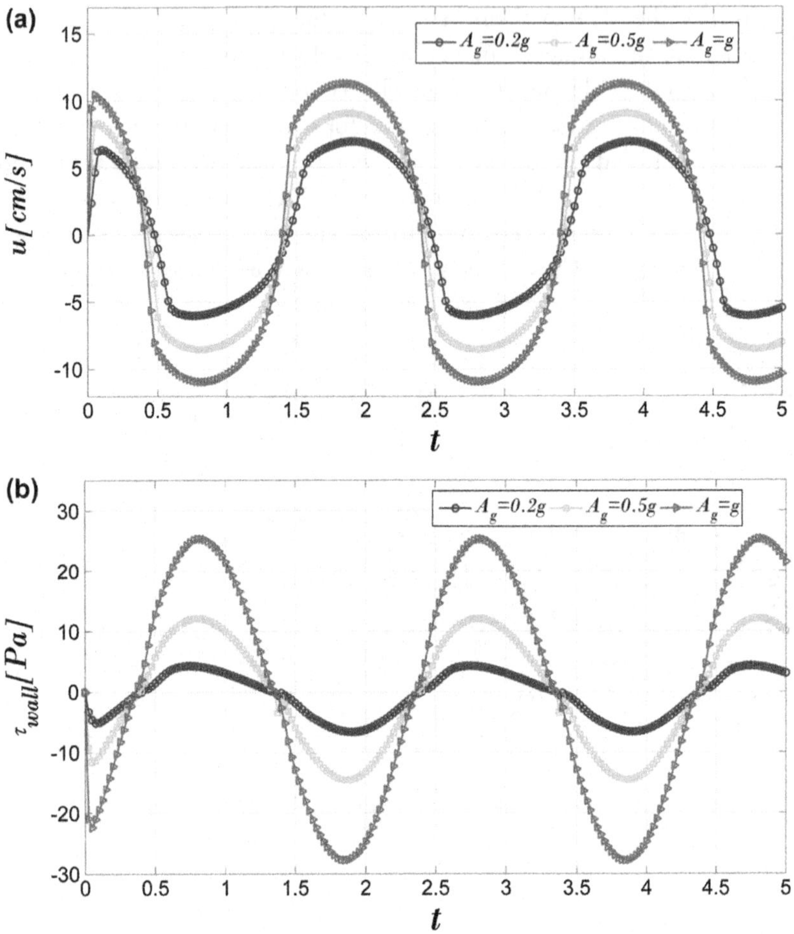

Figure 4.26 Effect of A_g for coronary artery on (a) center velocity profile and (b) wall shear stress values, when $\phi = \pi/6$, $e = 1$, and $f_b = 1.2$ Hz.

shown in Fig. 4.26 by the difference that they are shapely near to parabolic or convex/concave profiles.

Finally, the effects of ω_r on the center velocity profile and wall shear stress for femoral and coronary arteries are depicted in Figs. 4.27 and 4.28, respectively. As seen, changing this value makes a mutation in velocity profiles and completely changes the amplitude, shape, maximum values, etc. Reducing the ω_r from 1 to 0.5 makes a shift to right hand side in maximum points in both velocity and shear stress profiles. Also, it makes a

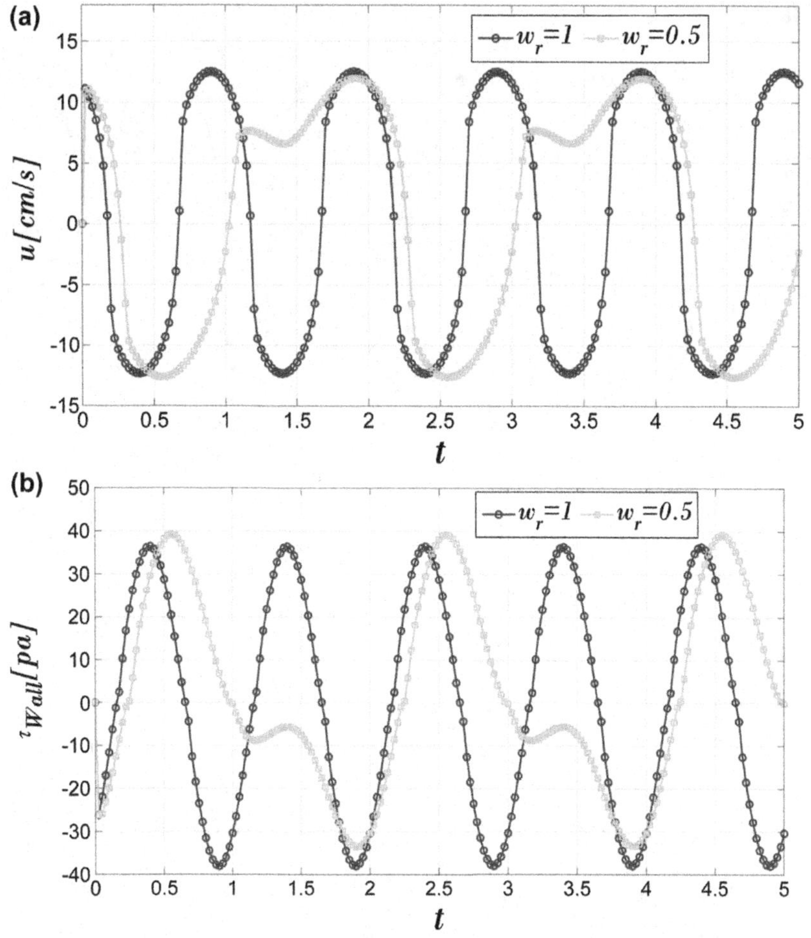

Figure 4.27 Effect of ω_r for femoral artery on (a) center velocity profile and (b) wall shear stress values, when $\phi = \pi/3$, $e = 20$, and $f_b = 1.2$ Hz.

reduction for maximum values of positive velocity and increase in its negative values. Its treatment for shear stress is completely vice versa (i.e., increase in its positive maximum values and decrease in its negative values). Furthermore, when $\omega_r = 0.5$ all the graphs have a relative maximum/minimum point and an absolute optimum point versus when $\omega_r = 1.0$.

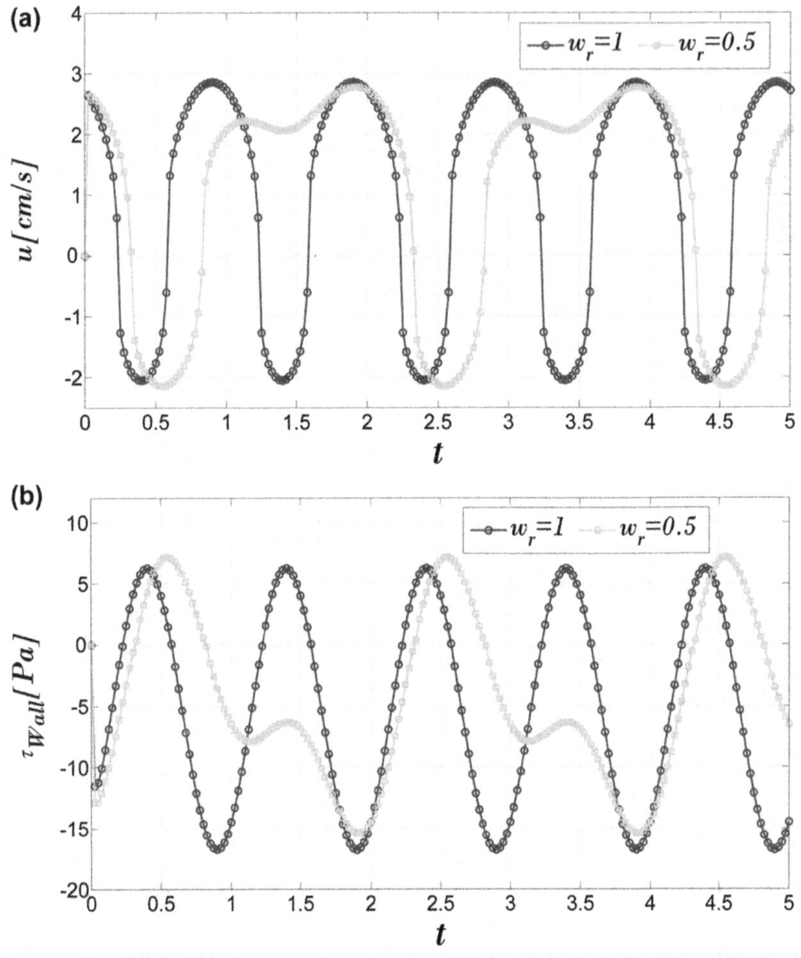

Figure 4.28 Effect of ω_r for coronary artery on (a) center velocity profile and (b) wall shear stress values, when $\phi = \pi/3$, $e = 1$, and $f_b = 1.2$ Hz.

REFERENCES

[1] Chang HN, Ha JS, Park JK, Kim IH, Shin HD. Velocity field of pulsatile flow in a porous tube. Journal of Biomechanics 1989;22:1257–62.
[2] Dinarvand S, Mehdi Rashidi M, Doosthoseini A. Analytical approximate solutions for two-dimensional viscous flow through expanding or contracting gaps with permeable walls. Central European Journal of Physics 2009;7(4):791–9.
[3] Rashidi MM. The modified differential transform method for solving MHD boundary-layer equations. Computer Physics Communications 2009;180(11):2210–7.

[4] Ahmadi AR, Zahmatkesh A, Hatami M, Ganji DD. A comprehensive analysis of the flow and heat transfer for a nanofluid over an unsteady stretching flat plate. Powder Technology 2014;258:125–33.
[5] Mandal PK, Chakravarty S, Mandal A, Amin N. Effect of body acceleration on unsteady pulsatile flow of non-Newtonian fluid through a stenosed artery. Applied Mathematics and Computation 2007;189(1):766–79.

CHAPTER 5

DTM for Nanofluids and Nanostructures Modeling

5.1 INTRODUCTION

Nanofluids are produced by dispersing the nanometer-scale solid particles into base liquids with low thermal conductivity such as water, ethylene glycol (EG), oils, etc. The term "nanofluid" was first coined by Choi [1] to describe this new class of fluids. The presence of the nanoparticles in the fluids noticeably increases the effective thermal conductivity of the fluid and consequently enhances the heat transfer characteristics. Therefore, numerous methods have been taken to improve the thermal conductivity of these fluids by suspending nano/micro-sized particle materials in liquids. Also nanostructures such as nanobeam, nanotube, etc., have many applications in mechanical engineering. This chapter focuses on the solving problems in this field and contains the following sections:

5.1 Introduction
5.2 Nanofluid in Divergent/Convergent Channels
5.3 MHD Couette Nanofluid Flow
5.4 Nanofluid Between Parallel Plates
5.5 Vibration Analysis of Nanobeams
5.6 Buckling Analysis of a Single-Walled Carbon Nanotube

5.2 NANOFLUID IN DIVERGENT/CONVERGENT CHANNELS

Consider a system of cylindrical polar coordinates (r, θ, z) which steady two-dimensional flow of an incompressible conducting viscous fluid from a source or sink at channel walls lie in planes, and intersect in the axis of z. Assuming purely radial motion which means that there is no change in the flow parameter along the z-direction. The flow depends on r and θ, and further assumes that there is no magnetic field in the z-direction (See Fig. 5.1). The reduced form of continuity, Navier—Stokes and Maxwell's equations are [2]:

$$\frac{\rho_{nf}}{r}\frac{\partial(ru(r,\theta))}{\partial r}(ru(r,\theta)) = 0 \tag{5.1}$$

Differential Transformation Method for Mechanical Engineering Problems
ISBN 978-0-12-805190-0
http://dx.doi.org/10.1016/B978-0-12-805190-0.00005-X

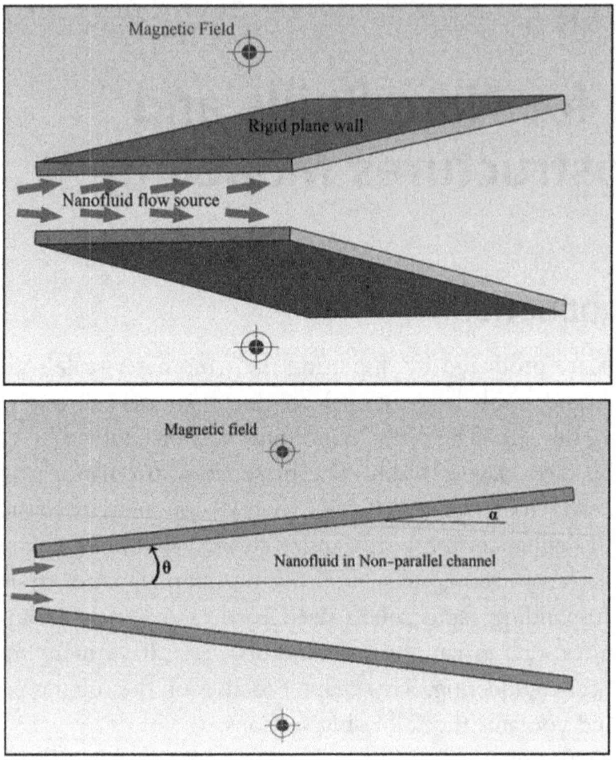

Figure 5.1 Schematic of the problem (MHD Jeffery–Hamel flow with nanofluid).

$$u(r,\theta)\frac{\partial u(r,\theta)}{\partial r} = -\frac{1}{\rho_{nf}}\frac{\partial P}{\partial r} + \upsilon_{nf}\left[\frac{\partial^2 u(r,\theta)}{\partial r^2} + \frac{1}{r}\frac{\partial u(r,\theta)}{\partial r} + \frac{1}{r^2}\frac{\partial^2 u(r,\theta)}{\partial \theta^2} - \frac{u(r,\theta)}{r^2}\right] - \frac{\sigma B_0^2}{\rho_{nf} r^2} u(r,\theta) \tag{5.2}$$

$$\frac{1}{\rho_{nf} r}\frac{\partial P}{\partial \theta} - \frac{2\upsilon_{nf}}{r^2}\frac{\partial u(r,\theta)}{\partial \theta} = 0 \tag{5.3}$$

where B_0 is the electromagnetic induction, σ_{nf} is the conductivity of the fluid, $u(r)$ is the velocity along radial direction, P is the fluid pressure, υ_{nf} is the coefficient of kinematic viscosity, and ρ_{nf} is the fluid density.

The effective density ρ_{nf}, the effective dynamic viscosity μ_{nf}, and kinematic viscosity υ_{nf} of the nanofluid are given as:

$$\rho_{nf} = \rho_f(1-\phi) + \rho_s\phi, \quad \mu_{nf} = \frac{\mu_f}{(1-\phi)^{2.5}}, \quad \upsilon_{nf} = \frac{\mu_f}{\rho_{nf}},$$
$$\frac{\sigma_{nf}}{\sigma_f} = 1 + \frac{3\left(\frac{\sigma_s}{\sigma_f} - 1\right)\phi}{\left(\frac{\sigma_s}{\sigma_f} + 2\right) - \left(\frac{\sigma_s}{\sigma_f} - 1\right)\phi} \tag{5.4}$$

Here, ϕ is the solid volume fraction. Considering $u_\theta = 0$ for purely radial flow, one can define the velocity parameter as:

$$f(\theta) = ru(r) \tag{5.5}$$

Introducing the $x = \frac{\theta}{\alpha}$ as the dimensionless degree, the dimensionless form of the velocity parameter can be obtained by dividing that to its maximum value as:

$$F(x) = \frac{f(\theta)}{f_{\max}} \tag{5.6}$$

Substituting Eq. (5.5) into Eqs. (5.2) and (5.3), and eliminating P, one can obtain the ordinary differential equation for the normalized function profile as [2]:

$$\begin{aligned} &F'''(x) + 2\alpha\,\mathrm{Re}\cdot A^*(1-\phi)^{2.5}F(x)F'(x) \\ &\quad + \left(4 - (1-\phi)^{1.25}B^*\,Ha\right)\alpha^2 F'(x) = 0 \end{aligned} \tag{5.7}$$

where A^* is a parameter. Reynolds number (Re) and Hartmann number (Ha) based on the electromagnetic parameter are introduced as following form:

$$A^* = (1-\phi) + \frac{\rho_s}{\rho_f}\phi$$
$$B^* = 1 + \frac{3\left(\frac{\sigma_s}{\sigma_f} - 1\right)\phi}{\left(\frac{\sigma_s}{\sigma_f} + 2\right) - \left(\frac{\sigma_s}{\sigma_f} - 1\right)\phi} \tag{5.8}$$

$$\mathrm{Re} = \frac{f_{\max}\alpha}{\upsilon_f} = \frac{U_{\max} r\alpha}{\upsilon_f}\begin{pmatrix} divergent - channel: \alpha > 0,\ f_{\max} > 0 \\ convergent - channel: \alpha < 0,\ f_{\max} < 0 \end{pmatrix} \tag{5.9}$$

$$Ha = \sqrt{\frac{\sigma_f B_0^2}{\rho_f \upsilon_f}} \tag{5.10}$$

With the following reduced form of boundary conditions

$$F(0) = 1,\ F'(0) = 0,\ F(1) = 0 \tag{5.11}$$

Physically these boundary conditions mean that maximum values of velocity are observed at centerline ($x = 0$) as shown in Fig. 5.1, and we consider fully developed velocity profile, thus the rate of velocity is zero at ($x = 0$). Also, in fluid dynamics, the no-slip condition for fluid states that at a solid boundary, the fluid will have zero velocity relative to the boundary. The fluid velocity at all fluid–solid boundaries is equal to that of the solid boundary, so we can see that the value of velocity is zero at ($x = 1$). Now we apply Differential Transformation Method (DTM) into Eq. (5.7). Taking the differential transform of Eq. (5.7) with respect to x and considering $H = 1$ gives:

$$\begin{aligned}&(k+1)(k+2)(k+3)F(k+3)\\&+ 2\alpha\mathrm{Re}\,A^*(1-\phi)^{2.5}\sum_{r=0}^{k}[(k-r+1)F(r)F(k-r+1)]\\&+\left(4 - Ha(1-\phi)^{1.25}\right)\alpha^2(k+1)F(k+1) = 0\end{aligned} \tag{5.12}$$

$$F(0) = 1,\ F(1) = 0\ ,\ F(2) = \beta, \tag{5.13}$$

where $F(k)$ is the differential transformation of $F(x)$, and β is a constant, which can be obtained through boundary condition, Eq. (5.11)

$$f(1) = 0 \text{ or } \sum_{k=0}^{N} F(k) = 0 \tag{5.14}$$

This problem can be solved for different values of H,

$$\begin{aligned}&F(0) = 1,\ F(1) = 0,\ F(2) = \beta,\ F(3) = 0\\&F(4) = \frac{-1}{6}\alpha\,\mathrm{Re}\,A^*(1-\phi)^{2.5}\beta - \frac{1}{3}\alpha^2\beta + \frac{1}{12}\alpha^2\beta Ha(1-\phi)^{1.25}\\&F(5) = 0\\&\ldots\end{aligned} \tag{5.15}$$

The above process is continuous. Substituting Eq. (5.15) into the main equation based on DTM. It can be obtained that the closed form of the solutions is:

$$F(x) = 1 + \beta x^2 + \left(\frac{-1}{6}\alpha\,\mathrm{Re}\,A^*(1-\phi)^{2.5}\beta - \frac{1}{3}\alpha^2\beta + \frac{1}{12}\alpha^2\beta Ha(1-\phi)^{1.25}\right)x^4 + \ldots \tag{5.16}$$

To obtain the value of β, we substitute the boundary condition from Eq. (5.11) into Eq. (5.16) in point $x = 1$. This value is too long that are not shown in this paper. By substituting obtained β into Eq. (5.16). We can find the expressions of $F(x)$. For example, when $\mathrm{Re} = 50$, $Ha = 2000$, $\phi = 0.04$, and $\alpha = 3$ degrees for a channel filled with Cu-water nanofluid, $F(x)$ function will be found as,

$$F(x) = 1 - 0.936173x^2 + 0.0298422x^4 - 0.0914287x^6 + 0.00373416x^8 - 0.00597469x^{10} \tag{5.17}$$

Subsequently, applying Padé approximation to Eq. (5.17) (for Padé [4,4] accuracy), we have,

$$\mathit{Padé}\ [4,4](F(x)) = \frac{0.999999996 - 0.9271132x^2 - 0.07601269x^4}{0.999999996 + 0.00905979x^2 - 0.09737341x^4} \tag{5.18}$$

In this example the accuracy of three analytical methods named DTM, DTM–Padé [4,4] compared with least square method (LSM) for obtaining the velocity profile of the MHD Jeffery–Hamel flow with nanofluid (Fig. 5.1) is investigated. Fig. 5.2 displays plots of $F(x)$ (nondimensional velocity profile for MHD Jeffery–Hamel flow) for different cases of α, Ha, and Re numbers for a divergent and convergent channel where Cu-water is selected as nanofluid from Table 5.1. This figure compared three described methods with those of the numerical method. The numerical solution that is applied to solve the present case is the fourth order Runge–Kutta procedure. As in the diagrams of Fig. 5.2, applied methods, specially DTM–Padé [4,4] and LSM, show a good agreement with the numerical solution. Tables 5.2 and 5.3 show the values of $F(x)$ when $\mathrm{Re} = 50$, $Ha = 2000$, $\phi = 0.04$, and $\alpha = 3$ degrees for a divergent and convergent channel respectively, which are derived from different applied methods for showing validity of them. From these two tables the absolute errors of methods were calculated and confirmed the accuracy and reliability of them. Also tables

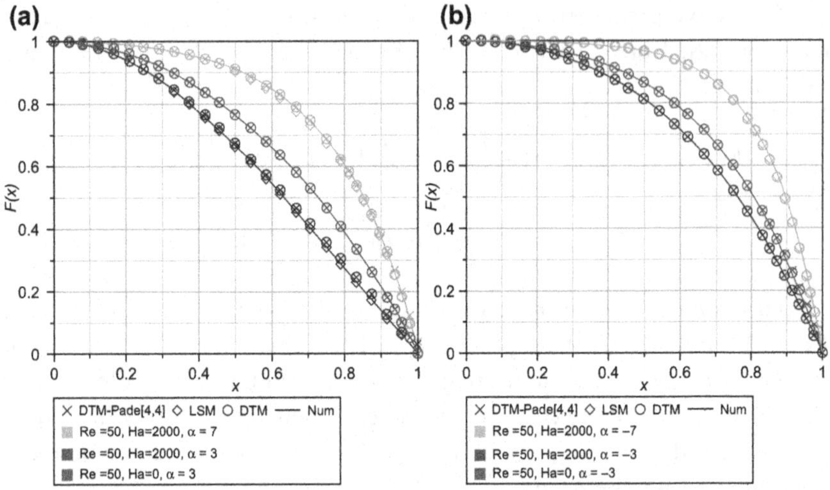

Figure 5.2 Comparison of differential transformation method (DTM)—Padé [4,4], least square method, DTM, and numerical results for nondimensional velocity in different values of parameters for (a) divergent channel and (b) convergent channel, filled with Cu-water nanofluid with $\phi = 0.04$.

Table 5.1 Thermophysical Properties of Nanofluids and Nanoparticles

Material	Density (kg/m^3)	Electrical Conductivity, σ ((Ω m)$^{-1}$)
Silver	10,500	6.30×10^7
Copper	8933	5.96×10^7
Ethylene glycol	1113.2	1.07×10^{-4}
Drinking water	997.1	0.05

reveal that the LSM has the lower error and higher accuracy among the other methods.

The effect of Hartmann number for a divergent and convergent channel is demonstrated in Fig. 5.3(a) and (b). Effect of Hartmann number on skin friction coefficient is presented as contour plot in Fig.5.3(c). The velocity curves show that the rate of transport is considerably reduced with the increase of Hartmann number. This clearly indicates that the transverse magnetic field opposes the transport phenomena. Because the variation of *Ha* leads to the variation of the Lorentz force due to magnetic field, and the Lorentz force produces more resistance to transport phenomena. As seen in this figure increasing the *Ha* makes an increase in velocity profile so, by the increasing of Hartmann number, the flow reversal disappears. Increasing Hartmann number leads to decrease in skin friction coefficient.

Table 5.2 Comparison of $F(x)$ Values and Errors (%) of Applied Methods for a Divergent Channel, When Re = 50, $Ha = 2000$, $\phi = 0.04$, and $\alpha = 3$ degrees With Cu–Water Nanofluid

	Num	DTM	DTM–Padé [4,4]	LSM	DTM	DTM–Padé [4,4]	LSM
x	$F(x)$ Values				Errors (%)		
0.0	1.00	1.00	1.00	1.00	0.00	0.00	0.00
0.1	0.9906413	0.9906411	0.9906411	0.9906669	2.02E-05	2.02E-05	0.00258
0.2	0.9625957	0.9625949	0.9625949	0.9627143	8.31E-05	8.31E-05	0.01232
0.3	0.9159215	0.9159197	0.9159196	0.9161175	0.000207	0.000197	0.0214
0.4	0.8506068	0.8506036	0.8506033	0.8507710	0.000411	0.000376	0.0193
0.5	0.7664072	0.7664020	0.7663992	0.7664192	0.001044	0.000678	0.00157
0.6	0.6626139	0.6626061	0.6625885	0.6624385	0.003833	0.001177	0.026471
0.7	0.5377426	0.5377303	0.5376469	0.5374677	0.017797	0.002287	0.051121
0.8	0.3891106	0.3890901	0.3887641	0.3888896	0.089049	0.005268	0.056796
0.9	0.2122444	0.2122145	0.2111039	0.2121622	0.537352	0.014088	0.038729
1.0	0.00	0.00	0.00	0.00	0.00	0.00	0.00

Table 5.3 Comparison of $F(x)$ Values and Errors (%) of Applied Methods for a Convergent Channel, When Re = 50, $Ha = 2000$, $\phi = 0.04$ and $\alpha = -3$ degrees With Cu–Water Nanofluid

	Num	DTM	DTM–Padé [4,4]	LSM	DTM	DTM–Padé [4,4]	LSM
x	***F*(*x*) Values**				**Errors (%)**		
0.0	1.00	1.00	1.00	1.00	0.00	0.00	0.00
0.1	0.9957323	0.9957536	0.9957536	0.9956519	0.00214	0.00214	0.008074
0.2	0.9824096	0.9824974	0.9824974	0.9820070	0.00894	0.00894	0.040981
0.3	0.9584170	0.9586246	0.9586246	0.9576683	0.02166	0.02166	0.078118
0.4	0.9208756	0.9212703	0.9212713	0.9200429	0.04297	0.04286	0.090425
0.5	0.8653567	0.8660262	0.8660365	0.8647706	0.07856	0.07737	0.067729
0.6	0.7854792	0.7865345	0.7866062	0.7852691	0.14348	0.13435	0.026748
0.7	0.6724157	0.6739705	0.6743519	0.6723980	0.28795	0.23123	0.002632
0.8	0.5143816	0.5164428	0.5181024	0.5142390	0.72335	0.40071	0.027723
0.9	0.2962961	0.2983765	0.3045470	0.2959939	2.78468	0.70214	0.101993
1.0	0.00	0.00	0.00	0.00	0.00	0.00	0.00

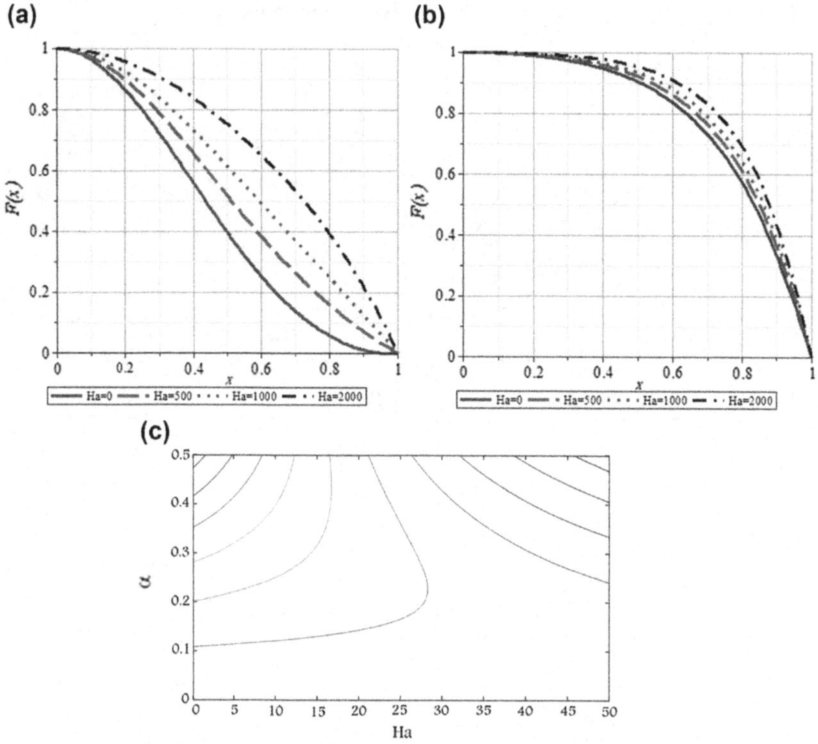

Figure 5.3 Effect of *Ha* on Cu–water velocity profile, when Re = 100, $\phi = 0.05$ for (a) divergent channel ($\alpha = 5$ degrees) and (b) convergent channel ($\alpha = -5$ degrees); (c) contour plot of $f'(0)$, when Pr = 6.2 (Cu–Water) and $\phi = 0.06$, Re = 5. *Ha*, Hartmann number; *Re*, Reynolds number.

Fig. 5.4(a) and (b) displays the effect of Reynolds number for a divergent and convergent channel with a slope of 5 degrees respectively. Fig. 5.4(c) and (d) shows the effect of Reynolds number on skin friction coefficient. These figures reveal that increasing Reynolds number makes a decrease in velocity profile in divergent channels, also for higher Reynolds number, the flow moves reversely and a region of backflow near the wall is observed (see Fig. 5.4(a) for Re = 200). As shown, in Fig. 5.4(b) for convergent channel, results were inversed and by increasing Reynolds number, velocity profiles were increased and no backflow was observed. Also for large Reynolds numbers velocity profile was approximately constant near the centerline and suddenly reached to zero near the wall. Skin friction coefficient increases with increase of Reynolds number.

Fig. 5.5 displays the effect of nanoparticles volume fraction, ϕ, when Re = 100, $Ha = 1000$ for a divergent and convergent channel with

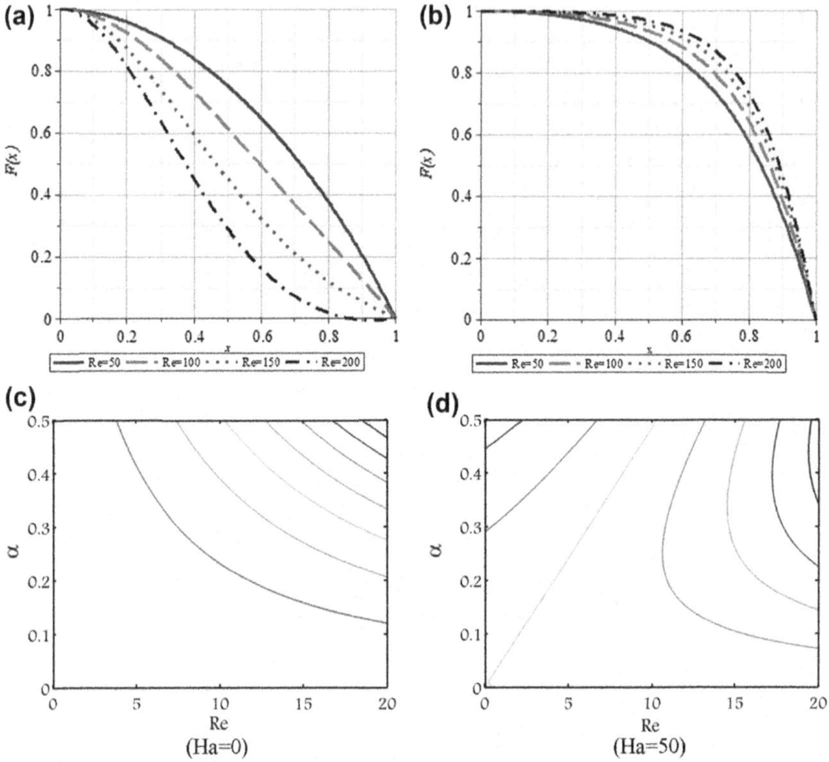

Figure 5.4 Effect of Re on Cu–water velocity profile, when $Ha = 1000$, $\phi = 0.05$ for (a) divergent channel ($\alpha = 5$ degrees) and (b) convergent channel ($\alpha = -5$ degrees); (c) and (d) contour plots of $f'(0)$, when Pr = 6.2(Cu–Water) and $\phi = 0.06$. *Ha*, Hartmann number; *Re*, Reynolds number.

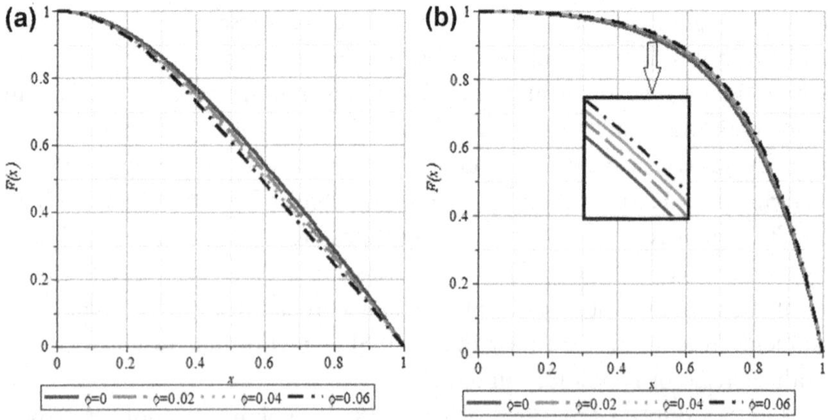

Figure 5.5 Effect of nanoparticles volume fraction, ϕ, on velocity profile, when Re = 100, $Ha = 1000$ for (a) divergent channel ($i = 5$ degrees) and (b) convergent channel ($\alpha = -5$ degrees). *Ha*, Hartmann number; *Re*, Reynolds number.

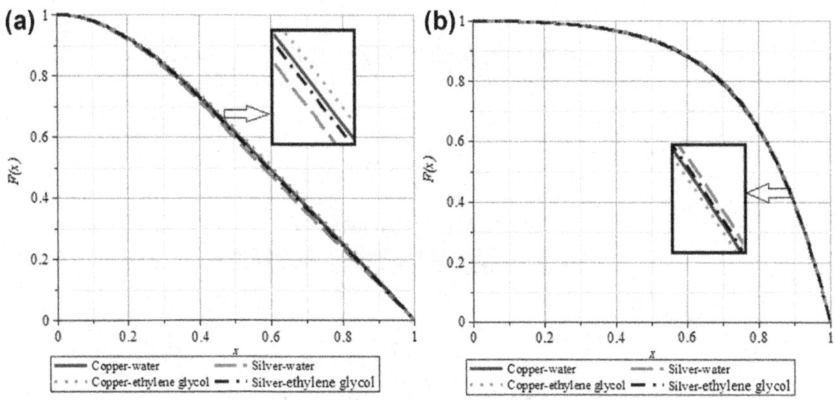

Figure 5.6 Effect of nanofluids structure on dimensionless velocity profile, when $\mathrm{Re} = 100$, $Ha = 1000$, $\phi = 0.05$ for (a) divergent channel ($\alpha = 5$ degrees) and (b) convergent channel ($\alpha = -5$ degrees). *Ha*, Hartmann number; *Re*, Reynolds number.

5 degrees slope. It is assumed that the base fluid and the nanoparticles (Cu-water) are in thermal equilibrium and no slip occurs between them. It can be seen that increasing nanoparticles volume fraction in divergent channel leads to decrease in velocity profile and the backflow may be started at high Re.

Finally we considered four different and common structures of nanofluid from Table 5.1 and their nondimensional velocity profiles, $F(x)$, are depicted in Figs. 5.6(a) and (b) for a divergent and convergent channel respectively. As seen in this figure, for a divergent channel, when nanofluid includes copper (as nanoparticles) or EG (as fluid phase) in its structure, $F(x)$ values are greater than the other structures, but this treatment of nanofluids structure is completely vice versa for convergent channels.

5.3 MHD COUETTE NANOFLUID FLOW

Turbulent CuO–water nanofluid flow and heat transfer between two infinite horizontal plates located at the $y = \pm h$ planes are investigated. The upper plate moves with a uniform velocity U_0 while the lower plate is kept stationary. The two plates are assumed to be electrically insulating and kept at two constant temperatures T_1 for the lower plate and T_2 for the upper plate with $T_2 > T_1$. A constant pressure gradient is applied in the x-direction. A uniform magnetic field B_0 is applied in the positive y-direction while the induced magnetic field is neglected by assuming a very small magnetic

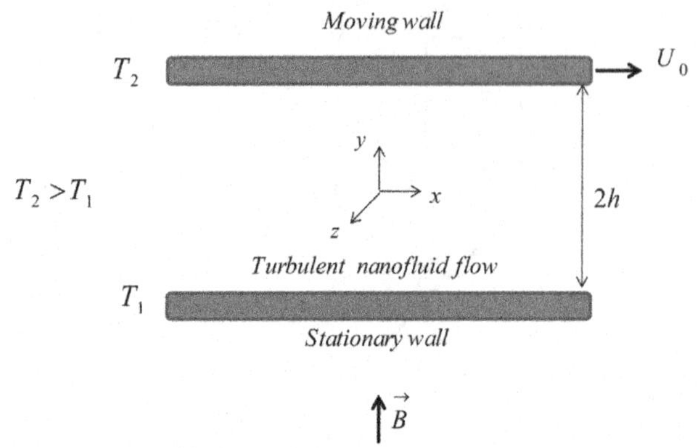

Figure 5.7 Schematic of the problem.

Reynolds number (see Fig. 5.7). The Hall effect is taken into consideration and consequently a z-component for the velocity is expected to arise. The uniform suction implies that the y-component of the velocity is constant. Thus, the nanofluid velocity vector is given by $\widehat{v}(y,t) = \widehat{u}(y,t)i + v_0 j + \widehat{w}(y,t)k$. The nanofluid motion starts from rest at $t = 0$, and the no-slip condition at the plates implies that the fluid velocity has neither a z- nor an x-component at $y = -h$ and $y = h$. The initial temperature of the fluid is assumed to be equal to T_1. The thermophysical properties of the nanofluid are given in Table 5.4 [3].

The flow of the nanofluid is governed by the Navier–Stokes equation, which has the two components

$$\rho_{nf}\left(\frac{\partial \widehat{u}}{\partial \widehat{t}} + v_0 \frac{\partial \widehat{u}}{\partial \widehat{y}}\right) = -\frac{d\widehat{P}}{d\widehat{x}} + (\mu_{nf} + \mu_t)\frac{\partial^2 \widehat{u}}{\partial \widehat{y}^2} - \frac{\sigma_{nf} B_0^2}{1 + m_{nf}{}^2}\left(\widehat{u} + m_{nf}\widehat{w}\right) \tag{5.19}$$

$$\rho_{nf}\left(\frac{\partial \widehat{w}}{\partial \widehat{t}} + v_0 \frac{\partial \widehat{w}}{\partial \widehat{y}}\right) = (\mu_{nf} + \mu_t)\frac{\partial^2 \widehat{w}}{\partial \widehat{y}^2} - \frac{\sigma_{nf} B_0^2}{1 + m_{nf}^2}\left(\widehat{w} - m_{nf}\widehat{u}\right) \tag{5.20}$$

Table 5.4 Thermophysical Properties of Water and Nanoparticles

	P (kg/m^3)	C_p (j/kgk)	K (W/m·k)	d_p (nm)	σ (Ω)
Pure water	997.1	4179	0.613	-	0.05
CuO	6500	540	18	29	1×10^{-12}

where m is the Hall parameter given by $m_{nf} = \sigma_{nf}\beta_1 B_0$ and β_1 is the Hall factor, and μ_t is turbulent viscosity given by $\mu_t = \rho_{nf}\ell_m^2\left|\frac{\partial \widehat{u}}{\partial \widehat{y}}\right|$. The energy equation describing the temperature distribution for the fluid is given by

$$(\rho c_p)_{nf}\left(\frac{\partial T}{\partial t} + v_0\frac{\partial T}{\partial y}\right) = k_{nf}\frac{\partial^2 T}{\partial \widehat{y}^2} + (\mu_{nf} + \mu_t)\left[\left(\frac{\partial \widehat{u}}{\partial \widehat{y}}\right)^2 + \left(\frac{\partial \widehat{w}}{\partial \widehat{y}}\right)^2\right] + \frac{\sigma_{nf} B_0^2}{\left(1 + m_{nf}^2\right)}\left(\widehat{u}^2 + \widehat{w}^2\right) \tag{5.21}$$

where T is the temperature of the fluid. The effective density (ρ_{nf}) and heat capacitance $(\rho C_p)_{nf}$ of the nanofluid are defined as,

$$\begin{aligned} \rho_{nf} &= \rho_f(1-\phi) + \rho_s\phi \\ (\rho C_p)_{nf} &= (\rho C_p)_f(1-\phi) + (\rho C_p)_s\phi \end{aligned} \tag{5.22}$$

Also effective electrical conductivity of nanofluid was presented by Maxwell as below:

$$\frac{\sigma_{nf}}{\sigma_f} = 1 + \frac{3(\sigma_s/\sigma_f - 1)\phi}{(\sigma_s/\sigma_f + 2) - (\sigma_s/\sigma_f - 1)\phi} \tag{5.23}$$

In this study Brownian motion impact on thermal conductivity (k_{nf}) and viscosity of nanofluid (μ_{nf}) has been considered

$$\begin{aligned} \mu_{nf} &= \mu_{static} + \mu_{Brownian} \\ \mu_{static} &= \frac{\mu_f}{(1-\phi)^{2.5}} \\ \mu_{Brownian} &= 5 \times 10^4 \phi\rho_f\beta\sqrt{\frac{\kappa_b T}{\rho_p d_p}} f(T,\phi) \\ f(T,\phi) &= (-6.04\phi + 0.4705)T + (1722.3\phi - 134.63), \\ \beta &= \begin{cases} 0.0137(100\phi)^{-0.8229} & \phi < 0.01 \\ 0.0011(100\phi)^{-0.7272} & \phi > 0.01 \end{cases} \end{aligned} \tag{5.24}$$

$$k_{eff} = k_{static} + k_{Brownian}$$

$$\frac{k_{static}}{k_f} = 1 + \frac{3\left(\frac{k_p}{k_f} - 1\right)\phi}{\left(\frac{k_p}{k_f} + 2\right) - \left(\frac{k_p}{k_f} - 1\right)\phi}$$

$$k_{Brownian} = 5 \times 10^4 \beta\phi(\rho C_p)_p \sqrt{\frac{\kappa_b T}{\rho_p d_p}} f(T, \phi), \quad k_b = 1.385 \times 10^{-23}$$

(5.25)

The problem is simplified by writing the equations in the nondimensional form. To achieve this define the following nondimensional quantities,

$$y = \frac{\widehat{y}}{h}, \; t = \frac{\widehat{t}\, U_0}{h}, \; P = \frac{\widehat{P}}{\rho_f U_0^2}, \; (u, w) = \frac{(\widehat{u}, \widehat{w})}{U_0}, \; \theta = \frac{T - T_1}{T_2 - T_1}, \; \alpha = -\frac{d\widehat{P}}{d\widehat{x}}$$

(5.26)

Prandtl number (Pr), Reynolds number (Re), Suction parameter (S), Hartmann number (Ha), Eckert number (Ec), Hall parameter (m), Turbulent parameter (L_t), and Turbulent Eckert number (Ec_t) for base fluid are introduced as follows:

$$\Pr = \frac{U_0 (\rho C_p)_f h}{\rho_f k_f}, \qquad \mathrm{Re} = \frac{\rho U_0 h}{\mu_f}, \qquad S = \frac{\rho v_0 h}{\mu_f}$$

$$Ha^2 = \frac{\sigma_f B_0^2 h^2}{\mu_f}, \qquad Ec = \frac{\rho_f U_0^2 h}{(\rho C_p)_f h (T_2 - T_1)}, \qquad m = \sigma_f \beta_1 B_0$$

$$L_t = \left(\frac{\ell_m}{h}\right)^2, \qquad Ec_t = \left(\frac{\ell_m}{h}\right)^2 \frac{\rho_f U_0^2}{(\rho C_p)_f (T_2 - T_1)}$$

(5.27)

In terms of the above nondimensional variables and parameters Eqs. (5.19)–(5.21) are,respectively, written as (where the hats are dropped for convenience):

$$\frac{\partial u}{\partial t} + \frac{S}{\mathrm{Re}} \frac{\partial u}{\partial y} = \alpha + \frac{1}{\mathrm{Re}} \frac{A_2}{A_1} \frac{\partial^2 u}{\partial y^2} + L_t \frac{\partial^2 u}{\partial y^2} \frac{\partial u}{\partial y} - \frac{A_3}{A_1} \frac{1}{\mathrm{Re}} \frac{Ha^2}{1 + (m\, A_3)^2} (u + m\, A_3 w)$$

(5.28)

$$\frac{\partial w}{\partial t} + \frac{S}{\mathrm{Re}} \frac{\partial w}{\partial y} = \frac{1}{\mathrm{Re}} \frac{A_2}{A_1} \frac{\partial^2 w}{\partial y^2} + L_t \frac{\partial^2 w}{\partial y^2} \frac{\partial u}{\partial y} - \frac{A_3}{A_1} \frac{1}{\mathrm{Re}} \frac{Ha^2}{1 + (m\, A_3)^2} (w - m\, A_3 u)$$

(5.29)

$$\frac{\partial\theta}{\partial t}+\frac{S}{\mathrm{Re}}\frac{\partial\theta}{\partial y}=\frac{1}{\Pr}\frac{A_5}{A_4}\frac{\partial^2\theta}{\partial y^2}+\frac{A_2}{A_4}Ec\left[\left(\frac{\partial u}{\partial y}\right)^2+\left(\frac{\partial w}{\partial y}\right)^2\right]$$
$$+\frac{A_3}{A_1}\frac{EcHa^2}{\mathrm{Re}\left(1+(m\,A_3)^2\right)}\left(u^2+w^2\right)+\frac{A_1}{A_4}\;Ec_t\left[\left(\frac{\partial u}{\partial y}\right)^3+\frac{\partial u}{\partial y}\left(\frac{\partial w}{\partial y}\right)^2\right] \tag{5.30}$$

where A_i $(i=1\ldots5)$ are defined as follows:

$$A_1=\frac{\rho_{nf}}{\rho_f},A_2=\frac{\mu_{nf}}{\mu_f},A_3=\frac{\sigma_{nf}}{\sigma_f},A_4=\frac{(\rho C_p)_{nf}}{(\rho C_p)_f},A_5=\frac{k_{nf}}{k_f} \tag{5.31}$$

The boundary and initial conditions for components of velocity and temperature are:

$$\begin{array}{lll} u=w=\theta=0 & & \text{for } t\le 0 \text{ and} \\ u=w=\theta=0 & \text{at } y=-1, & \text{for } t>0 \\ w=0,u=\theta=1 & \text{at } y=1 & \end{array} \tag{5.32}$$

To solve the coupled nonlinear partial equations (Eqs. 5.28–5.30) in the domain $t\in[0,\ T]$ and $y\in[-1,\ 1]$ using hybrid DTM and finite difference method, we apply finite difference approximation on y-direction and take DTM on t. The following finite difference scheme is used based on a uniform mesh. The length in direction of y is divided into N_y equal intervals. The y-coordinates of the grid points can be obtained by $y_j=j(\Delta y), j=0{:}N_y$, where Δy is the mesh size. After taking the second order accurate central finite difference approximation with respect to y and applying DTM on Eqs. (5.28)–(5.30) for time domain, the following recurrence relations can be obtained:

for $1\le j\le N_y$

$$U(j,k+1)=\frac{H}{k+1}\left\{-\frac{S}{\mathrm{Re}}\left(\frac{U(j+1,k)-U(j-1,k)}{2\Delta y}\right)+\alpha\delta(k)\right.$$
$$+\frac{1}{\mathrm{Re}}\frac{\overline{A}_2(j,k)}{A_1}\otimes\left(\frac{U(j+1,k)-2U(j,k)+U(j-1,k)}{\Delta y^2}\right)$$
$$+L_t\left(\frac{U(j+1,k)-2U(j,k)+U(j-1,k)}{\Delta y^2}\right)\cdot\left(\frac{U(j+1,k)-U(j-1,k)}{2\Delta y}\right)$$
$$\left.-\frac{A_3}{A_1}\frac{1}{\mathrm{Re}}\frac{Ha^2}{1+(mA_3)^2}\left(U(j,k)+mA_3W(j,k)\right)\right\} \tag{5.33}$$

$$W(j,k+1)=\frac{H}{k+1}\left\{-\frac{S}{\mathrm{Re}}\left(\frac{W(j+1,k)-W(j-1,k)}{2\Delta y}\right)\right.$$
$$+\frac{1}{\mathrm{Re}}\frac{\overline{A}_2(j,k)}{A_1}\otimes\left(\frac{W(j+1,k)-2W(j,k)+W(j-1,k)}{\Delta y^2}\right)$$
$$+L_t\left(\frac{W(j+1,k)-2W(j,k)+W(j-1,k)}{\Delta y^2}\right)\cdot\left(\frac{U(j+1,k)-U(j-1,k)}{2\Delta y}\right)$$
$$\left.-\frac{A_3}{A_1}\frac{1}{\mathrm{Re}}\frac{Ha^2}{1+(mA_3)^2}(W(j,k)-mA_3U(j,k))\right\} \tag{5.34}$$

$$\Theta(j,k+1)=\frac{H}{k+1}\left\{-\frac{S}{\mathrm{Re}}\left(\frac{\Theta(j+1,k)-\Theta(j-1,k)}{2\Delta y}\right)\right.$$
$$+\frac{1}{\mathrm{Pr}}\frac{\overline{A}_5(j,k)}{A_4}\otimes\left(\frac{\Theta(j+1,k)-2\Theta(j,k)+\Theta(j-1,k)}{\Delta y^2}\right)$$
$$+\frac{Ec}{A_4}\overline{A}_2(j,k)\otimes\left[\left(\frac{U(j+1,k)-U(j-1,k)}{2\Delta y}\right)\otimes\left(\frac{U(j+1,k)-U(j-1,k)}{2\Delta y}\right)\right.$$
$$\left.+\left(\frac{W(j+1,k)-W(j-1,k)}{2\Delta y}\right)\otimes\left(\frac{W(j+1,k)-W(j-1,k)}{2\Delta y}\right)\right]$$
$$+\frac{A_3}{A_1}\frac{1}{\mathrm{Re}}\frac{EcHa^2}{1+(mA_3)^2}(U(j,k)\otimes U(j,k)+W(j,k)\otimes W(j,k))$$
$$+Ec_t\frac{A_1}{A_4}\left[\left(\frac{U(j+1,k)-U(j-1,k)}{2\Delta y}\right)^3+\left(\frac{U(j+1,k)-U(j-1,k)}{2\Delta y}\right)\right.$$
$$\left.\left.\left(\frac{W(j+1,k)-W(j-1,k)}{2\Delta y}\right)^2\right]\right\} \tag{5.35}$$

where the symbol "⊗" denotes the convolution operation, $\overline{A}_2(j,k)$ and $\overline{A}_5(j,k)$ are the differential transform of the functions $A_2(\theta)$ and $A_5(\theta)$, respectively.

Applying DTM on initial conditions in Eq. (5.14), we have:

$$\begin{aligned}&\text{for } 0\le j\le N_y\\&U(j,0)=0,\quad W(j,0)=0,\quad \Theta(j,0)=0.\end{aligned} \tag{5.36}$$

The boundary conditions in Eq. (5.32) can be transformed as follows:

$$BC's \text{ for } u(y,t) \rightarrow \begin{cases} U(0,k) = 0, \quad k \geq 0 \\ U(N_y,0) = 1, U(N_y,k) = 0, \quad k \geq 1 \end{cases} \tag{5.37}$$

$$BC's \text{ for } w(y,t) \rightarrow \begin{cases} W(0,k) = 0, \quad k \geq 0 \\ W(N_y,k) = 0, k \geq 0 \end{cases} \tag{5.38}$$

$$BC's \text{ for } \theta(y,t) \rightarrow \begin{cases} \Theta(0,k) = 0, \quad k \geq 0 \\ \Theta(N_y,0) = 1, \Theta(N_y,k) = 0, \quad k \geq 1 \end{cases} \tag{5.39}$$

As shown in Fig. 5.8, they are in a very good agreement. After this validity, effects of nanoparticle volume fraction, Reynolds number, Hall parameter, Hartmann number, Eckert number, turbulent parameter, and turbulent Eckert number on flow and heat transfer characteristics are examined.

Fig. 5.9 shows the effect of volume fraction of nanofluid on the velocity and temperature profiles. The velocity components of nanofluid increase as a result of an increase in the energy transport in the fluid with the increasing

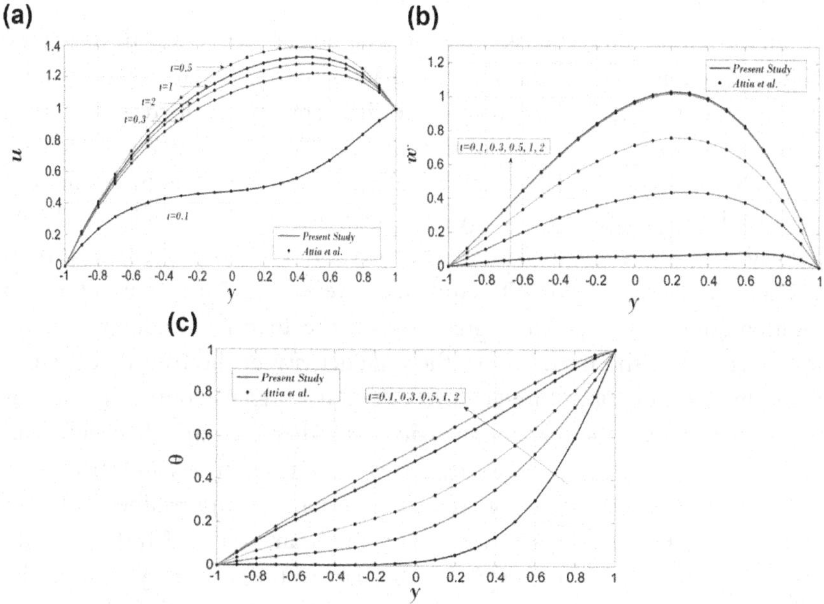

Figure 5.8 Comparison between Hybrid Differential Transformation Method in Ref. [3], when $\phi = 0.0$, $S = 1$, $\alpha = 5$, $\text{Re} = 1$, $Ec = 0.2$, $Ha = 3$, $m = 3$, $\text{Pr} = 1$, $Ec_t = 0.0$, $L_t = 0.0$.

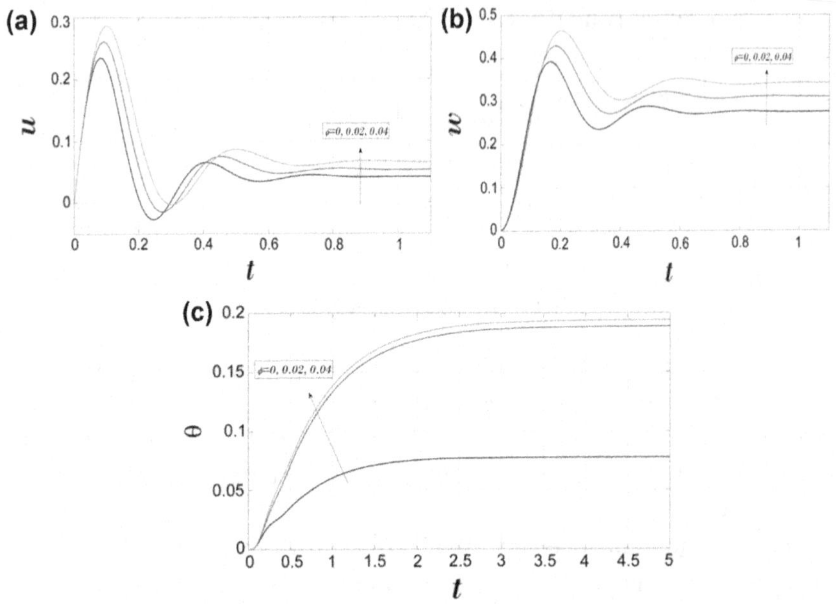

Figure 5.9 Effect of volume fraction on velocity and temperature profiles, when $S = 1$, $\alpha = 5$, $\Pr = 6.2$, $Ec = 0.2$, $\text{Re} = 1$, $Ha = 10$, $m = 5$, $Ec_t = 0.2$, $L_t = 0.1$.

volume fraction. Thus, the skin friction coefficient increases with increasing volume fraction of nanofluid. The sensitivity of thermal boundary layer thickness to volume fraction of nanoparticles is related to the increased thermal conductivity of the nanofluid. In fact, higher values of thermal conductivity are accompanied by higher values of thermal diffusivity. The high values of thermal diffusivity cause a drop in the temperature gradients and accordingly increase the boundary thickness. This increase in thermal boundary layer thickness reduces the Nusselt number; however, the Nusselt number is a multiplication of temperature gradient and the thermal conductivity ratio (conductivity of the nanofluid to the conductivity of the base fluid). Since the reduction in temperature gradient due to the presence of nanoparticles is much smaller than thermal conductivity ratio, therefore an enhancement in Nusselt is taken place by increasing the volume fraction of nanoparticles.

Effect of Hall parameter on velocity and temperature profiles is shown in Fig. 5.10. Hall parameter has a very important role in MHD Couette flows, because this parameter causes the start of the secondary flow in direction z. As Hall parameter increases, velocity profiles also increase thus skin friction coefficient increases with increase of m. Thermal boundary layer thickness decreases with the increase of Hall parameter.

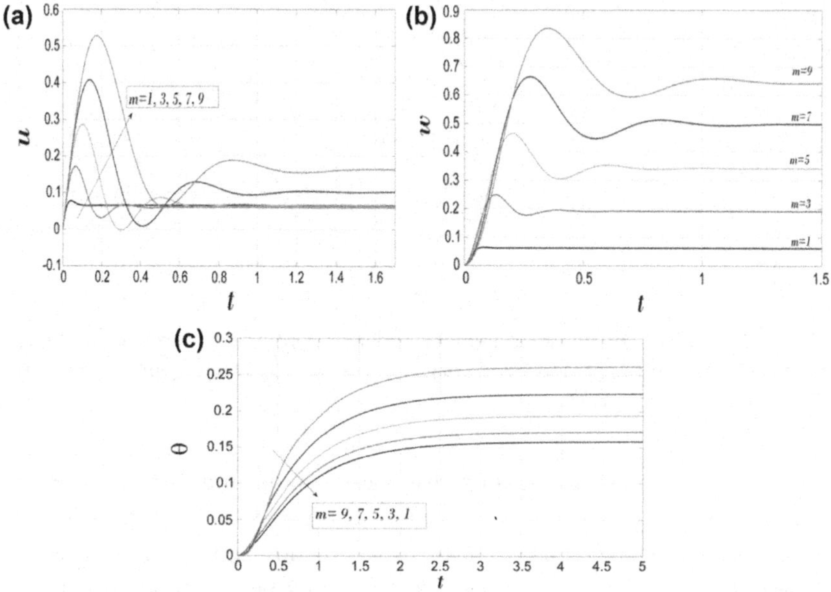

Figure 5.10 Effect of Hall parameter on velocity and temperature profiles, when $S = 1$, $\alpha = 5$, $\Pr = 6.2$, $Ec = 0.2$, $\mathrm{Re} = 1$, $Ha = 10$, $m = 5$, $Ec_t = 0.2$, $L_t = 0.1$.

Fig. 5.11 shows that the effect of nanoparticle volume fraction and Hartmann number on Nusselt number over the upper and lower plates. Nusselt number over the upper and lower plates increase with increase of nanoparticle volume fraction while they decrease with augment of Hartmann number. Effects of Reynolds number and turbulent Eckert number on Nusselt number over the upper and lower plates are shown in

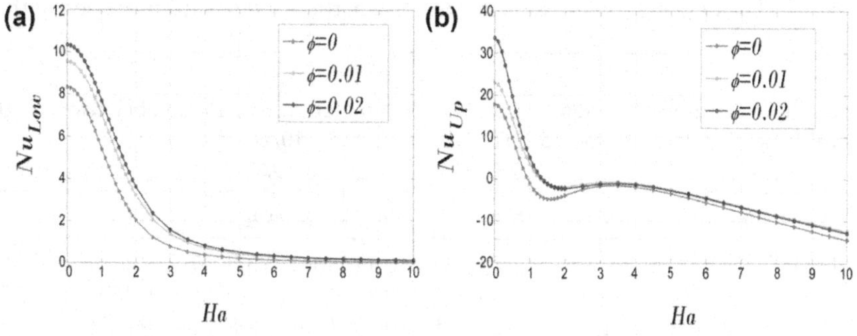

Figure 5.11 Effect of nanoparticle volume fraction and Hartmann number on Nusselt number over the upper and lower plates $S = 1$, $\alpha = 5$, $\Pr = 6.2$, $Ec = 0.2$, $\mathrm{Re} = 1$, $Ec_t = 0.2$, $L_t = 0.1$.

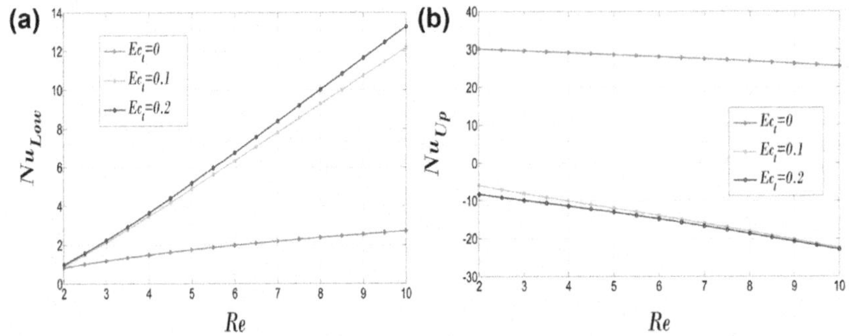

Figure 5.12 Effects of Reynolds number and turbulent Eckert number on Nusselt number over the upper and lower plates $S = 1$, $\alpha = 5$, $\Pr = 6.2$, $m = 1$, $Ec = 0.2$, $Ha = 10$, $L_t = 0.1$, $\phi = 0.02$.

Fig. 5.12. It is worth to mention that the Reynolds number indicates the relative significance of the inertia effect compared to the viscous effect. Thus, thermal boundary layer thickness over the lower plate decreases as Re increases and in turn increasing Reynolds number leads to increase in Nusselt number. But, an opposite trend is observed for thermal boundary layer thickness over the upper. Effects of turbulent Eckert number on Nusselt number over the upper and lower plates are similar to that of Reynolds number. In the case in which Nusselt number has negative values, temperature of the surface near the wall has higher than that of on the wall, so thermal gradient has a negative value. This phenomenon is due to increase of viscous dissipation with increase of turbulent Eckert number.

Table 5.5 illustrates the effects of turbulent parameter and Hall parameter on Nusselt number over the upper and lower plates. As turbulent parameter increases, absolute values of Nusselt number over the upper and lower plates decreases. Nusselt number over the lower plate increases with

Table 5.5 Effect of Turbulent Parameter and Hall Parameter on Nusselt Number Over the Upper and Lower Plates $S = 1$, $\alpha = 5$, $\Pr = 6.2$, $Ec = Ec_t = 0.2$, $Ha = 10$, $\phi = 0.02$

	$L_t = 0$		$L_t = 0.1$		$L_t = 0.2$		$L_t = 0.3$	
m	Nu_{Low}	Nu_{Up}	Nu_{Low}	Nu_{Up}	Nu_{Low}	Nu_{Up}	Nu_{Low}	Nu_{Up}
0	0.1151	−5.8986	0.1155	−5.8999	0.11580	−5.9047	0.1160	−5.9170
2	0.1566	−3.9305	0.1562	−3.4387	0.1557	−3.0751	0.1553	−2.7952
3	0.2040	−2.7836	0.2025	−2.2411	0.2011	−1.8450	0.1998	−1.539
4	0.2634	−1.9319	0.2604	−1.4038	0.2575	−1.0156	0.2548	−0.7142
5	0.3321	−1.3311	0.3268	−0.8318	0.3219	−0.4618	0.3175	−0.1711

increase of Hall parameter while absolute value of Nusselt number over the upper plate decreases with augment of Hall parameter.

5.4 NANOFLUID BETWEEN PARALLEL PLATES

We consider the flow and heat transfer analysis in the unsteady two-dimensional squeezing flow of an incompressible nanofluid between the infinite parallel plates. The two plates are placed at $z = \pm\ell(1 - \alpha t)^{1/2} = \pm h(t)$ (ℓ is distant of plate at $t = 0$ and α is squeezed parameter). For $\alpha > 0$, the two plates are squeezed until they touch $t = 1/\alpha$ and for $\alpha < 0$, the two plates are separated. The viscous dissipation effect, the generation of heat due to fraction caused by shear in the flow, is neglected. Further the symmetric nature of the flow is adopted [4] and shown in Fig. 5.13.

It is also assumed that the time variable magnetic field $\left(\vec{B} = B\vec{e_y},\, B = B_0(1 - \alpha t)\right)$ is applied, where $\vec{e_y}$ is unit vector in the Cartesian coordinate system. The electric current J and the electromagnetic force F are defined by $J = \sigma\left(\vec{V} \times \vec{B}\right)$ and $F = \sigma\left(\vec{V} \times \vec{B}\right) \times \vec{B}$, respectively. Also a heat source ($Q = Q_0/(1 - \alpha t)$) is applied between two plates.

The nanofluid is a two component mixture with the following assumptions: incompressible; no-chemical reaction; negligible radiative heat transfer; nanosolid particles and the base fluid are in thermal equilibrium

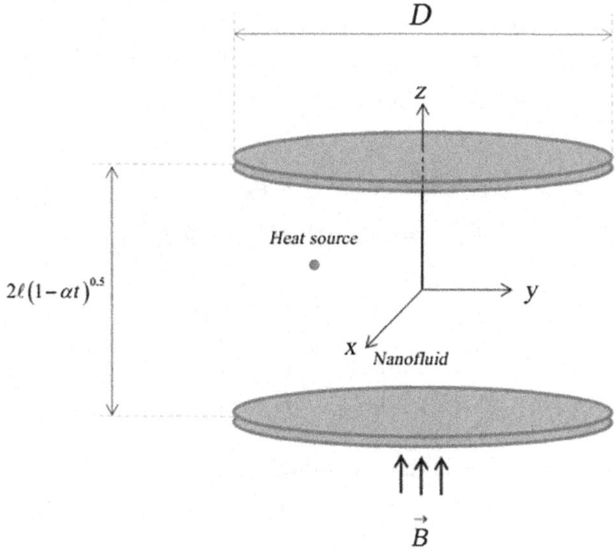

Figure 5.13 Geometry of problem.

Table 5.6 Thermophysical Properties of Water and Nanoparticles at Room Temperature

	P (kg/m^3)	C_p (j/kgk)	K (W/m·k)	d_p (nm)
Pure water	997.1	4179	0.613	—
Al_2O_3	3970	765	25	47
CuO	6500	540	18	29

and no slip occurs between them. The thermophysical properties of the nanofluid are given in Table 5.6 [4].

The governing equations for conservative momentum and energy in unsteady two-dimensional flow of a nanofluid are:

$$\frac{\partial u}{\partial x}+\frac{\partial v}{\partial y}=0 \tag{5.40}$$

$$\rho_{nf}\left(\frac{\partial u}{\partial t}+u\frac{\partial u}{\partial x}+v\frac{\partial u}{\partial y}\right)=-\frac{\partial p}{\partial x}+\mu_{nf}\left(\frac{\partial^2 u}{\partial x^2}+\frac{\partial^2 u}{\partial y^2}\right)-\sigma_{nf}\,B^2 u \tag{5.41}$$

$$\rho_{nf}\left(\frac{\partial v}{\partial t}+u\frac{\partial v}{\partial x}+v\frac{\partial v}{\partial y}\right)=-\frac{\partial p}{\partial y}+\mu_{nf}\left(\frac{\partial^2 v}{\partial x^2}+\frac{\partial^2 v}{\partial y^2}\right) \tag{5.42}$$

$$\frac{\partial T}{\partial t}+u\frac{\partial T}{\partial x}+v\frac{\partial T}{\partial y}=\frac{k_{nf}}{(\rho C_p)_{nf}}\left(\frac{\partial^2 T}{\partial x^2}+\frac{\partial^2 T}{\partial y^2}\right)+\frac{Q}{(\rho C_p)_{nf}}T \tag{5.43}$$

Here u and v are the velocities in the x and y directions respectively, T is the temperature, P is the pressure, effective density (ρ_{nf}), the effective heat capacity $(\rho C_p)_{nf}$, and electrical conductivity (σ_{nf}) of the nanofluid are defined as:

$$\begin{gathered}\rho_{nf}=(1-\phi)\,\rho_f+\phi\,\rho_p,\\ (\rho C_p)_{nf}=(1-\phi)(\rho C_p)_f+\phi(\rho C_p)_p\\ \frac{\sigma_{nf}}{\sigma_f}=1+\frac{3\left(\frac{\sigma_p}{\sigma_f}-1\right)\phi}{\left(\frac{\sigma_p}{\sigma_f}+2\right)-\left(\frac{\sigma_p}{\sigma_f}-1\right)\phi}\end{gathered} \tag{5.44}$$

The Brownian motion has a significant impact on the effective thermal conductivity. The effective thermal conductivity is

$$k_{eff}=k_{static}+k_{Brownian} \tag{5.45}$$

$$\frac{k_{static}}{k_f} = 1 + \frac{3\left(\frac{k_p}{k_f} - 1\right)\phi}{\left(\frac{k_p}{k_f} + 2\right) - \left(\frac{k_p}{k_f} - 1\right)\phi} \tag{5.46}$$

where, k_{static} is the static thermal conductivity based on Maxwell classical correlation. The enhanced thermal conductivity component generated by microscale convective heat transfer of a particle's Brownian motion and affected by ambient fluid motion is obtained via simulating Stokes' flow around a sphere (nanoparticle). By introducing two empirical functions (β and f), the interaction between nanoparticles in addition to the temperature effect in the model, leading to:

$$k_{Brownian} = 5 \times 10^4 \beta \phi \rho_f c_{p,f} \sqrt{\frac{\kappa_b T}{\rho_p d_p}} f(T, \phi) \tag{5.47}$$

By introducing a thermal interfacial resistance $R_f = 4 \times 10^{-8}\ \text{km}^2/\text{W}$, the original k_p in Eq. (5.46) was replaced by a new $k_{p,eff}$ in the form:

$$R_f + \frac{d_p}{k_p} = \frac{d_p}{k_{p,eff}} \tag{5.48}$$

For different based fluids and different nanoparticles, the function should be different. Only water-based nanofluids are considered in the current study. For Al_2O_3—water nanofluids and CuO—water nanofluids, this function follows the format:

$$g'(T, \phi, d_p) = \left(a_1 + a_2 \ln(d_p) + a_3 \ln(\phi) + a_4 \ln(\phi)\ln(d_p) + a_5 \ln(d_p)^2\right)\ln(T) + \left(a_6 + a_7 \ln(d_p) + a_8 \ln(\phi) + a_9 \ln(\phi)\ln(d_p) + a_{10} \ln(d_p)^2\right) \tag{5.49}$$

with the coefficients a_i ($i = 0\ldots10$) are based on the type of nanopartices and also with these coefficients, Al_2O_3—water nanofluids and CuO—water nanofluids have an R^2 of 96% and 98%, respectively (Table 5.7). Finally, the KKL (Koo—Kleinstreuer—Li) correlation is written as:

$$k_{Brownian} = 5 \times 10^4 \phi \rho_f c_{p,f} \sqrt{\frac{\kappa_b T}{\rho_p d_p}} g'(T, \phi, d_p) \tag{5.50}$$

For the *effective viscosity* due to micromixing in suspensions, its proposed:

$$\mu_{eff} = \mu_{static} + \mu_{Brownian} = \mu_{static} + \frac{k_{Brownian}}{k_f} \times \frac{\mu_f}{\Pr_f} \tag{5.51}$$

Table 5.7 The Coefficient Values of Al_2O_3–Water Nanofluids and CuO–Water Nanofluids

Coefficient Values	Al_2O_3–Water	CuO–Water
a_1	52.813488759	−26.593310846
a_2	6.115637295	−0.403818333
a_3	0.6955745084	−33.3516805
a_4	4.17455552786E-02	−1.915825591
a_5	0.176919300241	6.42185846658E-02
a_6	−298.19819084	48.40336955
a_7	−34.532716906	−9.787756683
a_8	−3.9225289283	190.245610009
a_9	−0.2354329626	10.9285386565
a_{10}	−0.999063481	−0.72009983664

where $\mu_{static} = \frac{\mu_f}{(1-\phi)^{2.5}}$ is viscosity of the nanofluid, as given originally by Brinkman.

The relevant boundary conditions are:

$$\begin{aligned} &v = v_w = dh/dt, \quad T = T_H \quad \text{at } y = h(t), \\ &v = \partial u/\partial y = \partial T/\partial y = 0 \quad \text{at } y = 0. \end{aligned} \tag{5.52}$$

We introduce these parameters:

$$\begin{aligned} &\eta = \frac{y}{\left[l(1-\alpha t)^{1/2}\right]}, \quad u = \frac{\alpha x}{[2(1-\alpha t)]} f'(\eta), \\ &v = -\frac{\alpha l}{\left[2(1-\alpha t)^{1/2}\right]} f(\eta), \quad \theta = \frac{T}{T_H}. \end{aligned} \tag{5.53}$$

Substituting the above variables into Eqs. (5.41) and (5.42), and then eliminating the pressure gradient from the resulting equations gives:

$$f^{iv} - S(A_1/A_4)\left(\eta f''' + 3f'' + f'f'' - ff'''\right) - Ha^2(A_5/A_4)f'' = 0, \tag{5.54}$$

Using Eq. (5.53), Eq. (5.43) reduces to the following differential equations:

$$\theta'' + \Pr S\left(\frac{A_2}{A_3}\right)(f\theta' - \eta\theta') + \frac{Hs}{A_3}\theta = 0 \tag{5.55}$$

Here A_1, A_2, A_3, A_4, and A_5 are dimensionless constants given by:

$$A_1 = \frac{\rho_{nf}}{\rho_f}, \quad A_2 = \frac{(\rho C_p)_{nf}}{(\rho C_p)_f}, \quad A_3 = \frac{k_{eff}}{k_f}, \quad A_4 = \frac{\mu_{eff}}{\mu_f}, \quad A_5 = \frac{\sigma_{nf}}{\sigma_f} \tag{5.56}$$

With these boundary conditions:

$$\begin{aligned} &f(0)=0, \quad f''(0)=0, \\ &f(1)=1, \quad f'(1)=0, \\ &\theta'(0)=0, \quad \theta(1)=1. \end{aligned} \tag{5.57}$$

where S is the squeeze number, Pr is the Prandtl number, Ha is the Hartmann number, and Hs is the heat source parameter, which are defined as:

$$S=\frac{\alpha\ell^2}{2\upsilon_f}, \quad \Pr=\frac{\mu_f(\rho C_p)_f}{\rho_f\, k_f}, \quad Ha=\ell B_0\sqrt{\frac{\sigma_f}{\mu_f}}, \quad Hs=\frac{Q_0\ell^2}{k_f} \tag{5.58}$$

Physical quantities of interest are the skin fraction coefficient and Nusselt number, which are defined as:

$$Cf^*=\frac{\mu_{nf}\left(\frac{\partial u}{\partial y}\right)_{y=h(t)}}{\rho_{nf}v_w^2}, \quad Nu^*=\frac{-lk_{nf}\left(\frac{\partial T}{\partial y}\right)_{y=h(t)}}{k_f T_H}. \tag{5.59}$$

In terms of (5.56), we obtain

$$\begin{aligned} &C_f=\left|(A_1/A_4)f''(1)\right|, \\ &Nu=\left|A_3\,\theta'(1)\right|. \end{aligned} \tag{5.60}$$

Now DTM has been applied into governing equations (Eqs. 5.54 and 5.55). Taking the differential transforms of Eqs. (5.54) and (5.55) with respect to χ and considering $H=1$ gives:

$$\begin{aligned} &(k+1)(k+2)(k+3)(k+4)F[k+4]+S(A_1/A_4)\sum_{m=0}^{k} \\ &\quad(\Delta[k-m-1](m+1)(m+2)(m+3)F[m+3]) \\ &\quad-3S(A_1/A_4)(k+1)(k+2)F[k+2]-S(A_1/A_4)\sum_{m=0}^{k} \\ &\quad((k-m+1)F[k-m+1](m+1)(m+2)F[m+2]) \\ &\quad+S(A_1/A_4)\sum_{m=0}^{k}(F[k-m](m+1)(m+2)(m+3)F[m+3]) \\ &\quad-Ha^2(A_5/A_4)(k+1)(k+2)F[k+2]=0, \\ &\Delta[m]=\begin{cases}1 & m=1\\ 0 & m\neq 1\end{cases} \end{aligned} \tag{5.61}$$

$$F[0] = 0,\ \ F[1] = a_1,\ \ F[2] = 0,\ \ F[3] = a_2 \tag{5.62}$$

$$(k+1)(k+2)\Theta[k+2] + \Pr\cdot S\cdot\left(\frac{A_2}{A_3}\right)\sum_{m=0}^{k}(F[k-m](m+1)\Theta[m+1])$$

$$-\Pr\cdot S\cdot\left(\frac{A_2}{A_3}\right)\sum_{m=0}^{k}(\Delta[k-m](m+1)\Theta[m+1])$$

$$+\frac{Hs}{A_3}\Theta[k] = 0,$$

$$\Delta[m] = \begin{cases} 1 & m = 1 \\ 0 & m \neq 1 \end{cases}$$

(5.63)

$$\Theta[0] = a_3, \Theta[1] = 0 \tag{5.64}$$

where $F[k]$ and $\Theta[k]$ are the differential transforms of $f(\eta)$, $\theta(\eta)$, and a_1, a_2, a_3 are constants which can be obtained through boundary condition. This problem can be solved as followed:

$$F[0] = 0,\ \ F[1] = a_1,\ \ F[2] = 0,\ \ F[3] = a_2, F[4] = 0$$

$$F[5] = \frac{3}{20}S\ a_2\ (A_1/A_2) + \frac{1}{20}Sa_1a_2(A_1/A_2) + \frac{1}{20}a_1a_2 + \frac{1}{20}Ha^2a_2, \ldots \tag{5.65}$$

$$\begin{aligned}&\Theta[0] = a_3,\ \Theta[1] = 0, \Theta[2] = -\frac{1}{2}\frac{\mathrm{Hs}}{\mathrm{A}_3}\mathrm{a}_3, \Theta[3] = 0.0,\\ &\Theta[4] = \frac{1}{12}\mathrm{Pr}S\left(\frac{A_2}{A_3}\right)\frac{\mathrm{Hs}}{A_3}a_3a_1, \Theta[5] = 0, \ldots\end{aligned} \tag{5.66}$$

The above process is continuous. By substituting Eqs. (5.64) and (5.66) into the main Eq. (5.61) based on DTM, it can be obtained that the closed form of the solutions is:

$$\begin{aligned}F(\eta) = a_1\eta + a_2\eta^3 + \Bigg(&\frac{3}{20}S\ a_2(A_1/A_2) + \frac{1}{20}Sa_1a_2(A_1/A_2)\\ &+\frac{1}{20}a_1a_2 + \frac{1}{20}Ha^2a_2\Bigg)\eta^4 + \ldots\end{aligned} \tag{5.67}$$

$$\theta(\eta) = a_3 + \left(-\frac{1}{2}\frac{Hs}{A_3}a_3\right)\eta^2 + \left(\frac{1}{12}Pr\ S\left(\frac{A_2}{A_3}\right)\frac{Hs}{A_3}\ a_3a_1\right)\eta^4 + \ldots \tag{5.68}$$

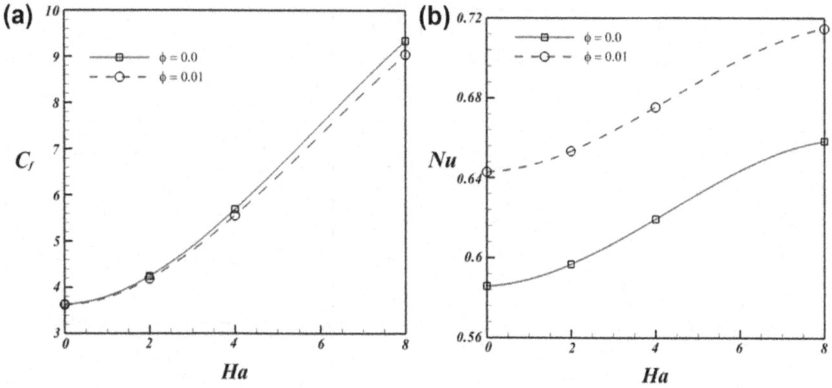

Figure 5.14 Effect of volume fraction of nanofluid on skin friction coefficient and Nusselt number, when $S = 1$, $Hs = -1$, $\Pr = 6.2$ (CuO–water).

by substituting the boundary condition from Eq. (5.53) into Eqs. (5.67) and (5.68) in point $\eta = 1$, it can be obtained the values of a_1, a_2, a_3. By substituting obtained a_1, a_2, a_3 into Eqs. (5.67) and (5.68), it can be obtained the expression of $F(\eta)$ and $\Theta(\eta)$.

The results of this method are compared with previous work in Fig. 5.14. This companion indicates that DTM has a good accuracy to solve such problem. The kind of nanoparticle is a key factor for heat transfer enhancement. Fig. 5.14 shows the effect of nanofluid volume fraction on skin friction coefficient and Nusselt number. As nanofluid volume fraction increase, Nusselt number also increases while skin friction coefficient decreases.

Fig. 5.15 shows the effect of the squeeze number on the velocity and temperature profiles. It is important to note that the squeeze number (S) describes the movement of the plates ($S > 0$ corresponds to the plates moving apart, while $S < 0$ corresponds to the plates moving together the so-called squeezing flow). Vertical velocity decreases with increase of squeeze number while horizontal velocity has different behavior. It means that horizontal velocity decrease with increase of S when $\eta < 0.5$ while opposite trend is observed for $\eta > 0.5$. Thermal boundary layer thickness increases with increase of squeeze number. Effect of the Hartmann number on velocity and temperature profiles is shown in Fig. 5.16. Effects of Hartmann number on velocity profiles are similar to that of squeeze number. While increasing Hartmann number leads to decrease in thermal boundary layer thickness.

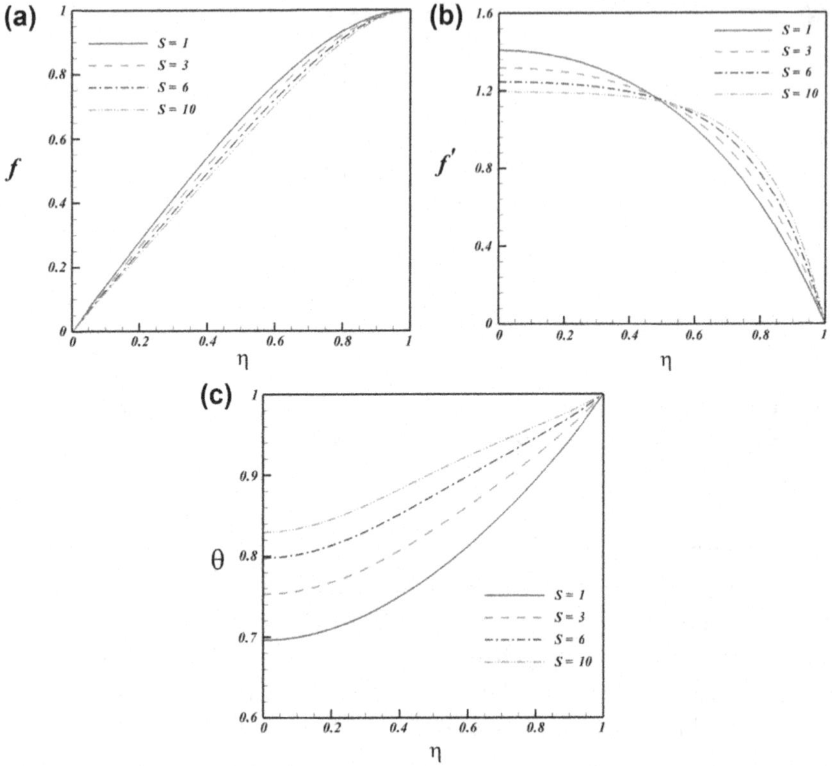

Figure 5.15 Effect of the squeeze number on velocity and temperature profiles, when $Ha = 1$, $Hs = -1$, $\Pr = 6.2$ (CuO—Water).

Fig. 5.17 depicts the effects of Hartmann number and squeeze number on skin friction coefficient and Nusselt number. This figure shows that Hartmann number has direct relationship with both of the skin friction coefficient and Nusselt number. While squeeze number has direct relationship with skin friction coefficient and reverse relationship with Nusselt number. Effects of heat source parameter on temperature profile and Nusselt number is shown in Fig. 5.18. As heat source parameter increases, temperature boundary layer thickness decrease and in turn Nusselt number increases.

5.5 VIBRATION ANALYSIS OF NANOBEAMS

In this example, nonlocal Euler—Bernoulli beam theory is employed for vibration analysis of functionally graded (FG) size-dependent nanobeams. Fig. 5.19 shows the coordinate system for an FG nanobeam with length L,

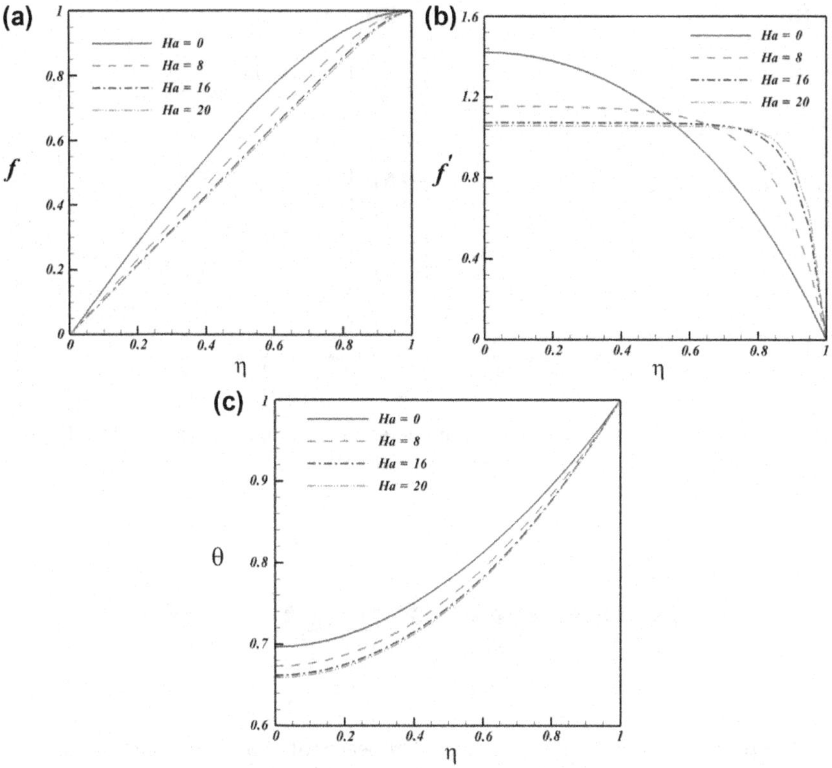

Figure 5.16 Effect of the Hartmann number on velocity and temperature profiles, when $S = 1$, $Hs = -1$, $\Pr = 6.2$ (CuO–Water).

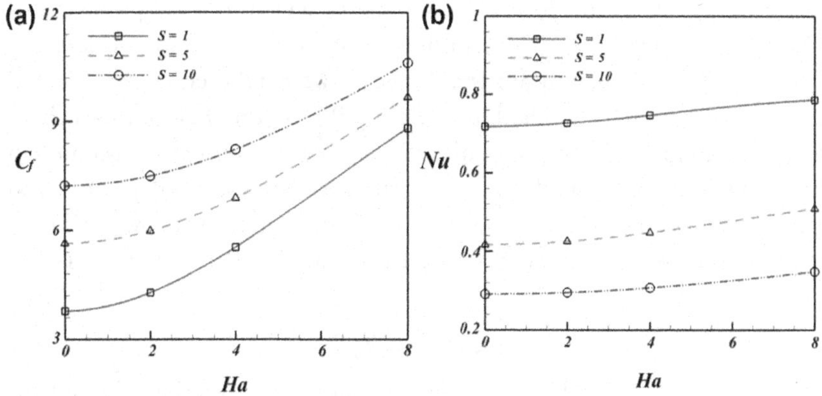

Figure 5.17 Effects of Hartmann number and squeeze number on skin friction coefficient and Nusselt number, when $\phi = 0.04$, $Hs = -1$, $\Pr = 6.2$ (CuO–Water).

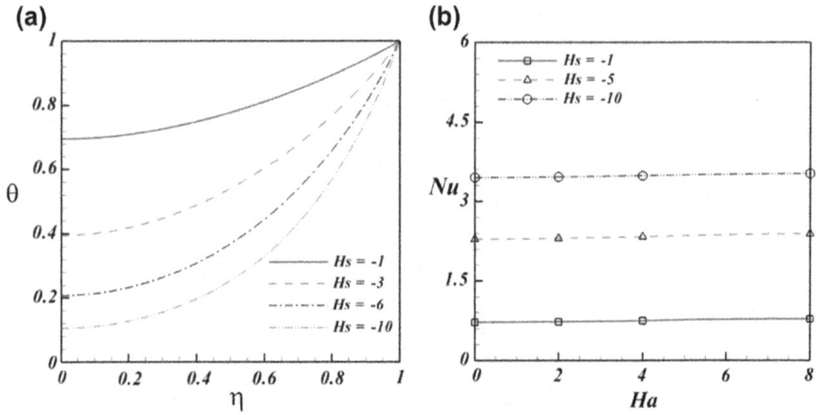

Figure 5.18 Effects of heat source parameter on temperature profile and Nusselt number, when $\phi = 0.04$, $Hs = -1$, $Pr = 6.2$ (CuO–Water).

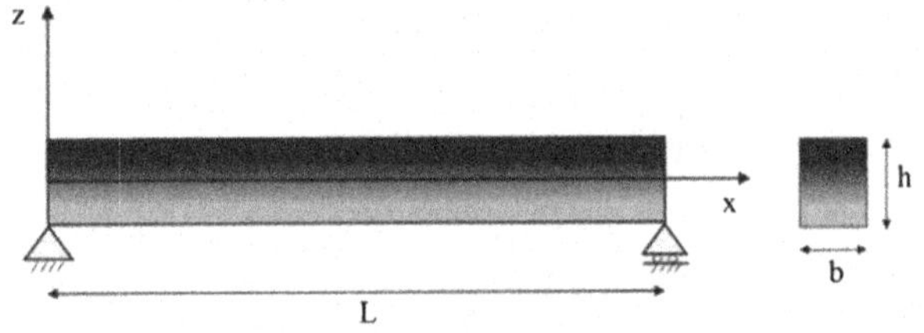

Figure 5.19 Typical functionally graded beam with Cartesian coordinates.

width b, and thickness h. It is assumed that bottom surface ($z = -h/2$) of functionally graded beam is pure metal, whereas the top surface ($z = h/2$) is pure ceramics. One of the most favorable models for functionally graded materials (FGMs) is the power-law model, in which material properties of FGMs are assumed to vary according to a power law about spatial coordinates. The FG nanobeam is assumed to be composed of ceramic and metal, and effective material properties (P_f) of the FG beam such as Young's modulus E_f, shear modulus G_f, and mass density ρ_f are assumed to vary continuously in the thickness direction (z-axis direction) according to a power function of the volume fractions of the constituents. According to the rule of mixture, the effective material properties, P, can be expressed as [5],

$$P_f = P_c V_c + P_m V_m \tag{5.69}$$

where P_m, P_c, V_m, and V_c are the material properties, and the volume fractions of the metal and the ceramic constituents related by:

$$V_c + V_m = 1 \tag{5.70a}$$

The volume fraction of the ceramic and metal constituents of the beam in both power law and Mori–Tanaka models are assumed to be given by:

$$V_c = \left(\frac{z}{h}+\frac{1}{2}\right)^p,\ V_m = 1 - \left(\frac{z}{h}+\frac{1}{2}\right)^p \tag{5.70b}$$

Here p is the nonnegative variable parameter (power-law exponent), which determines the material distribution through the thickness of the beam. Therefore, from Eqs. (5.69) and (5.70), the effective material properties of the FG nanobeam can be expressed as follows:

$$P_f(z) = (P_c - P_m)\left(\frac{z}{h}+\frac{1}{2}\right)^p + P_m \tag{5.71}$$

Additionally, in this study, Mori–Tanaka homogenization technique is also employed to model the effective material properties of the FG nanobeam. According to Mori–Tanaka homogenization technique the local effective material properties of the FG nanobeam such as effective local bulk modulus K_e and shear modulus μ_e can be calculated by:

$$\frac{K_e - K_m}{K_c - K_m} = \frac{V_c}{1 + V_m(K_c - K_m)/(K_m + 4\mu_m/3)} \tag{5.72a}$$

$$\frac{\mu_e - \mu_m}{\mu_c - \mu_m} = \frac{V_c}{1 + V_m(\mu_c - \mu_m)/[\mu_m + \mu_m(9K_m + 8\mu_m)/(6(K_m + 2\mu_m))]} \tag{5.72b}$$

Therefore from Eq. (5.72), the effective Young's modulus (E), Poisson's ratio (υ), and mass density (ρ) based on Mori–Tanaka scheme can be expressed by:

$$E(z) = \frac{9K_e\mu_e}{3K_e + \mu_e} \tag{5.73a}$$

$$\upsilon(z) = \frac{3K_e - 2\mu_e}{6K_e + 2\mu_e} \tag{5.73b}$$

$$\rho(z) = \rho_c V_c + \rho_m V_m \tag{5.73c}$$

It is worth mentioning that the chosen material gradations demonstrate the principle and are employed for demonstration purposes. Upon the

Euler–Bernoulli beam model, the displacement field at any point of the beam can be written as:

$$u_x(x,z,t) = u(x,t) - z\frac{\partial w(x,t)}{\partial x} \tag{5.74a}$$

$$u_z(x,z,t) = w(x,t) \tag{5.74b}$$

where t is time, u and w are displacement components of the midplane along x and z directions, respectively. By assuming the small deformations, the only nonzero strain of the Euler–Bernoulli beam theory is:

$$\varepsilon_{xx} = \varepsilon_{xx}^0 - zk^0,\ \varepsilon_{xx}^0 = \frac{\partial u(x,t)}{\partial x},\ k^0 = \frac{\partial^2 w(x,t)}{\partial x^2} \tag{5.75}$$

where ε_{xx}^0 is the extensional strain and k^0 is the bending strain. Based on the Hamilton's principle, which states that, the motion of an elastic structure during the time interval $t_1 < t < t_2$ is such that the time integral of the total dynamics potential is extremum:

$$\int_0^t \delta(U - T + W_{ext})dt = 0 \tag{5.76}$$

Here U is strain energy, T is kinetic energy, and W_{ext} is work done by external forces. The virtual strain energy can be calculated as:

$$\delta U = \int_v \sigma_{ij}\delta\varepsilon_{ij}dV = \int_v (\sigma_{xx}\delta\varepsilon_{xx})dV \tag{5.77}$$

Substituting Eq. (5.75) into Eq. (5.77) yields:

$$\delta U = \int_0^L \left(N\left(\delta\varepsilon_{xx}^0\right) - M(\delta k^0)\right)dx \tag{5.78}$$

In which N, M are the axial force and bending moment respectively. These stress resultants used in Eq. (5.78) are defined as:

$$N = \int_A \sigma_{xx}dA,\ M = \int_A \sigma_{xx}z dA \tag{5.79}$$

The kinetic energy for Euler–Bernoulli beam can be written as:

$$T = \frac{1}{2}\int_0^L \int_A \rho(z)\left(\left(\frac{\partial x_x}{\partial t}\right)^2 + \left(\frac{\partial u_z}{\partial t}\right)^2\right)dAdx \tag{5.80}$$

Also the virtual kinetic energy is:

$$\delta T = \int_0^L \left[I_0 \left(\frac{\partial u}{\partial t}\frac{\partial \delta u}{\partial t} + \frac{\partial w}{\partial t}\frac{\partial \delta w}{\partial t} \right) - I_1 \left(\frac{\partial u}{\partial t}\frac{\partial^2 \delta w}{\partial t \partial x} + \frac{\partial \delta u}{\partial t}\frac{\partial^2 w}{\partial t \partial x} \right) + I_2 \frac{\partial^2 w}{\partial t \partial x}\frac{\partial^2 \delta w}{\partial t \partial x} \right] dx \tag{5.81}$$

Where (I_0, I_1, I_2) are the mass moment of inertias, defined as follows:

$$(I_0, I_1, I_2) = \int_A \rho(z)(1, z, z^2) dA \tag{5.82}$$

The first variation of external forces work of the beam can be written in the form:

$$\delta W_{ext} = \int_0^L (f(x)\delta u + q(x)\delta w) dx \tag{5.83}$$

where $f(x)$ and $q(x)$ are external axial and transverse loads distribution along length of beam, respectively. By Substituting Eqs. (5.78), (5.81), and (5.83) into Eq. (5.76), and setting the coefficients of δu, δw, and $\delta \partial w/\partial x$ to zero, the following Euler–Lagrange equation can be obtained:

$$\frac{\partial N}{\partial x} + f = I_0 \frac{\partial^2 u}{\partial t^2} - I_1 \frac{\partial^3 w}{\partial x \partial t^2} \tag{5.84a}$$

$$\frac{\partial^2 M}{\partial x^2} + q = I_0 \frac{\partial^2 w}{\partial t^2} + I_1 \frac{\partial^3 u}{\partial x \partial t^2} - I_2 \frac{\partial^4 w}{\partial x^2 \partial t^2} \tag{5.84b}$$

Under the following boundary conditions:

$$N = 0 \text{ or } u = 0 \text{ at } x = 0 \text{ and } x = L \tag{5.85a}$$

$$\frac{\partial M}{\partial x} - I_1 \frac{\partial^2 u}{\partial t^2} + I_2 \frac{\partial^3 w}{\partial x \partial t^2} = 0 \text{ or } w = 0 \text{ at } x = 0 \text{ and } x = L \tag{5.85b}$$

$$M = 0 \text{ or } \frac{\partial w}{\partial x} = 0 \text{ at } x = 0 \text{ and } x = L \tag{5.85c}$$

By using nonlocal elasticity model for FG nanobeam the explicit relation of the nonlocal normal force can be derived by [5].

$$N = A_{xx} \frac{\partial u}{\partial x} - B_{xx} \frac{\partial^2 w}{\partial x^2} + \mu \left(I_0 \frac{\partial^3 u}{\partial x \partial t^2} - I_1 \frac{\partial^4 w}{\partial x^2 \partial t^2} - \frac{\partial f}{\partial x} \right) \tag{5.86}$$

Also the explicit relation of the nonlocal bending moment can be derived by [5].

$$M = B_{xx}\frac{\partial u}{\partial x} - C_{xx}\frac{\partial^2 w}{\partial x^2} + \mu\left(I_0\frac{\partial^3 w}{\partial t^2} + I_1\frac{\partial^3 u}{\partial x \partial t^2} - I_2\frac{\partial^4 w}{\partial x^2 \partial t^2} - q\right) \tag{5.87}$$

The nonlocal governing equations of Euler—Bernoulli FG nanobeam in terms of the displacement can be derived by substituting for N and M from Eqs. (5.86) and (5.87), respectively, into Eq. (5.84) as follows:

$$\begin{aligned} A_{xx}\frac{\partial^2 u}{\partial x^2} - B_{xx}\frac{\partial^3 w}{\partial x^3} + \mu\left(I_0\frac{\partial^4 u}{\partial x^2 \partial t^2} - I_1\frac{\partial^5 w}{\partial x^3 \partial t^2} - \frac{\partial^2 f}{\partial x^2}\right) \\ - I_0\frac{\partial^2 u}{\partial t^2} + I_1\frac{\partial^3 w}{\partial t^2 \partial x} + f = 0 \end{aligned} \tag{5.88a}$$

$$\begin{aligned} B_{xx}\frac{\partial^3 u}{\partial x^3} - C_{xx}\frac{\partial^4 w}{\partial x^4} + \mu\left(I_0\frac{\partial^4 w}{\partial t^2 \partial x^2} + I_1\frac{\partial^5 u}{\partial t^2 \partial x^3} - I_2\frac{\partial^6 w}{\partial t^2 \partial x^4} - \frac{\partial^2 q}{\partial x^2}\right) \\ - I_0\frac{\partial^2 w}{\partial t^2} - I_1\frac{\partial^3 u}{\partial t^2 \partial x} + I_2\frac{\partial^4 w}{\partial t^2 \partial x^2} + q = 0 \end{aligned} \tag{5.88b}$$

In this section to solve the free vibration problem of the FG nanobeam with various boundary conditions, the DTM is employed. A sinusoidal variation of u $(x,\ t)$ and w $(x,\ t)$ with a circular natural frequency ω is assumed and the functions are approximated as,

$$u(x,t) = u(x)e^{i\omega t} \tag{5.89a}$$

$$w(x,t) = w(x)e^{i\omega t} \tag{5.89b}$$

Substituting Eqs. (5.89a) and (5.89b) into Eqs. (5.88a) and (5.88b), equations of motion can be rewritten as follows:

$$\begin{aligned} A_{xx}\frac{\partial^2 u}{\partial x^2} - B_{xx}\frac{\partial^3 w}{\partial x^3} + \mu\left(-I_0\omega^2\frac{\partial^2 u}{\partial x^2} + I_1\omega^2\frac{\partial^3 w}{\partial x^3} - \frac{\partial^2 f}{\partial x^2}\right) \\ + I_0\omega^2 u - I_1\omega^2\frac{\partial w}{\partial x} + f = 0 \end{aligned} \tag{5.90a}$$

$$\begin{aligned} B_{xx}\frac{\partial^3 u}{\partial x^3} - C_{xx}\frac{\partial^4 w}{\partial x^4} + \mu\left(-I_0\omega^2\frac{\partial^2 w}{\partial x^2} - I_1\omega^2\frac{\partial^3 u}{\partial x^3} + I_2\omega^2\frac{\partial^4 w}{\partial x^4} - \frac{\partial^2 q}{\partial x^2}\right) \\ + I_0\omega^2 w + I_1\omega^2\frac{\partial u}{\partial x} - I_2\omega^2\frac{\partial^2 w}{\partial x^2} + q = 0 \end{aligned} \tag{5.90b}$$

According to the basic transformation operations introduced in Chapter 1, the transformed form of the governing Eqs. (5.90a) and (5.90b) around $x_0 = 0$ may be obtained as:

$$
\begin{aligned}
&A_{xx}(k+1)(k+2)U[k+2] - B_{xx}(k+1)(k+2)(k+3)W[k+3] \\
&- I_0\omega^2(-U[k] + \mu(k+1)(k+2)U[k+2]) \\
&- I_1\omega^2(-\mu(k+1)(k+2)(k+3)W[k+3] + (k+1)W[k+1]) = 0
\end{aligned}
\tag{5.91}
$$

$$
\begin{aligned}
&B_{xx}(k+1)(k+2)(k+3)U[k+3] \\
&- C_{xx}(k+1)(k+2)(k+3)(k+4)W[k+4] \\
&- I_0\omega^2(-W[k] + \mu(k+1)(k+2)W[k+2]) \\
&- I_1\omega^2(-(k+1)U[k+1] + \mu(k+1)(k+2)(k+3)U[k+3]) \\
&- I_2\omega^2(-\mu(k+1)(k+2)(k+3)(k+4)W[k+4] \\
&+ (k+1)(k+2)W[k+2]) = 0
\end{aligned}
\tag{5.92}
$$

where $U[k\cdot]$and $W\,[k\cdot]$ are the transformed functions of u and w respectively. Additionally, applying DTM to Eqs. (5.85a)–(5.85c), the various boundary conditions are given as follows:

Simply supported–simply supported:

$$
\begin{aligned}
&W[0] = 0,\ W[2] = 0,\ U[0] = 0 \\
&\sum_{k=0}^{\infty} W[k] = 0,\ \sum_{k=0}^{\infty} k(k-1)W[k] = 0,\ \sum_{k=0}^{\infty} kU[k] = 0
\end{aligned}
\tag{5.93a}
$$

Clamped–clamped:

$$
\begin{aligned}
&W[0] = 0, W[1] = 0, U[0] = 0 \\
&\sum_{k=0}^{\infty} W[k] = 0, \sum_{k=0}^{\infty} kW[k] = 0, \sum_{k=0}^{\infty} U[k] = 0
\end{aligned}
\tag{5.93b}
$$

Clamped–simply supported:

$$
\begin{aligned}
&W[0] = 0, W[1] = 0, U[0] = 0 \\
&\sum_{k=0}^{\infty} W[k] = 0, \sum_{k=0}^{\infty} k(k-1)W[k] = 0, \sum_{k=0}^{\infty} kU[k] = 0
\end{aligned}
\tag{5.93c}
$$

Clamped-Free:

$$
\begin{aligned}
&W[0] = 0, W[1] = 0, U[0] = 0 \\
&\sum_{k=0}^{\infty} k(k-1)W[k] = 0, \sum_{k=0}^{\infty} k(k-1)(k-2)W[k] = 0, \sum_{k=0}^{\infty} kU[k] = 0
\end{aligned}
\tag{5.93d}
$$

By using Eqs. (5.91) and (5.92) together with the transformed boundary conditions, one arrives at the following eigenvalue problem:

$$\begin{bmatrix} A_{11}(\omega) & A_{12}(\omega) & A_{13}(\omega) \\ A_{21}(\omega) & A_{22}(\omega) & A_{23}(\omega) \\ A_{31}(\omega) & A_{32}(\omega) & A_{33}(\omega) \end{bmatrix} [C] = 0 \tag{5.94a}$$

where $[C]$ corresponds to the missing boundary conditions at $x =$. For the nontrivial solutions of Eq. (5.94a), it is necessary that the determinant of the coefficient matrix is equal to zero:

$$\begin{bmatrix} A_{11}(\omega) & A_{12}(\omega) & A_{13}(\omega) \\ A_{21}(\omega) & A_{22}(\omega) & A_{23}(\omega) \\ A_{31}(\omega) & A_{32}(\omega) & A_{33}(\omega) \end{bmatrix} = 0 \tag{5.94b}$$

Solution of Eq. (5.94b) is simply a polynomial root finding problem. In this example, the Newton–Raphson method is used to solve the governing equation of the nondimensional natural frequencies.

Solving Eq. (5.94b), the ith estimated eigenvalue for nth iteration $\left(\omega = \omega_i^{(n)}\right)$ may be obtained and the total number of iterations is related to the accuracy of calculations, which can be determined by the following equation:

$$\left|\omega_i^{(n)} - \omega_i^{(n-1)}\right| \langle \varepsilon \tag{5.95}$$

In this study $\varepsilon = 0.0001$ considered in procedure of finding eigenvalues which results in four digit precision in estimated eigenvalues. Further a Matlab program has been developed according to DTM rule stated above, to find eigenvalues. As mentioned before, DTM implies an iterative procedure to obtain the high-order Taylor series solution of differential equations. The Taylor series method requires a long computational time for large orders, whereas one advantage of employing DTM in solving differential equations is a fast convergence rate and a small calculation error. To show the results, functionally graded nanobeam is composed of steel and alumina (Al_2O_3), where its properties are given in [5]. The bottom surface of the beam is pure steel, whereas the top surface of the beam is pure alumina. The beam geometry has the following dimensions: L (length) = 10,000 nm, b (width) = 1000 nm, and h (thickness) = 100 nm. Relation described in Eq. (5.96) is performed to calculate the nondimensional natural frequencies.

$$\overline{\omega} = \omega L^2 \sqrt{\rho A / EI} \tag{5.96}$$

where $I = \text{bh}^3/12$ is the moment of inertia of the cross section of the beam.

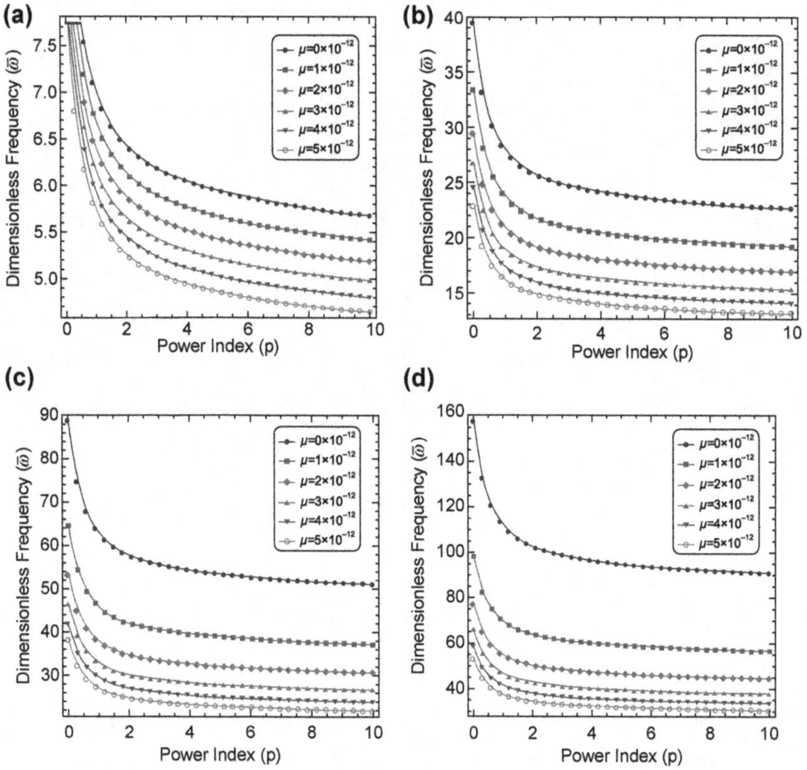

Figure 5.20 The variation of the (a) first, (b) second, (c) third, and (d) fourth dimensionless frequency of simply supported FG nanobeam with material graduation for different nonlocality parameter ($L/h = 100$).

Fig. 5.20 demonstrates the variation of first four fundamental frequencies of power law FG nanobeam with varying material distribution and nonlocality parameter at $L/h = 100$. As can be noted, the first four dimensionless frequency of simply supported FG nanobeam decrease acutely as the material index parameter increases from 0 to 10. It can be observed that, the first and second frequencies reduce with a high rate, where the power exponent in range from 0 to 5 than the power exponent in range between 5 and 10. While the third and fourth frequencies reduce and they have high rate in range from 0 to 2. Figs. 5.21 and 5.22 demonstrate the variation of mode number with changing of the nonlocality parameter at constant slenderness ratio ($L/h = 50$) of FG nanobeam with simply supported edge conditions and different material distribution for power law and Mori–Tanaka models, respectively. As presented, the influence of

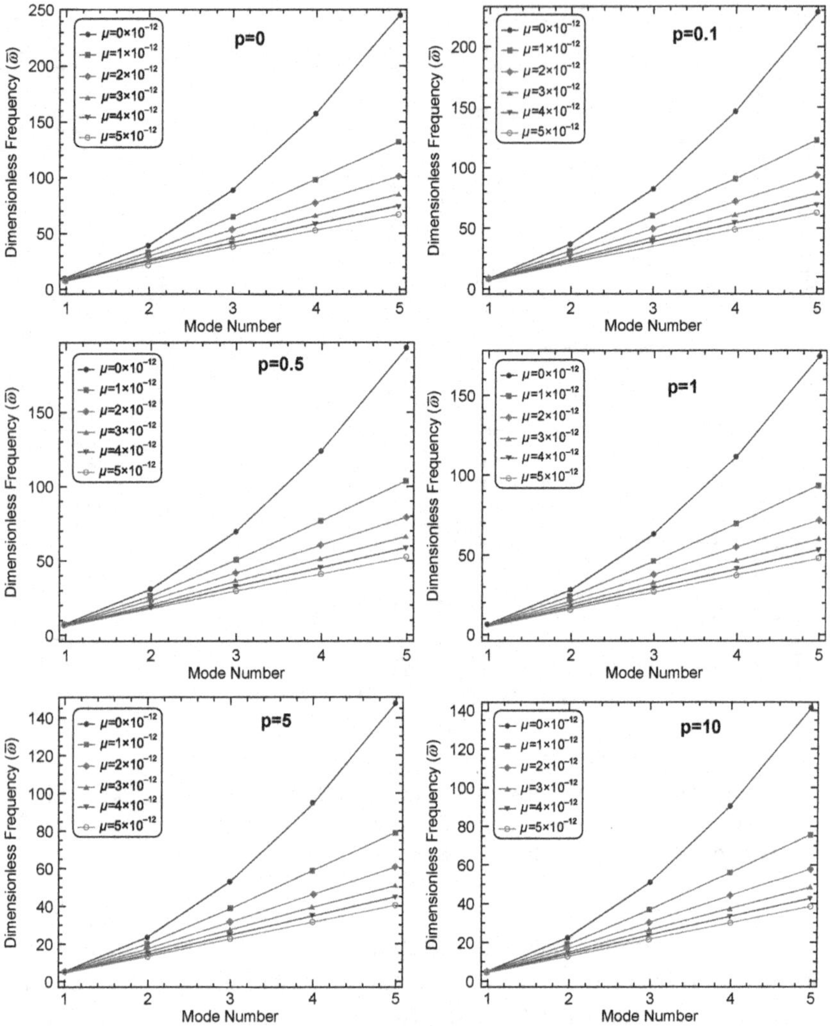

Figure 5.21 The effect of nonlocality parameter on dimensionless frequency of power law FG nanobeam for various mode numbers and with different material graduation indexes ($p = 0, 0.1, 0.5, 1, 5, 10$) ($L/h = 50$).

nonlocality parameter on the nondimensional frequency increased by the growing in mode number. Also, it can be deduced that, the influence of nonlocality parameter on the frequencies is unaffected with the material distribution. More results and discussion on accuracy of method can be found in [5].

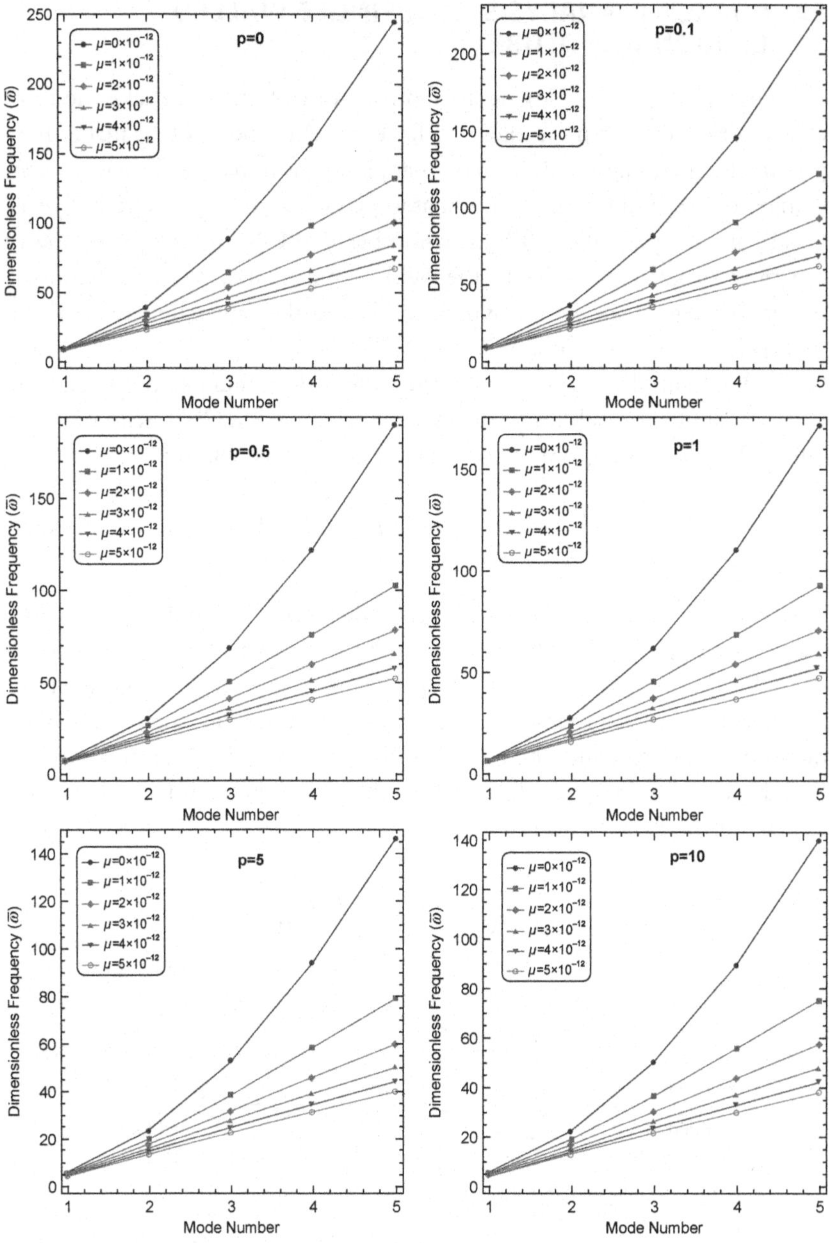

Figure 5.22 The effect of nonlocality parameter on dimensionless frequency of Mori–Tanaka FG nanobeam for various mode numbers and with different material graduation indexes ($p = 0$, 0.1, 0.5, 1, 5, 10) ($L/h = 50$).

5.6 BUCKLING ANALYSIS OF A SINGLE-WALLED CARBON NANOTUBE

Carbon nanotubes (CNTs) are an allotrope of carbon. They take the form of cylindrical carbon molecules and have novel properties that make them potentially useful in a wide variety of applications in nanotechnology, electronics, optics, and other fields of materials science. In the recent few years, CNTs and single-walled carbon nanotubes (SWCNTs) have been one of the most promising studies in the field of mechanics, physics, chemistry, materials science, and so on, which motivated the researchers to work on [6–8]. Fig. 5.23 show a sample of CNT.

In this example, we considered the Euler–Bernoulli beam model using stress gradient approach for the buckling analysis of SWCNT with nonlocal effect. The governing differential equation for the buckling is

$$EI\frac{d^4w}{dx^4} + P\frac{d^2}{dx^2}\left(w - (e_0a)^2\frac{d^2w}{dx^2}\right) = 0 \tag{5.97}$$

where $w = (x \cdot t)$ is the transverse beam deflection, t, x, are the spatial coordinate and the time; E is the Young modulus of elasticity; P is the buckling load, I is the moment of inertia of the beam cross section, e_0 is a constant appropriate to each material, a is an internal characteristic length. The e_0a is determined by matching the dispersion curves based on the atomic models. For the buckling analysis, Eq. (5.97) can be nondimensionalised using L (length of the beam) and rewritten as

$$\frac{d^4w}{dx^4} + \frac{PL^2}{EI}\frac{d^2}{dx^2}\left(w - (e_0a/L)^2\frac{d^2w}{dx^2}\right) = 0 \tag{5.98}$$

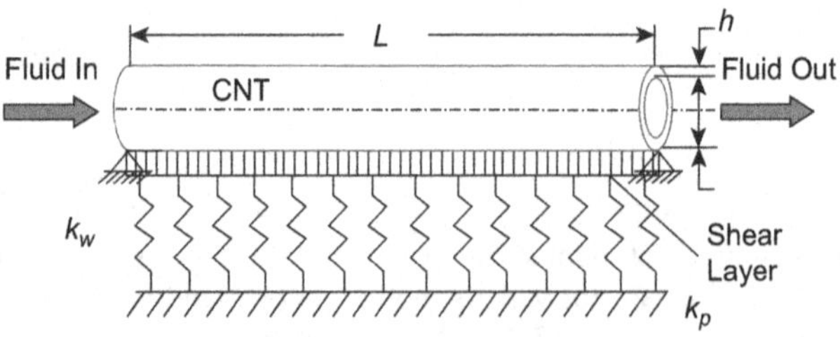

Figure 5.23 Single-walled carbon nanotube embedded in a Pasternak-type foundation model.

By applying DTM to above equation, following expression can be written easily

$$W(k+4) = \frac{-rW(k+2)}{(k+3)(k+4)} \tag{5.99}$$

where

$$r = \frac{PL^2/EI}{\left[1-(e_0 a)^2(P/EI)\right]} \tag{5.100}$$

The boundary conditions are different for different conditions.

A. Clamped–clamped

In this case, consider the SWCNT supported by clamped at both the ends and the boundary conditions defined as

$$w(0)=0, w'(0)=0, w(L)=0, w'(L)=0 \tag{5.101}$$

And DTM transformed shape is

$$W(0)=0,\ W(1)=0, \sum_{k=0}^{\infty} W(k)=0, \sum_{k=0}^{\infty} kW(k)=0 \tag{5.102}$$

B. Simply supported

The boundary conditions for the case of simply supported SWCNT at both the ends are defined as

$$w(0)=0, w''(0)=0, w(L)=0, w''(L)=0 \tag{5.103}$$

And DTM transformed shape is

$$W(0)=0,\ W(2)=0, \sum_{k=0}^{\infty} W(k)=0, \sum_{k=0}^{\infty} k(k-1)W(k)=0 \tag{5.104}$$

C. Clamped–simply supported

For the SWCNT supported by clamped at one end and simply supported in the other end, the boundary conditions are defined as

$$w(0)=0, w'(0)=0, w(L)=0, w''(L)=0 \tag{5.105}$$

And DTM transformed shape is

$$W(0)=0,\ W(1)=0, \sum_{k=0}^{\infty} W(k)=0, \sum_{k=0}^{\infty} k(k-1)W(k)=0 \tag{5.106}$$

D. Clamped—free SWCNT

For the SWCNT supported by clamped at one end and free in the other end, the boundary conditions are defined as

$$
\begin{aligned}
& w(0)=0, w'(0)=0, M_{NL}(L)=0,\ V_{NL}(L)=0 \\
& M_{NL}=-EI\left(\frac{d^2 w}{dx^2}+(e_0 a)^2 \frac{d^4 w}{dx^4}\right) \\
& V_{NL}=-EI\left(\frac{d^3 w}{dx^3}+(e_0 a)^2 \frac{d^5 w}{dx^5}\right)-P\frac{dw}{dx}
\end{aligned}
\tag{5.107}
$$

where M_{NL} and V_{NL} are the nonlocal moment and nonlocal shear force. And DTM transformed shape is

$$
\begin{aligned}
& W(0)=0, W(3)=0 \\
& \sum_{k=0}^{\infty} k(k-1)W(k)\left[1+(e_0 a)^2(k-2)(k-3)\right]=0 \\
& \sum_{k=0}^{\infty} k(k-1)(k-2)W(k)\left[1+(e_0 a)^2(k-3)(k-4)\right]+(r)(k)W(k)=0
\end{aligned}
\tag{5.108}
$$

REFERENCES

[1] Choi SUS. Enhancing thermal conductivity of fluids with nanoparticles. In: Siginer DA, Wang HP, editors. Developments and applications of non-Newtonian flows, FED-vol. 231/MD-vol. 66. New York: ASME; 1995. p. 99—105.

[2] Hatami M, Sheikholeslami M, Hosseini M, Ganji DD. Analytical investigation of MHD nanofluid flow in non-parallel walls. Journal of Molecular Liquids 2014;194:251—9.

[3] Mosayebidorcheh S, Sheikholeslami M, Hatami M, Ganji DD. Analysis of turbulent MHD Couette nanofluid flow and heat transfer using hybrid DTM—FDM. Particuology 2016;26:95—101.

[4] Sheikholeslami M, Ganji DD. Nanofluid flow and heat transfer between parallel plates considering Brownian motion using DTM. Computer Methods in Applied Mechanics and Engineering 2015;283:651—63.

[5] Ebrahimi F, Salari E. Size-dependent free flexural vibrational behavior of functionally graded nanobeams using semi-analytical differential transform method. Composites Part B: Engineering 2015;79:156—69.

[6] Senthilkumar V. Buckling analysis of a single-walled carbon nanotube with nonlocal continuum elasticity by using differential transform method. Advanced Science Letters September 2010;3(3):337—40 (4).

[7] Valipour P, Ghasemi SE, Khosravani MR, Ganji DD. Theoretical analysis on nonlinear vibration of fluid flow in single-walled carbon nanotube. Journal of Theoretical and Applied Physics. http://dx.doi.org/10.1007/s40094-016-0217-9.

[8] Ahmadi Asoor AA, Valipour P, Ghasemi SE. Investigation on vibration of single-walled carbon nanotubes by variational iteration method. Applied Nanoscience 2016;6:243—9.

CHAPTER 6

DTM for Magnetohydrodynamic (MHD) and Porous Medium Flows

6.1 INTRODUCTION

In recent years the effect of magnetic field and porous medium in different engineering applications such as the cooling of reactors and many metallurgical processes involves the cooling of continuous tiles has been more considerable. And also, in several engineering processes, materials manufactured by extrusion processes and heat treated materials traveling between a feed roll and a wind up roll on convey belts possess the characteristics of a moving continuous surface, oil industry and combustion, penetration etc., you can find some applications of Magnetohydrodynamic (MHD) and porous medium [1]. Although in previous chapters some examples were presented, which contained these two important topics, but in this separate chapter some other examples are discussed due to its importance in the following sections:

6.1 Introduction
6.2 Magnetohydrodynamic Couette Fluid Flow Between Parallel Plates
6.3 Micropolar Fluid in a Porous Channel
6.4 Magnetohydrodynamic Viscous Flow Between Porous Surfaces

6.2 MAGNETOHYDRODYNAMIC COUETTE FLUID FLOW BETWEEN PARALLEL PLATES

The fluid is assumed to be flowing between two infinite horizontal plates located at the $y = \pm h$ planes. The upper plate moves with a uniform velocity U_0 while the lower plate is kept stationary. The two plates are assumed to be electrically insulating and kept at two constant temperatures T_1 for the lower plate and T_2 for the upper plate with $T_2 > T_1$. A constant pressure gradient is applied in the x-direction. A uniform magnetic field $\boldsymbol{B}_0$ is applied in the positive y-direction while the induced magnetic field is

Differential Transformation Method for Mechanical Engineering Problems
ISBN 978-0-12-805190-0
http://dx.doi.org/10.1016/B978-0-12-805190-0.00006-1

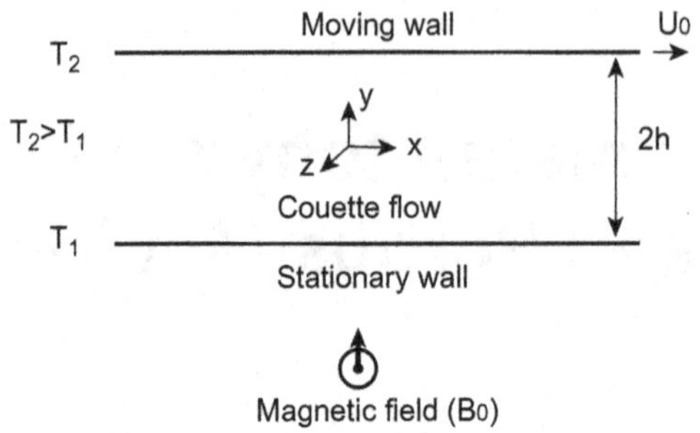

Figure 6.1 Schematic of the problem.

neglected by assuming a very small magnetic Reynolds number (See Fig. 6.1). The Hall effect is taken into consideration and consequently a z-component for the velocity is expected to arise. The fluid motion starts from rest at $t = 0$, and the no-slip condition at the plates implies that the fluid velocity has neither a z nor an x-component at $y = -h$ and $y = h$. The initial temperature of the fluid is assumed to be equal to T_1. Since the plates are infinite in the x and z directions, the physical quantities do not change in these directions and the problem is essentially one dimensional [2].

The flow of the fluid is governed by the Navier–Stokes equation which has the two components.

$$\rho\frac{\partial u}{\partial t} = -\frac{dP}{dx} + \mu\frac{\partial^2 u}{\partial y^2} + \frac{\partial \mu}{\partial y}\frac{\partial u}{\partial y} - \frac{\sigma B_0^2}{1+m^2}(u+mw) \tag{6.1}$$

$$\rho\frac{\partial w}{\partial t} = \mu\frac{\partial^2 w}{\partial y^2} + \frac{\partial \mu}{\partial y}\frac{\partial w}{\partial y} - \frac{\sigma B_0^2}{1+m^2}(w-mu) \tag{6.2}$$

where ρ is the density of the fluid, μ is the viscosity of the fluid, v is the velocity vector of the fluid $= u(y, t)i + w(y, t)j$, σ is the electric conductivity of the fluid, m is the Hall parameter given by $m = \sigma\beta B_0$, and β is the Hall factor. The energy equation describing the temperature distribution for the fluid is given by Ref. [2].

$$\rho c_p\frac{\partial T}{\partial t} = \frac{\partial}{\partial y}\left(k\frac{\partial T}{\partial y}\right) + \mu\left[\left(\frac{\partial u}{\partial y}\right)^2 + \left(\frac{\partial w}{\partial y}\right)^2\right] + \frac{\sigma B_0^2}{(1+m^2)}(u^2+w^2) \tag{6.3}$$

where T is the temperature of the fluid, c_p is the specific heat at constant pressure of the fluid, and k is thermal conductivity of the fluid. The viscosity of the fluid is assumed to vary exponentially with temperature and is defined as $\mu = \mu_0 f(T) = \mu_0\exp(-a(T - T_1))$. Also the thermal conductivity of the fluid is varying linearly with temperature as $k = k_0 g(T) = k_0[1 + b(T - T_1)]$. The problem is simplified by writing the equations in the nondimensional form. To achieve this define the following nondimensional quantities,

$$\widehat{y} = \frac{y}{h}, \widehat{t} = \frac{tU_0}{h}, \widehat{P} = \frac{P}{\rho U_0^2}, (\widehat{u}, \widehat{w}) = \frac{(u, w)}{U_0}, \theta = \frac{T - T_1}{T_2 - T_1}, \alpha = -\frac{d\widehat{P}}{d\widehat{x}} \tag{6.4}$$

$f(\theta) = e^{-a\theta}$, a is the viscosity parameter. $g(\theta) = 1 + b\theta$ b is the thermal conductivity parameter. $Re = \frac{\rho U_0 h}{\mu_0}$ is the Reynolds number. $Ha^2 = \frac{\sigma B_0^2 h^2}{U_0^2 \mu_0}$, Ha is Hartmann number. $Pr = \frac{\mu_0 c_p}{k_0}$ is Prandtl number, and $Ec = \frac{U_0^2}{c_p(T_2 - T_1)}$ is Eckert number. In terms of the above nondimensional quantities the velocity and energy Eqs. (6.1)–(6.3) read,

$$\frac{\partial u}{\partial t} = \alpha + \frac{1}{Re} f(\theta) \frac{\partial^2 u}{\partial y^2} + \frac{1}{Re} \frac{\partial f(\theta)}{\partial y} \frac{\partial u}{\partial y} - \frac{1}{Re} \frac{Ha^2}{1 + m^2}(u + mw) \tag{6.5}$$

$$\frac{\partial w}{\partial t} = \frac{1}{Re} f(\theta) \frac{\partial^2 w}{\partial y^2} + \frac{1}{Re} \frac{\partial f(\theta)}{\partial y} \frac{\partial w}{\partial y} - \frac{1}{Re} \frac{Ha^2}{1 + m^2}(w - mu) \tag{6.6}$$

$$\frac{\partial \theta}{\partial t} = \frac{1}{RePr} g(\theta) \frac{\partial^2 \theta}{\partial y^2} + \frac{1}{RePr} \left(\frac{\partial g(\theta)}{\partial y}\right)\left(\frac{\partial \theta}{\partial y}\right) + \frac{Ec}{Re} f(\theta)\left[\left(\frac{\partial u}{\partial y}\right)^2 + \left(\frac{\partial w}{\partial y}\right)^2\right] + \frac{EcHa^2}{Re(1 + m^2)}(u^2 + w^2) \tag{6.7}$$

The boundary and initial conditions for components of velocity and temperature are,

$$IC's \rightarrow \begin{cases} u(y, 0) = 0 \\ w(y, 0) = 0 \\ \theta(y, 0) = 0 \end{cases} \qquad BC's \rightarrow \begin{cases} u(-1, t) = 0, \; u(1, t) = 1 \\ w(-1, t) = 0, \; w(1, t) = 0 \\ \theta(-1, t) = 0, \; \theta(1, t) = 1 \end{cases} \tag{6.8}$$

Once the values of the velocities and temperature are obtained, the friction coefficient and Nusselt number will be determined. The local skin friction coefficient at the lower wall is [2],

$$C_f = \frac{2}{Re}\frac{\partial U}{\partial y}\bigg|_{y=-1} \tag{6.9}$$

And the local Nusselt number for lower wall is defined as,

$$Nu_x = \frac{\partial \theta}{\partial y}\bigg|_{y=-1} \tag{6.10}$$

To solve the partial differential equation $u(y, t)$ in the domain $t \in [0, T]$ and $y \in [y_{first}, y_{end}]$ using hybrid modified differential transformation method and finite difference method (FDM), we apply finite difference approximate on y-direction and take modified differential transformation method (MDTM) on t. The time domain is divided to N_t sections. We suppose the time subdomains are equal and length of each subdomain is $H = T/N_t$. So there is a separate function for every subdomain as follows:

$$u_i(y,t) = \begin{cases} u_1(j,t), \ t \in [t_1, t_2], 1 \le j \le N_y + 1 \\ \vdots \\ u_i(j,t), \ t \in [t_i, t_{i+1}], 1 \le j \le N_y + 1 \\ \vdots \\ u_{N_t}(j,t), \ t \in [t_N, t_{N_t+1}], 1 \le j \le N_y + 1 \end{cases} \tag{6.11}$$

where $t_i = (i-1)H$ and N_y is the number of cells in y-direction. The solution of the system of Eqs. (6.5) to (6.8) can be assumed as the following form:

$$\begin{aligned} &\text{for } 1 \le i \le N_t, \quad 1 \le j \le N_y + 1 \\ &u_i(j,t) = \sum_{k=0}^{m} U_i(j,k)\left(\frac{t-t_i}{H}\right)^k \quad t \in [t_i, t_{i+1}] \\ &w_i(j,t) = \sum_{k=0}^{m} W_i(j,k)\left(\frac{t-t_i}{H}\right)^k \quad t \in [t_i, t_{i+1}] \\ &\theta_i(j,t) = \sum_{k=0}^{m} \Theta_i(j,k)\left(\frac{t-t_i}{H}\right)^k \quad t \in [t_i, t_{i+1}] \end{aligned} \tag{6.12}$$

After taking second-order accurate central finite difference approximation with respect to y and applying MDTM on Eqs. (6.5) to (6.8), the following recurrence relations can be obtained:

for $1 \le i \le N_t,\ 1 \le j \le N_y - 1$

$$U_i(j,k+1) = \frac{H}{k+1}\left\{\alpha\delta(k) + \frac{1}{Re\Delta y^2}\sum_{r=0}^{k} F_i(j,k-r)(U_i(j+1,r) - 2U_i(j,r) + U_i(j-1,r)) + \frac{1}{4Re\Delta y^2}\sum_{r=0}^{k}(F_i(j+1,k-r) - F_i(j-1,k-r))(U_i(j+1,r) - U_i(j-1,r)) - \frac{Ha^2}{Re(1+m^2)}(U_i(j,k) + mW_i(j,k))\right\} \quad (6.13)$$

$$W_i(j,k+1) = \frac{H}{k+1}\left\{\frac{1}{Re\Delta y^2}\sum_{r=0}^{k} F_i(j,k-r)(W_i(j+1,r) - 2W_i(j,r) + W_i(j-1,r)) + \frac{1}{4Re\Delta y^2}\sum_{r=0}^{k}(F_i(j+1,k-r) - F_i(j-1,\ k-r))(W_i(j+1,r) - W_i(j-1,r)) - \frac{Ha^2}{Re(1+m^2)}(W_i(j,k) - mU_i(j,k))\right\} \quad (6.14)$$

$$\Theta_i(j,k+1) = \frac{H}{k+1}\left\{\frac{1}{RePr\Delta y^2}(\Theta_i(j+1,k) - 2\Theta_i(j,k) + \Theta_i(j-1,k)) + \frac{b}{RePr\Delta y^2}\sum_{r=0}^{k}\Theta_i(j,k-r)(\Theta_i(j+1,r) - 2\Theta_i(j,r) + \Theta_i(j-1,r)) + \frac{1}{4RePr\Delta y^2}\sum_{r=0}^{k}(\Theta_i(j+1,k-r) - \Theta_i(j-1,k-r)) (\Theta_i(j+1,r) - \Theta_i(j-1,r)) + \frac{Ec}{4Re\Delta y^2}\sum_{r=0}^{k}\sum_{s=0}^{r} F_i(j,s) [(U_i(j+1,r-s) - U_i(j-1,r-s))(U_i(j+1,k-r) - U_i(j-1,k-r)) (W_i(j+1,r-s) - W_i(j-1,r-s))(W_i(j+1,k-r) - W_i(j-1,k-r))] + \frac{EcHa^2}{Re(1+m^2)}\sum_{r=0}^{k}[U_i(j,r)U_i(j,k-r) + W_i(j,r)W_i(j,k-r)]\right\} \quad (6.15)$$

where $F_i(j, k)$ and $G_i(j, k)$ are the differential transform of the functions $f(\theta)$ and $g(\theta)$, respectively. Applying MDTM on initial conditions in Eq. (6.8),

$$\begin{aligned}&\text{for } 1 \leq j \leq N_y + 1\\ &U_1(j,0) = 0,\ W_1(j,0) = 0,\ T_1(j,0) = 0\end{aligned} \tag{6.16}$$

The boundary conditions in Eq. (6.8) can be transformed as follows:

for $1 \leq j \leq N_t$

$$BC\text{'s for } u(y,t) \rightarrow \begin{cases} U_i(1,k) = 0,\ k \geq 0 \\ U_i(N_y+1,0) = 1,\ U_i(N_y+1,k) = 0,\ k \geq 1 \end{cases} \tag{6.17}$$

$$BC\text{'s for } w(y,t) \rightarrow \begin{cases} W_i(1,k) = 0,\ k \geq 0 \\ W_i(N_y+1,k) = 0,\ k \geq 0 \end{cases} \tag{6.18}$$

$$BC\text{'s for } \theta(y,t) \rightarrow \begin{cases} \Theta_i(1,k) = 0,\ k \geq 0 \\ \Theta_i(N_y+1,0) = 1,\ \Theta_i(N_y+1,k) = 0,\ k \geq 1 \end{cases} \tag{6.19}$$

For solving the problem in whole of the time subdomains, we must use the continuity condition in each time subdomain. These conditions can be expressed as,

$$\begin{aligned}&\text{for } 2 \leq j \leq N_y\,,\ 2 \leq i \leq N_t\\ &U_i(j,0) = \sum_{k=0}^{m} U_{i-1}(j,k),\quad W_i(j,0) = \sum_{k=0}^{m} W_{i-1}(j,k)\\ &\Theta_i(j,0) = \sum_{k=0}^{m} \Theta_{i-1}(j,k)\end{aligned} \tag{6.20}$$

In this study all calculations are based on main data for $\alpha = 5$, $Pr = 1$, $Re = 1$, $Ec = 0.2$, $Ha = 1$, $m = 3$, $a = 0.5$, $b = 0.5$ for Eq. (6.7), and in the parametric study for effect of each parameter, other parameters are assumed to be constant and the effect of the main parameter in an acceptable range is investigated. In the first step, the accuracy of hybrid differential transformation method (DTM) is investigated comparing to FDM. As seen in Fig. 6.2, which presents the velocity and temperature distributions as functions of y for various values of time (0.1, 0.5, and 2 s), hybrid-DTM is completely accurate and efficient. This figure reveals that temperature in Couette flow can exceed from hot plate temperature in large times (Fig. 6.2(c), $t = 2$s). As seen hybrid-DTM has an excellent agreement with

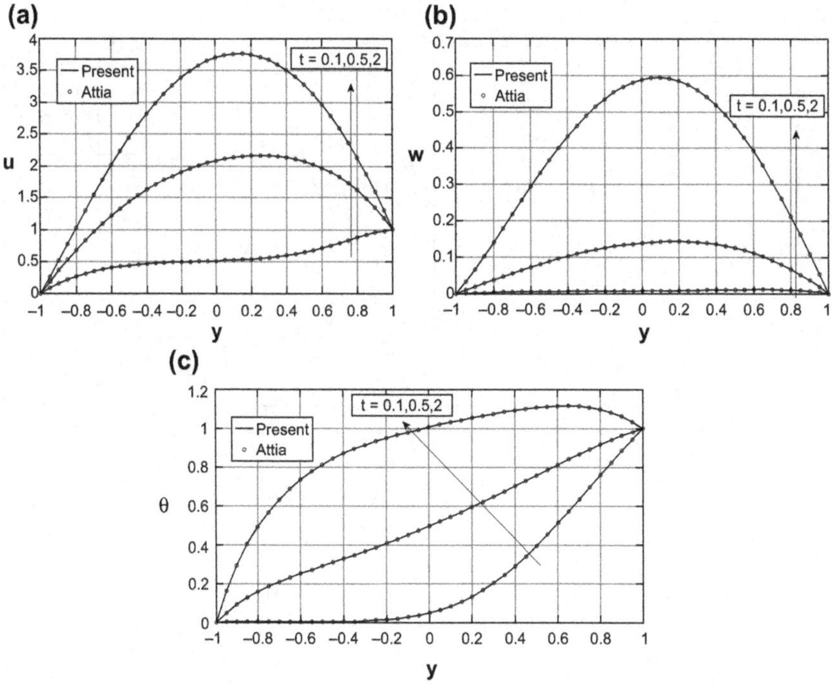

Figure 6.2 Comparison between hybrid differential transformation method, when $\alpha = 5$, $Pr = 1$, $Re = 1$, $Ec = 0.2$, $Ha = 1$, $m = 3$, $a = 0.5$, $b = 0.5$. (a) u velocity, (b) w velocity and, (c) temperature.

FDM numerical technique, furthermore its solution is obtained using a simple iterative procedure, also it reduces the execution time and memory requirements for large scale computations. Following a parametric study for showing the effect of constant numbers appeared in the mathematical equations is investigated by hybrid-DTM.

Fig. 6.3 shows the effect of Hartmann number (*Ha*) on velocities and temperature versus time in the center of channel ($y = 0$). Fig. 6.3(a) confirms that increasing *Ha* decreases u as it increases the damping force on u. Fig. 6.3(b) presents an interesting phenomenon, which is, increasing *Ha* increases w for small t and decreases w for large t and steady state condition. Fig. 6.3(c) shows that the effect of *Ha* on the temperature θ depends on t. If *Ha* is small $0 < Ha < 1$, then increasing *Ha* increases θ as a result of increasing the Joule dissipation. However, for large values of t, increasing *Ha* decreases θ due to the reduction in the Joule and viscous dissipations.

Fig. 6.4 demonstrates the effect of Reynolds number (*Re*) on velocities and temperature distribution in the center of channel. These figures reveal

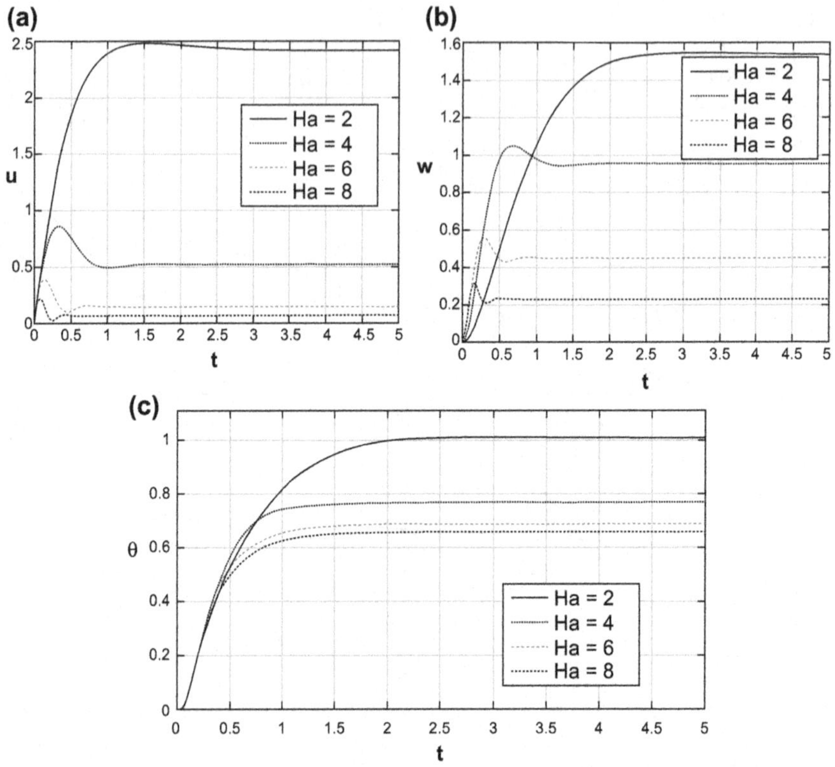

Figure 6.3 Effect of Hartmann number (*Ha*) in center ($y = 0$), when $\alpha = 5$, $Pr = 1$, $Re = 1$, $Ec = 0.2$, $m = 3$, $a = 0.5$, $b = 0.5$. (a) *u* velocity, (b) *w* velocity and, (c) temperature.

that the velocity components, *u and w*, and temperature θ reach steady state quickly for low *Re* numbers but their steady states value is great for large *Re* numbers. Also these figures confirm that *u* reaches steady state faster than *w*. This is expected as *u* is the source of *w*.

Effect of Hall parameter (*m*) is depicted in Fig. 6.5. It is evident that *u* increases with *m* for all times (See Fig. 6.5(a)). This is due to the fact that an increase in *m* decreases the effective conductivity, and hence the magnetic damping decreases which increases *u*. Fig. 6.5(b) illustrates effect of *m* on *w*, velocity component. The effect of *m* on θ is shown in Fig. 6.5(c). For all values for *m*, increasing *m* decreases θ slightly. This is because an increase in *m* results in an increase in *u* but a decrease in *w*, so the Joule dissipation that is proportional also to $(1/(1 + \mathrm{m}^2))$ decreases.

According to Eqs. (6.9) and (6.10), local Nusselt number and skin friction coefficient for lower plate are calculated and results are presented

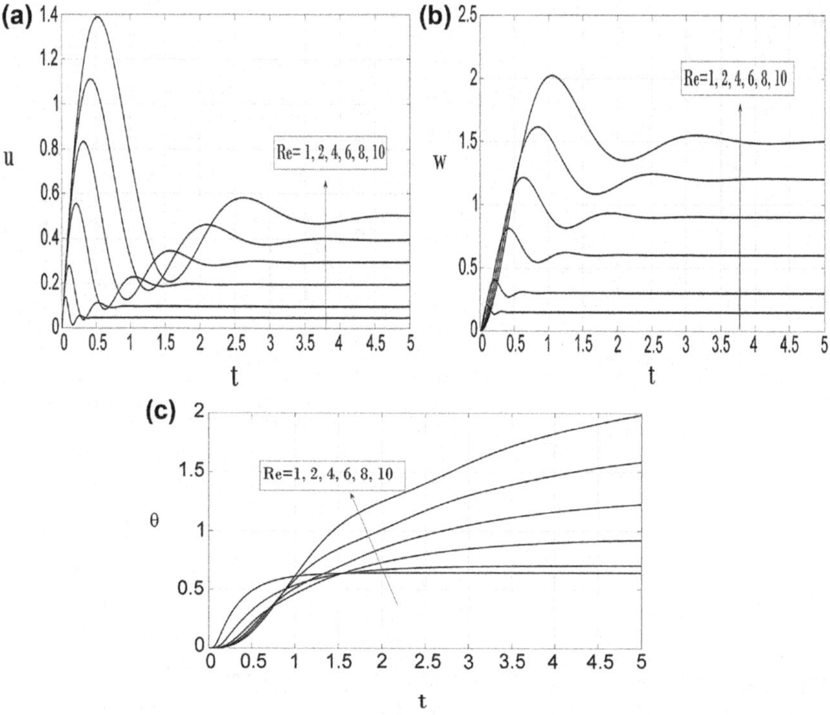

Figure 6.4 Effect of Reynolds number (*Re*) in center ($y = 0$), when $\alpha = 5$, $Pr = 1$, $Ha = 1$, $Ec = 0.2$, $m = 3$, $a = 0.5$, $b = 0.5$. (a) *u* velocity, (b) *w* velocity and, (c) temperature.

through Figs. 6.6 and 6.7 respectively. As seen in Fig. 6.7, increasing in Reynolds number and Hall parameter makes high Nusselt number, but increasing the Hartmann number decreases Nusselt number and heat transfer. Fig. 6.7 reveals that increasing the Hall parameter, Hartmann and Reynolds number have a similar behavior on skin friction coefficient compared to Nusselt number in Fig. 6.6, which is due to relevancy between temperature and velocity gradients.

6.3 MICROPOLAR FLUID IN A POROUS CHANNEL

A micropolar fluid is the fluid with internal structures in which coupling between the spin of each particle and the macroscopic velocity field is taken into account. We consider the steady laminar flow of a micropolar fluid along a two-dimensional channel with porous walls through which fluid is uniformly injected or removed with speed v_0. The lower channel wall has a

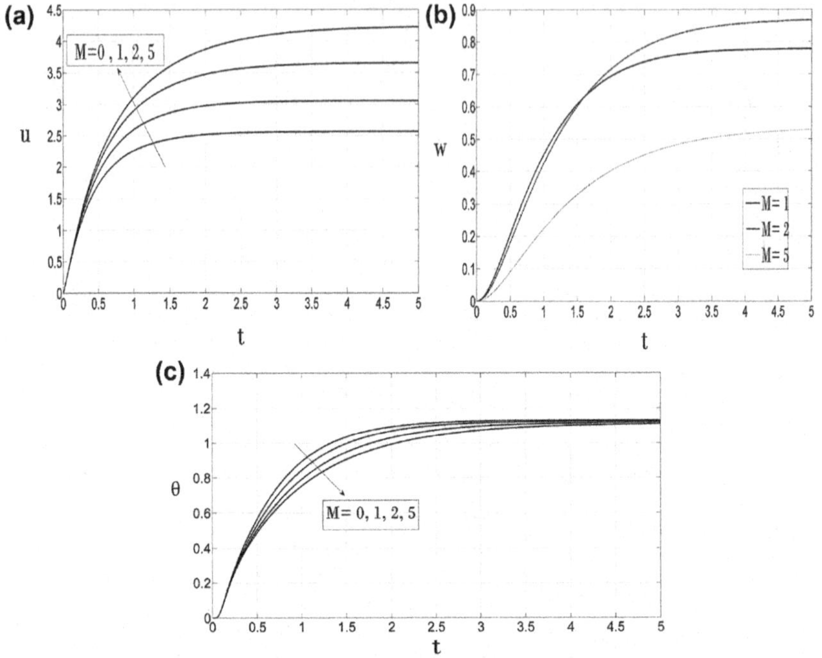

Figure 6.5 Effect of Hall parameter (*m*) in center ($y = 0$), when $\alpha = 5$, $Pr = 1$, $Ha = 1$, $Ec = 0.2$, $R = 1$, $a = 0.5$, $b = 0.5$. (a) *u* velocity, (b) *w* velocity and, (c) temperature.

solute concentration C_1 and temperature T_1 while the upper wall has solute concentration C_2 and temperature T_2 as shown in Fig. 6.8. Using Cartesian coordinates, the channel walls are parallel to the x-axis and located at $y = \pm h$, where $2h$ is the channel width. The relevant equations governing the flow are [3]:

$$\frac{\partial u}{\partial x} + \frac{\partial v}{\partial y} = 0, \tag{6.21}$$

$$\rho\left(u\frac{\partial u}{\partial x} + v\frac{\partial u}{\partial y}\right) = -\frac{\partial P}{\partial x} + (\mu + \kappa)\left(\frac{\partial^2 u}{\partial x^2} + \frac{\partial^2 u}{\partial y^2}\right) + \kappa\frac{\partial N}{\partial y}, \tag{6.22}$$

$$\rho\left(u\frac{\partial v}{\partial x} + v\frac{\partial v}{\partial y}\right) = -\frac{\partial P}{\partial y} + (\mu + \kappa)\left(\frac{\partial^2 v}{\partial x^2} + \frac{\partial^2 v}{\partial y^2}\right) - \kappa\frac{\partial N}{\partial x}, \tag{6.23}$$

$$\rho\left(u\frac{\partial N}{\partial x} + v\frac{\partial N}{\partial y}\right) = -\frac{\kappa}{j}\left(2N + \frac{\partial u}{\partial y} - \frac{\partial v}{\partial x}\right) + \left(\frac{\mu_s}{j}\right)\left(\frac{\partial^2 N}{\partial x^2} + \frac{\partial^2 N}{\partial y^2}\right), \tag{6.24}$$

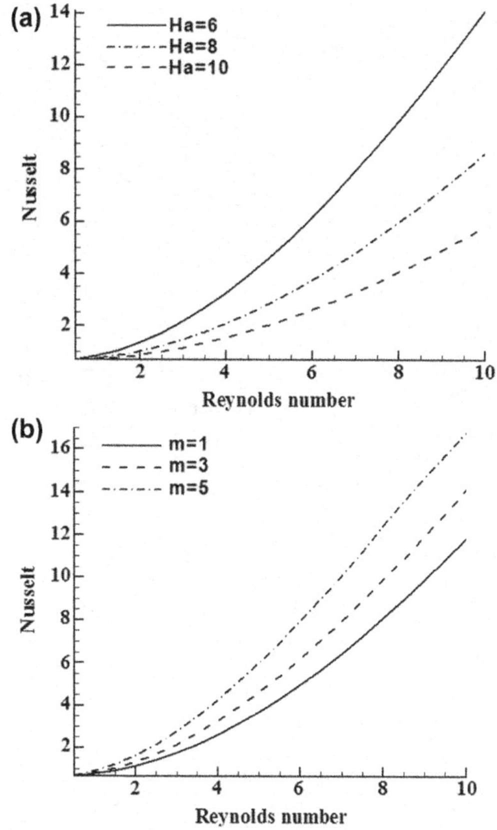

Figure 6.6 Nusselt number variations versus Reynolds number and (a) Hartmann number (b) Hall parameter (m), when $\alpha = 5$, $Pr = 1$, $Ec = 0.2$, $a = 0.5$ and $b = 0.5$ for steady state condition.

$$\rho\left(u\frac{\partial T}{\partial x} + v\frac{\partial T}{\partial y}\right) = \frac{k_1}{c_p}\frac{\partial^2 T}{\partial y^2}, \tag{6.25}$$

$$\rho\left(u\frac{\partial C}{\partial x} + v\frac{\partial C}{\partial y}\right) = D^*\frac{\partial^2 C}{\partial y^2}. \tag{6.26}$$

where u and v are the velocity components along the x- and y-axis respectively, ρ is the fluid density, μ is the dynamic viscosity, N is the angular or microrotation velocity, P is the fluid pressure, T and c_p are the fluid temperature and specific heat at constant pressure respectively, C is the species concentration, k_1 and D^* are the thermal conductivity and molecular

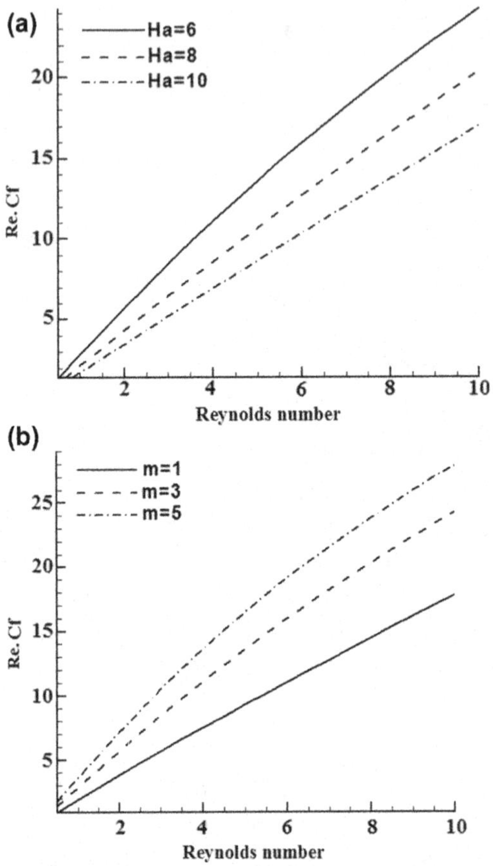

Figure 6.7 Skin friction coefficient variations versus Reynolds number and (a) Hartmann number (b) Hall parameter (m), when $\alpha = 5$, $Pr = 1$, $Ec = 0.2$, $a = 0.5$ and $b = 0.5$ for steady state condition.

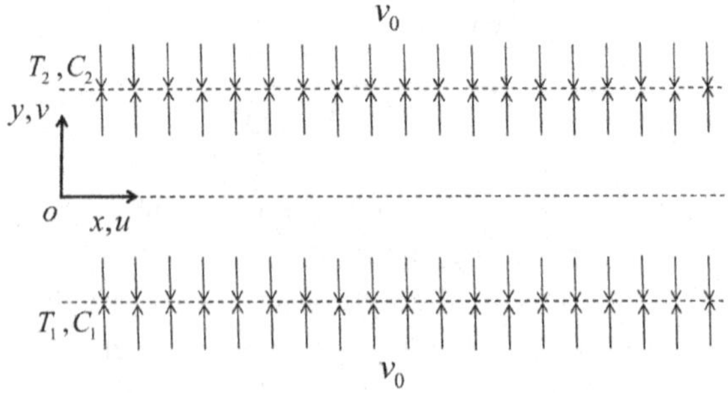

Figure 6.8 Geometry of problem.

diffusivity respectively, j is the microinertia density, k is a material parameter, and $v_s = \left(\mu + \frac{k}{2}\right)j$ is the micro rotation viscosity.

The appropriate boundary conditions are:

$$
\begin{aligned}
y = -h&: v = u = 0, N = -s\frac{\partial u}{\partial y},\\
y = +h&: v = 0, u = \frac{v_0 x}{h}, N = \frac{v_0 x}{h^2}.
\end{aligned}
\tag{6.27}
$$

where s is a boundary parameter and indicates the degree to which the microelements are free to rotate near the channel walls. The case $s = 0$ represents concentrated particle flows in which microelements close to the wall are unable to rotate. Other interesting particular cases that have been considered in the literature include $s = 1/2$, which represents weak concentrations and the vanishing of the antisymmetric part of the stress tensor, and $s = 1$, which represents turbulent flow. We introduce the following dimensionless variables:

$$
\begin{aligned}
&\eta = \frac{y}{h},\ \psi = -v_0 x f(\eta),\ N = \frac{v_0 x}{h^2} g(\eta),\\
&\theta(\eta) = \frac{T - T_2}{T_1 - T_2},\ \phi(\eta) = \frac{C - C_2}{C_1 - C_2},
\end{aligned}
\tag{6.28}
$$

where $T_2 = T_1 - Ax$, $C_2 = C_1 - Bx$ with A and B as constants. The stream function is defined in the usual way; $u = \frac{\partial \psi}{\partial y}, v = -\frac{\partial \psi}{\partial x}$.

Eqs. (6.21)–(6.27) reduce to the coupled system of nonlinear differential equations:

$$(1 + N_1) f^{IV} - N_1 g - Re\left(ff''' - f'f''\right) = 0, \tag{6.29}$$

$$N_2 g'' + N_1 (f'' - 2g) - N_3 Re\,(fg' - f'g) = 0, \tag{6.30}$$

$$\theta'' + Pe_h f'\theta - Pe_h f\theta' = 0, \tag{6.31}$$

$$\phi'' + Pe_m f'\phi - Pe_m f\phi' = 0, \tag{6.32}$$

subject to the boundary conditions:

$$
\begin{aligned}
\eta &= -1 : f = f' = g = 0,\ \theta = \phi = 1,\\
\eta &= +1 : f = \theta = \phi = 0, f' = -1,\ g = 1.
\end{aligned}
\tag{6.33}
$$

The parameters are the buoyancy ratio N, the Peclet numbers for the diffusion of heat Pe_h and mass Pe_m, respectively, the Reynolds number Re

where for suction $Re > 0$ and for injection $Re < 0$ and the Grashof number Gr given by:

$$N_1 = \frac{\kappa}{\mu}, N_2 = \frac{\nu_s}{\mu h^2}, N_3 = \frac{j}{h^2}, Re = \frac{v_0}{\upsilon} h,$$
$$Pr = \frac{\upsilon \rho c_p}{k_1}, Sc = \frac{\upsilon}{D^*}, Gr = \frac{g\beta_T A h^4}{\upsilon^2}, \quad (6.34)$$
$$Pe_h = PrRe, Pe_m = ScRe,$$

where Pr is the Prandtl number, Sc is the generalized Schmidt number, N_1 is the coupling parameter, and N_2 is the spin-gradient viscosity parameter. In technological processes, the parameters of particular interest are the local Nusselt and Sherwood numbers. These are defined as follows:

$$Nu_x = \frac{q''_{y=-h} x}{(T_1 - T_2)k_1} = -\theta'(-1), \quad (6.35)$$

$$Sh_x = \frac{m''_{y=-h} x}{(C_1 - C_2)D^*} = -\phi'(-1), \quad (6.36)$$

where q'' and m'' are local heat flux and mass flux respectively. Now DTM into governing equations has been applied. Taking the differential transform of Eqs. (6.29) and (6.32) with respect to χ, and considering $H = 1$ gives:

$$(1 + N_1)(k+1)(k+2)(k+3)(k+4)F[k+4] - N_1 G[k]$$
$$-Re\sum_{m=0}^{k}(F[k-m](m+1)(m+2)(m+3)F[m+3])$$
$$+Re\sum_{m=0}^{k}((k-m+1)F[k-m+1](m+1)(m+2)F[m+2]) = 0, \quad (6.37)$$

$$F[0] = a_0, F[1] = a_1, F[2] = a_2, F[3] = a_3, \quad (6.38)$$

$$N_2(k+1)(k+2)G[k+2] + N_1(k+1)(k+2)F[k+2] - 2N_1 G[k]$$
$$-N_3 Re\sum_{m=0}^{k}((m+1)G[m+1]F[k-m])$$
$$+N_3 Re\sum_{m=0}^{k}((m+1)F[m+1]Gk-m]) = 0, \quad (6.39)$$

$$G[0] = b_0, G[1] = b_1, \quad (6.40)$$

$$
\begin{aligned}
&(k+1)(k+2)\vartheta[k+2] + Pe_h \sum_{m=0}^{k}((m+1)F[m+1]\vartheta[k-m]) \\
&-Pe_h \sum_{m=0}^{k}((m+1)\vartheta[m+1]F[k-m]) = 0,
\end{aligned}
\tag{6.41}
$$

$$
\vartheta[0] = c_0,\ \vartheta[1] = c_1, \tag{6.42}
$$

$$
\begin{aligned}
&(k+1)(k+2)\varphi[k+2] + Pe_m \sum_{m=0}^{k}((m+1)F[m+1]\varphi[k-m]) \\
&-Pe_m \sum_{m=0}^{k}((m+1)\varphi[m+1]F[k-m]) = 0,
\end{aligned}
\tag{6.43}
$$

$$
\varphi[0] = d_0, \varphi[1] = d_1, \tag{6.44}
$$

where $F(k), G(k), \vartheta(k)$, and $\varphi(k)$ are the differential transforms of $f(\eta), g(\eta), \theta(\eta)$, and $\phi(\eta)$, and a_0, a_1, a_2, a_3, b_0, b_1, c_0, c_1, d_0, d_1 are constants which can be obtained through boundary condition. Using Eqs. (6.29) and (6.32) this problem can be solved as followed:

$$
\begin{aligned}
&F[0] = a_0,\ F[1] = a_1,\ F[2] = a_2,\ F[3] = a_3, \\
&F[4] = \frac{1}{12(1+N_1)}\left(N_1 \frac{1}{2N_2}(-\ N_3 Re b_0 a_1 - 2N_1 a_2 + 2N_1 b_0 \right. \\
&\qquad\qquad \left. +\ N_3 Re a_0\ b_1) + 3ReN_1N_4 - ReN_2N_3\right),
\end{aligned}
\tag{6.45}
$$

...

$$
\begin{aligned}
&G[0] = b_0, G[1] = b_1, \\
&G[2] = \frac{1}{2N_2}(-\ N_3 Re b_0 a_1 - 2N_1 a_2 + 2N_1 b_0 + N_3 Re a_0 b_1), \\
&G[3] = \frac{1}{6b^2}(-\ 2N_3 Re N_2 b_0 a_2 - 6N_1 a_3 N_2 + 2N_1 N_2 b_1 - N_3^2 Re^2 a_0 a_1 b_0 \\
&\qquad -2N_3 Re N_1 a_0 a_2 + 2N_3 N_1 Re a_0 b_0 + N_3^2 Re^2 a_0^2 b_1),
\end{aligned}
\tag{6.46}
$$

...

$$
\begin{aligned}
&\vartheta[0] = c_0, \vartheta[1] = c_1, \\
&\vartheta[2] = -\frac{1}{2} Pe_h a_1 c_0 + \frac{1}{2} Pe_h c_1 a_0, \\
&\vartheta[3] = -\frac{1}{3} Pe_h a_2 c_0 - \frac{1}{6} a_0 Pe_h^2 a_1 c_0 + \frac{1}{6} Pe_h^2 a_0^2 c_1,
\end{aligned}
\tag{6.47}
$$

$$\begin{aligned}
&\ldots\\
&\varphi[0] = d_0, \varphi[1] = d_1,\\
&\varphi[2] = -\frac{1}{2}Pe_m a_1 d_0 + \frac{1}{2}Pe_m d_1 a_0,\\
&\varphi[3] = -\frac{1}{3}Pe_m a_2 d_0 - \frac{1}{6}a_0 Pe_m^2 a_1 d_0 + \frac{1}{6}Pe_m^2 a_0^2 d_1.\\
&\ldots
\end{aligned} \tag{6.48}$$

The above process is continuous. By substituting Eqs. (6.45) to (6.48) into the main equation based on DTM, it can be obtained that the closed form of the solutions is:

$$\begin{aligned}
F(\eta) = a_0 + a_1\eta + a_3\eta^2 + &\left(\frac{1}{12(1+N_1)}\left(N_1\frac{1}{2N_2}\left(-N_3Reb_0a_1 - 2N_1a_2\right.\right.\right.\\
&\left.\left.\left.+2N_1b_0 + N_3Rea_0\,b_1\right) + 3ReN_1N_4 - ReN_2N_3\right)\eta^3 + \cdots,
\end{aligned} \tag{6.49}$$

$$\begin{aligned}
G(\eta) = b_0 + b_1\eta + &\frac{1}{2N_2}(-N_3Reb_0a_1 - 2N_1a_2 + 2N_1b_0 + N_3Rea_0b_1)\eta^2\\
&+\left(\frac{1}{6b^2}\left(-2N_3ReN_2b_0a_2 - 6N_1a_3N_2 + 2N_1N_2b_1 - N_3^2Re^2a_0a_1b_0\right.\right.\\
&\left.\left.-2N_3ReN_1a_0a_2 + 2N_3N_1Rea_0b_0 + N_3^2Re^2a_0^2b_1\right)\eta^3\cdots,\right.
\end{aligned} \tag{6.50}$$

$$\begin{aligned}
\vartheta(\eta) = c_0 + c_1\eta + &\left(-\frac{1}{2}Pe_h a_1 c_0 + \frac{1}{2}Pe_h c_1 a_0\right)\eta^2\\
&+\left(-\frac{1}{3}Pe_h a_2 c_0 - \frac{1}{6}a_0 Pe_h^2 a_1 c_0 + \frac{1}{6}Pe_h^2 a_0^2 c_1\right)\eta^3\cdots,
\end{aligned} \tag{6.51}$$

$$\begin{aligned}
\varphi(\eta) = d_0 + d_1\eta + &\left(-\frac{1}{2}Pe_m a_1 d_0 + \frac{1}{2}Pe_m d_1 a_0\right)\eta^2\\
&\left(-\frac{1}{3}Pe_m a_2 d_0 - \frac{1}{6}a_0 Pe_m^2 a_1 d_0 + \frac{1}{6}Pe_m^2 a_0^2 d_1\right)\eta^3\cdots,
\end{aligned} \tag{6.52}$$

By substituting the boundary conditions into Eqs. (6.49) to (6.52) in point $\eta = -1$ and $\eta = 1$ the values of a_0, a_1, a_2, a_3, b_0, b_1, c_0, c_1, d_0, d_1 can be obtained.

$$\begin{aligned} &F(-1) = F'(-1) = G(-1) = 0, \ \vartheta(-1) = \phi(-1) = 1, \\ &F(1) = \vartheta(1) = \phi(1) = 0, \ F'(-1) = -1, G(1) = 1. \end{aligned} \tag{6.53}$$

By solving Eq. (6.53) the values of a_0, a_1, a_2, a_3, b_0, b_1, c_0, c_1, d_0, d_1 will be given. By substituting obtained a_0, a_1, a_2, a_3, b_0, b_1, c_0, c_1, d_0, d_1 into Eqs. (6.50)–(6.53), the expression of $F(\eta), G(\eta), \vartheta(\eta)$, and $\varphi(\eta)$ can be obtained. For DTM solving, constant values with different nondimensional parameters are shown in Table 6.1. $\theta(\eta)$ values in different steps of DTM at $N_1 = N_2 = N_3 = Re = 1$ and $Pe_h = Pe_m = 1$ are shown in Table 6.2. This table shows that this method is converged in step 7 and error has been minimized. In this table, %error is introduced as followed:

$$\%\text{Error} = \left| \frac{f(\eta)_{\text{NM}} - f(\eta)_{\text{DTM}}}{f(\eta)_{\text{NM}}} \right| \tag{6.54}$$

Also, it can also be seen that the maximum error for the DTM occurs near the middle of the interval $\eta = 0$. The results that obtained by DTM were well matched with the results carried out by the numerical solution obtained using four-order Runge–kutta method as shown in Fig. 6.9. After this validity, the influence of significant parameters such as Reynolds numbers, microrotation/angular velocity, and Peclet number on the flow and heat transfer characteristics is discussed. Fig. 6.10 shows the effects of moderate Reynolds numbers on the f and g, when $N_1 = N_2 = N_3 = Pe_h = Pe_m = 1$.

It is worth to mention that the Reynolds number indicates the relative significance of the inertia effect compared to the viscous effect. Thus, velocity boundary layer thickness decreases as Re increases. The Reynolds number has little effect on the temperature and concentration fields. In a general manner, there is an increment in velocity profiles from suction to injection. At higher Reynolds numbers the maximum velocity (f) point shifts to the upper wall where shear stress becomes larger as the Reynolds number grows. Also from this figure we observe that with an increase in the value of the Reynolds number the point at which minimum rotation (g) occurs moves away from the origin of the channel to upper wall.

Fig. 6.11 depicts the effect of coupling parameter (N_1) on the (a) f (b) g, when $N_2 = N_3 = Re = Pe_h = Pe_m = 1$. The values of velocity profile (f) increases with increase in N_1, but the values of microrotation profile (g) decreases with increasing N_1. Fig. 6.12 displays the effects of spin-gradient viscosity parameter (N_2) on the (a) f (b) g, when $N_1 = N_3 = Re = Pe_h = Pe_m = 1$. The values of velocity profile decreases

Table 6.1 Constant Values With Different Nondimensional Parameters

$N_1 = N_2 = N_3$	Re	$Pe_h = Pe_m$	a_0	a_1	a_2	a_3	b_0	b_1	c_0	c_1	d_0	d_1
0.5	1	1	0.252	0.250	−0.254	−0.250	0.072	0.161	0.619	−0.540	0.619	−0.540
1	1	1	0.262	0.250	−0.274	−0.250	0.076	0.173	0.623	−0.544	0.076	−0.544
1	0.5	1	0.262	0.250	−0.274	−0.250	0.076	0.173	0.623	−0.544	0.623	−0.544
1	1	0.5	0.259	0.250	−0.268	−0.250	0.067	0.160	0.557	−0.521	0.557	−0.522

Table 6.2 %Error of $\theta(\eta)$ Values in Different Steps of Differential Transformation Method at: $N_1 = N_2 = N_3 = Re = Pe_h = Pe_m = 1$

η	Step = 4	Step = 5	Step = 6	Step = 7
−1	0	0	0	0
−0.9	0.084731	0.086267	0.003871	0.000468
−0.8	0.168765	0.17176	0.005964	0.001576
−0.7	0.251416	0.255721	0.006769	0.002984
−0.6	0.332018	0.337421	0.006684	0.004466
−0.5	0.409935	0.416176	0.006017	0.005876
−0.4	0.484572	0.491359	0.005001	0.007128
−0.3	0.555383	0.562408	0.003801	0.008171
−0.2	0.621875	0.628838	0.002522	0.00898
−0.1	0.683621	0.690244	0.001226	0.009539
0	0.740259	0.746307	6.02E-05	0.009838
0.1	0.791502	0.796792	0.00133	0.009865
0.2	0.837137	0.841553	0.002582	0.009604
0.3	0.87703	0.880522	0.003805	0.00904
0.4	0.911122	0.91371	0.004962	0.008158
0.5	0.939433	0.9412	0.005968	0.00696
0.6	0.962059	0.963137	0.006676	0.005476
0.7	0.979164	0.979721	0.006846	0.003789
0.8	0.990982	0.991199	0.006135	0.002072
0.9	0.997811	0.997854	0.004063	0.000637
1	1	1	0	0

with increase in N_2. Similarly, with the range $N_2 > 1$, the angular velocity increases with N_2. However, when $N_2 < 1$ the behavior of the angular velocity is oscillatory and irregular. The parameter N_3 was found to have an effect only on the angular velocity (g) and as shown in Fig. 6.13, increasing N_3 leads to decrease in the angular velocity.

The topographical effects of the Peclet number on the fluid temperature and solute concentration are shown in Fig. 6.14. The fluid temperature and concentration increases with increasing Peclet number. The maximum value of temperature profile occurred in the middle of the channel. However, the Peclet number was found to have no effect on the velocity and the microrotation vectors. Effects of Re and $Pe_h = Pe_m$ on Nusselt number and Sherwood number when $N_1 = N_2 = N_3 = 1$ are shown in Figs. 6.15 and 6.16, respectively. For both suction and injection it can be found that Reynolds number has direct relationship with Nusselt number and Sherwood number, but Peclet number has reverse relationship with them.

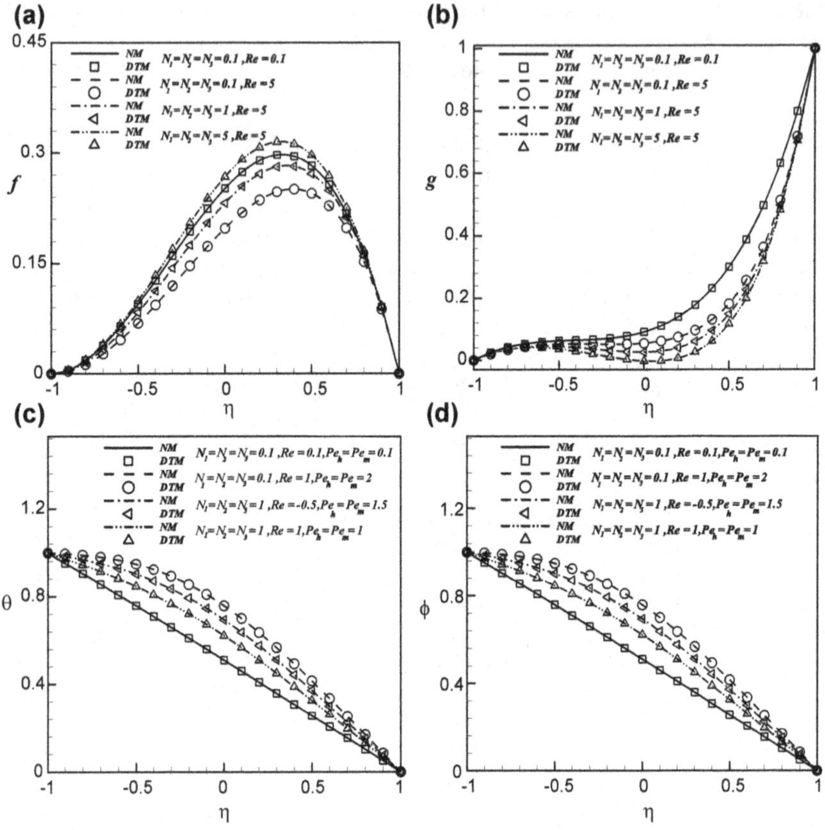

Figure 6.9 Comparison between numerical and differential transformation method solution results.

6.4 MAGNETOHYDRODYNAMIC VISCOUS FLOW BETWEEN POROUS SURFACES

Consider the steady flow of an electrically conducting fluid between two horizontal parallel plates when the fluid and the plates rotate together around the axis, which is normal to the plates with an angular velocity [4]. A Cartesian coordinate system is considered as followed: the x-axis is along the plate, the y-axis is perpendicular to it, and the z-axis is normal to the xy plane (see Fig. 6.17). The origin is located on the lower plate, and the plates are located at $y = 0$ and h. The lower plate is being stretched by two equal and opposite forces so that the position of the point (0,0,0) remains unchanged. A uniform magnetic flux with density B_0 is acting along y-axis about which the system is rotating. The upper plate is subjected to a

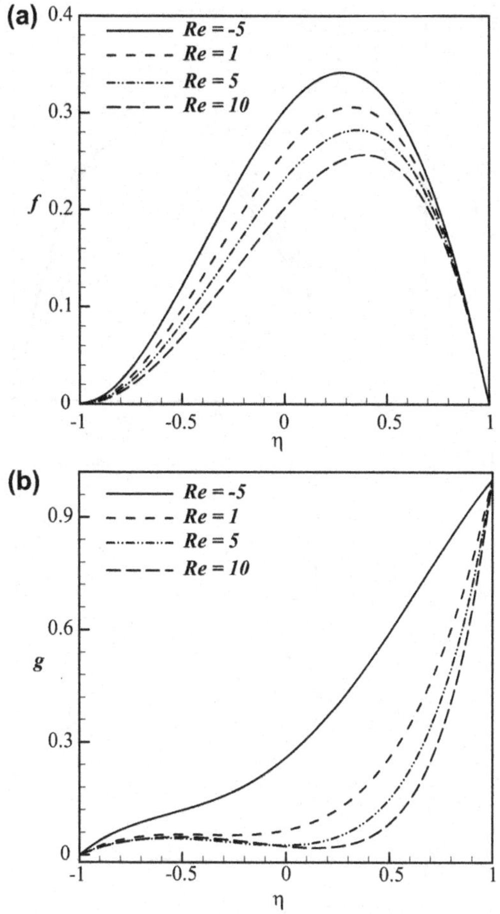

Figure 6.10 Effects of moderate Reynolds number on the (a) f (b) g, when $N_1 = N_2 = N_3 = 1$.

constant wall injection with velocity v_0. The governing equations of motion in a rotating frame of reference are:

$$\frac{\partial u}{\partial x} + \frac{\partial v}{\partial y} + \frac{\partial w}{\partial z} = 0 \tag{6.55}$$

$$u\frac{\partial u}{\partial x} + v\frac{\partial u}{\partial y} + 2\Omega w = -\frac{1}{\rho}\frac{\partial p^*}{\partial x} + \upsilon\left[\frac{\partial^2 u}{\partial x^2} + \frac{\partial^2 u}{\partial y^2}\right] - \frac{\sigma B_0^2}{\rho}u, \tag{6.56}$$

$$u\frac{\partial v}{\partial y} = -\frac{1}{\rho}\frac{\partial p^*}{\partial y} + \upsilon\left[\frac{\partial^2 v}{\partial x^2} + \frac{\partial^2 v}{\partial y^2}\right], \tag{6.57}$$

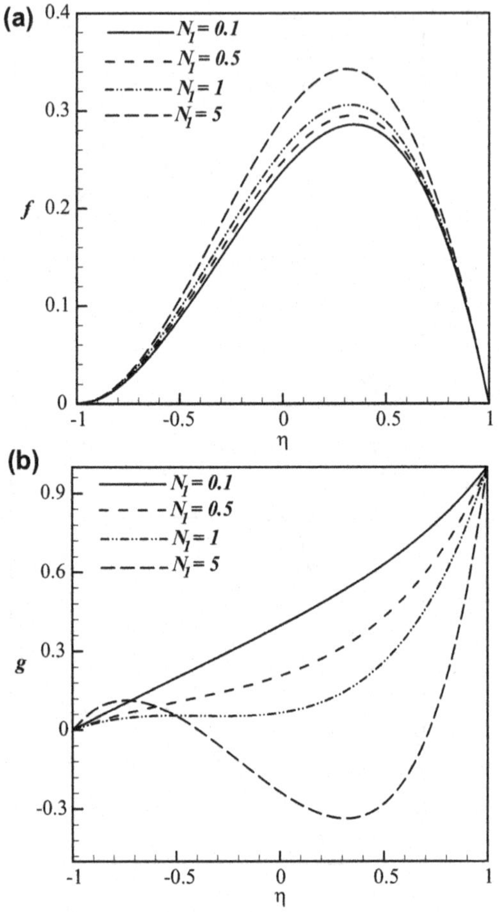

Figure 6.11 Effects of N_1 on the (a) f (b) g, when $N_2 = N_3 = Re = Pe_h = Pe_m = 1$.

$$u\frac{\partial w}{\partial x} + v\frac{\partial w}{\partial y} - 2\Omega w = \upsilon\left[\frac{\partial^2 w}{\partial x^2} + \frac{\partial^2 w}{\partial y^2}\right] - \frac{\sigma B_0^2}{\rho} w, \tag{6.58}$$

where u, v, and w denote the fluid velocity components along the x, y, and z directions, υ is the kinematic coefficient of viscosity, ρ is the fluid density, and p^* is the modified fluid pressure. The absence of $\frac{\partial p^*}{\partial z}$ in Eq. (6.58) implies that there is a net cross-flow along the z-axis.

The boundary conditions are:

$$u = ax, \ v = 0, \ w = 0 \ \ at \ \ y = 0 \tag{6.59}$$

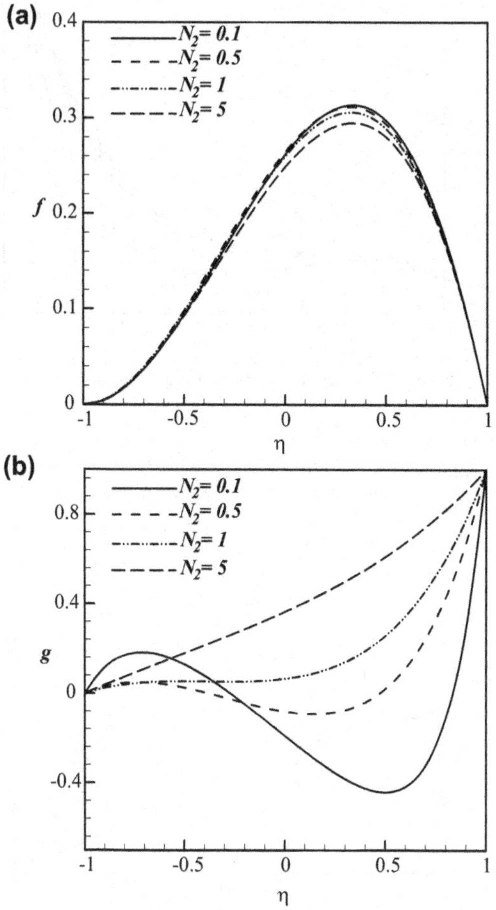

Figure 6.12 Effects of N_2 on the (a) f (b) g, when $N_1 = N_3 = Re = Pe_h = Pe_m = 1$.

$$u = 0,\ v = -v_0,\ w = 0\ \ at\ \ y = +h$$

Nondimensional variables are introduced as followed:

$$\eta = \frac{y}{h},\ u = axf'(\eta),\ v = -ah\,f(\eta),\ w = ax\,g(\eta) \tag{6.60}$$

where a prime denotes differentiation with respect to η.

Substituting Eq. (6.60) in Eqs. (6.55)–(6.58), we have:

$$-\frac{1}{\rho}\frac{\partial p^*}{\partial y} = a^2x\left[f' - ff'' - \frac{f'''}{R} + \frac{M}{R} + \frac{2K_r}{R}g\right], \tag{6.61}$$

$$-\frac{1}{\rho h}\frac{\partial p^*}{\partial \eta} = a^2h\left[ff' + \frac{1}{R}f''\right], \tag{6.62}$$

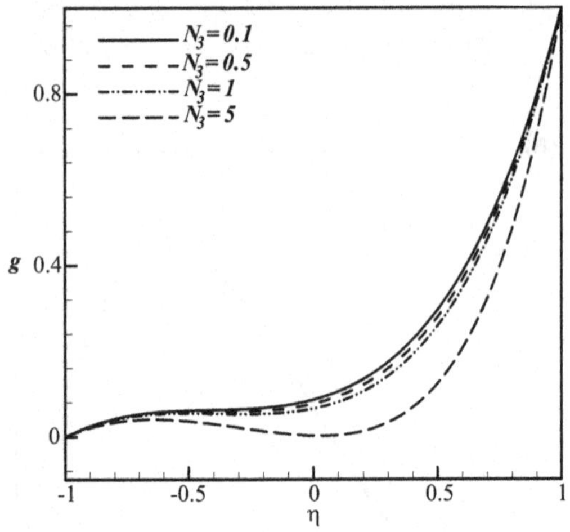

Figure 6.13 Effects of N_3 on the g, when $N_1 = N_2 = Re = Pe_h = Pe_m = 1$.

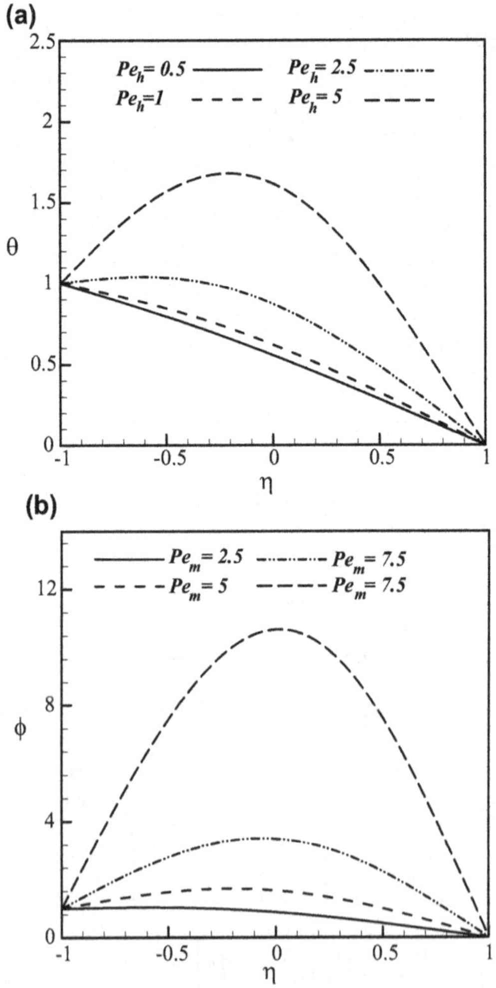

Figure 6.14 Effects of (a) Pe_h on temperature profile at $Pe_m = 1$ (b) Pe_m on concentration profile at $Pe_h = 1$, when $N_1 = N_2 = N_3 = Re = 1$.

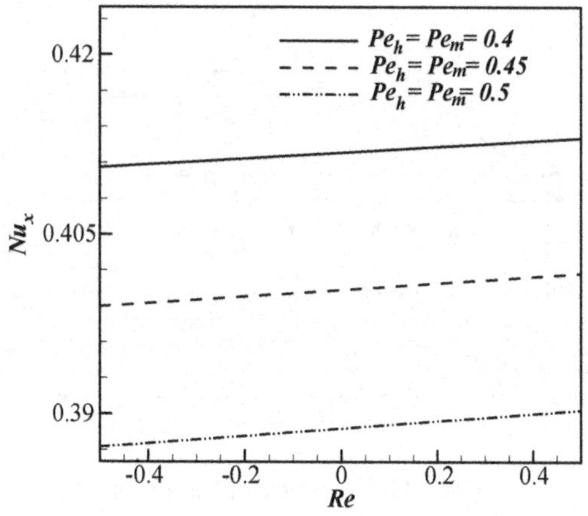

Figure 6.15 Effects of Re and $Pe_h = Pe_m$ on Nusselt number, when $N_1 = N_2 = N_3 = 1$.

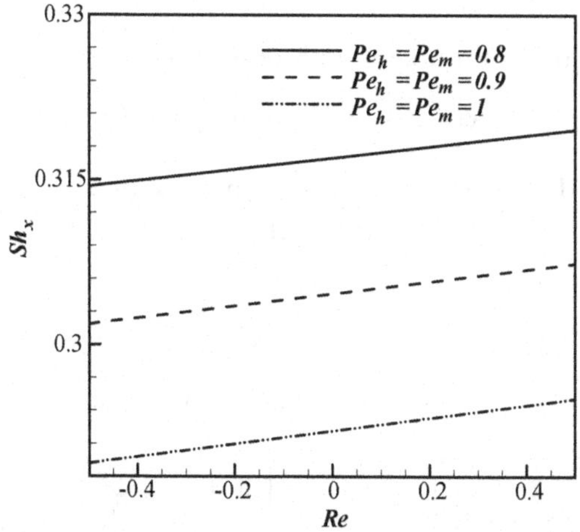

Figure 6.16 Effects of Re and $Pe_h = Pe_m$ on Sherwood number, when $N_1 = N_2 = N_3 = 1$.

$$g'' - R(f'g - fg') + 2K_r f' - Mg = 0 \tag{6.63}$$

and the nondimensional quantities are defined through in which R is the viscosity parameter, M is the magnetic parameter, and K_r is the rotation parameter.

$$R = \frac{ah^2}{\upsilon}, M = \frac{\sigma B_0^2 h^2}{\rho \upsilon}, K_r = \frac{\Omega h^2}{\upsilon} \tag{6.64}$$

Eq. (6.61) with the help of Eq. (6.62) can be written as,

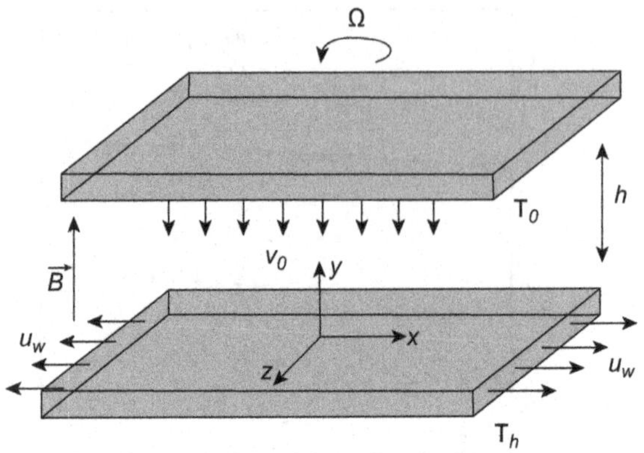

Figure 6.17 Geometry of the problem.

$$f''' - R[f'^2 - ff''] - 2K_r^2 g - M^2 f' = A \tag{6.65}$$

Differentiation of Eq. (6.65) with respect to η gives,

$$f^{iv} - R(f' f'' - f f'') - 2K_r g' - M f'' = 0 \tag{6.66}$$

Therefore, the governing equations and boundary conditions for this case in nondimensional form are given by:

$$f^{iv} - R(f' f'' - f f'') - 2K_r g' - M f'' = 0 \tag{6.67}$$

$$u\frac{\partial T}{\partial x} + v\frac{\partial T}{\partial y} + w\frac{\partial T}{\partial z} = \frac{\bar{k}}{\rho\, c_p}\left(\frac{\partial^2 T}{\partial x^2} + \frac{\partial^2 T}{\partial y^2} + \frac{\partial^2 T}{\partial z^2}\right) + \mu\varphi \tag{6.68}$$

$$\phi = 2\left[\left(\frac{\partial u}{\partial x}\right)^2 + \left(\frac{\partial v}{\partial y}\right)^2 + \left(\frac{\partial w}{\partial z}\right)^2\right] + \left(\frac{\partial v}{\partial x} + \frac{\partial u}{\partial y}\right)^2 + \left(\frac{\partial w}{\partial y} + \frac{\partial v}{\partial z}\right)^2$$
$$+ \left(\frac{\partial w}{\partial x} + \frac{\partial u}{\partial z}\right)^2 - \frac{2}{3}\left(\frac{\partial u}{\partial x} + \frac{\partial v}{\partial y} + \frac{\partial w}{\partial z}\right)^2 \tag{6.69}$$

$$g'' - R(f'g - fg') + 2K_r f' - Mg = 0 \tag{6.70}$$

subject to the boundary conditions:

$$f = 0,\ f' = 1,\ g = 0 \ \text{at}\ \eta = 0$$
$$f = \lambda,\ f' = 0,\ g = 0 \ \text{at}\ \eta = 1 \ \ \lambda = \frac{v_0}{ah} \tag{6.71}$$

With replacing nondimensional variables and using similarity solution method, by neglecting last term of viscous dissipation in the energy equation, we have the following energy equation:

$$\theta'' + Pr[Rf\,\theta' + Ec(4f'^2 + g^2) + Ec_x(f''^2 + g'^2)] = 0 \qquad (6.72)$$

subject to the boundary conditions

$$\theta(0) = 1,\ \theta(1) = 0 \qquad (6.73)$$

where $Pr = \dfrac{\mu C_p}{\overline{k}}$ is the Prandtl number, $Ec = \dfrac{a^2h^2}{C_p(\theta_0 - \theta_h)}$ is the Eckert number, $Ec_x = \dfrac{a^2x^2}{C_p(\theta_0 - \theta_h)}$ is the local Eckert number, and the nondimensional temperature is defined as,

$$\theta(\eta) = \frac{T - T_h}{T_0 - T_h} \qquad (6.74)$$

where T_0 and T_h are temperatures at the lower and upper plates.

Now we apply DTM into governing equations. Taking the differential transform of these equations with respect to χ and considering $H = 1$ gives,

$$(k+1)(k+2)(k+3)(k+4)F[k+4]$$

$$-R\sum_{m=0}^{k}((k-m+1)F[k-m+1](m+1)(m+2)F[m+2])$$

$$+R\sum_{m=0}^{k}(F[k-m](m+1)(m+2)(m+3)F[m+3]) - 2K_r(k+1)F[k+2]$$

$$-M(k+1)(k+2)F[k+2] = 0 \qquad (6.75)$$

$$F[0] = 0, F[1] = 1, F[2] = \alpha, F[3] = \beta \qquad (6.76)$$

$$(k+1)(k+2)\vartheta[k+2] + PrR\sum_{m=0}^{k}(F[k-m](m+1)\vartheta[m+1])$$

$$+\,4PrEc\sum_{m=0}^{k}((k-m+1)F[k-m+1](m+1)F[m+1])$$

$$+\,PrEc\sum_{m=0}^{k}(G[k-m]G[m])$$

$$+\,PrEc_x\sum_{m=0}^{k}((k-m+1)(k-m+2)F[k-m+2](m+1)(m+2)F[m+2])$$

$$+\,PrEc_x\sum_{m=0}^{k}((k-m+1)G[k-m+1](m+1)G[m+1]) = 0 \qquad (6.77)$$

$$G[0] = 0, G[1] = \gamma \qquad (6.78)$$

$$(k+1)(k+2)\vartheta[k+2] + PrR\sum_{m=0}^{k}(F[k-m](m+1)\vartheta[m+1])$$

$$+4PrEc\sum_{m=0}^{k}((k-m+1)F[k-m+1](m+1)F[m+1])$$

$$+PrEc\sum_{m=0}^{k}(G[k-m]G[m])$$

$$+PrEc_x\sum_{m=0}^{k}((k-m+1)(k-m+2)F[k-m+2](m+1)(m+2)F[m+2])$$

$$+PrEc_x\sum_{m=0}^{k}((k-m+1)G[k-m+1](m+1)G[m+1]) = 0 \tag{6.79}$$

$$\vartheta[0] = 1, \vartheta[1] = \xi \tag{6.80}$$

where $F(k)$, $G(k)$ and $\vartheta(k)$ are the differential transforms of $f(\eta)$, $g(\eta)$, and $\theta(\eta)$ and α, β, γ and ξ are constants, which can be obtained through boundary conditions.

This problem can be solved as followed:

$$F[0] = 0, F[1] = 1$$

$$F[2] = \alpha, F[3] = \beta$$

$$F[4] = \frac{1}{12}R\alpha + \frac{1}{12}K_r + \frac{1}{12}M\alpha$$

$$F[5] = \frac{1}{20}M\beta + \frac{1}{30}R\alpha^2 + \frac{1}{30}K_r\alpha$$

$$F[6] = \frac{1}{360}MK_r + \frac{1}{360}M^2\alpha + \frac{1}{30}R\beta\alpha - \frac{1}{360}R^2\alpha - \frac{1}{300}R\alpha + \frac{1}{60}\alpha\beta \tag{6.81}$$

$$G[0] = 0$$

$$G[1] = \gamma$$

$$G[2] = -K_r$$

$$G[3] = -\frac{2}{3}K_r\alpha + \frac{1}{6}M\gamma$$

$$G[4] = -\frac{1}{12}MK_r + \frac{1}{12}RK_r + \frac{1}{12}R\alpha\gamma - \frac{1}{2}\beta K_r$$

$$G[5] = -\frac{1}{15}M\alpha K_r + \frac{1}{120}M^2\gamma + \frac{1}{30}R\alpha K_r - \frac{1}{60}RM\gamma + \frac{1}{10}R\beta K_r - \frac{1}{30}K_r^2$$

$$G[6] = -\frac{1}{360}M^2K_r + \frac{1}{90}M\alpha K_r + \frac{1}{180}MR\alpha\gamma - \frac{1}{30}MK_r\beta$$
$$-\frac{1}{120}R^2K_r + \frac{1}{60}RK_r\beta + \frac{1}{90}RK_r\alpha^2 + \frac{1}{120}RK_r\alpha - \frac{1}{90}K_r^2\alpha \tag{6.82}$$

$$\vartheta[0] = 1$$

$$\vartheta[1] = \xi$$

$$\vartheta[2] = -2PrEc - 2PrEc_x\alpha^2 - \frac{1}{2}PrEc_x\gamma^2$$

$$\vartheta[3] = \frac{2}{3}PrEc_x\gamma K_r - \frac{1}{6}PrR\xi - \frac{8}{3}\alpha PrEc - 4PrEc_x\beta\alpha$$

$$\begin{aligned}\vartheta[4] = &-\frac{1}{3}PrEc_xR\alpha^2 - \frac{1}{3}PrEc_xK_r\alpha - \frac{1}{3}PrEc_xM\alpha^2 - 3PrEc_x\beta^2 - \frac{1}{12}PrR\alpha\xi \\ &+\frac{1}{3}Pr^2REc + \frac{1}{3}Pr^2Ec_x\alpha^2 + \frac{1}{12}PrEc_xM\gamma^2 - \frac{4}{3}PrEc\alpha^2 - 2PrEc\beta \\ &-\frac{1}{12}PrEc\gamma^2 + \frac{1}{3}PrEc_x\gamma\alpha K_r - \frac{1}{12}PrEc_xM\gamma^2 - \frac{1}{3}PrEc_xK_r^2\end{aligned}$$

$$\begin{aligned}\vartheta[5] = &-\frac{4}{5}PrEc_x\alpha M\beta - \frac{2}{15}PrEc_xR\alpha^3 - \frac{2}{15}PrEc_xK_r\alpha^2 - \frac{3}{5}PrEc_xR\beta\alpha \\ &-\frac{3}{5}PrEc_xK_r\beta - \frac{1}{20}PrR\beta\xi + \frac{3}{5}Pr^2RE\alpha + \frac{1}{5}Pr^2R\alpha^3Ec_x + \frac{1}{20}Pr^2\gamma^2R\alpha Ec_x \\ &-\frac{1}{10}pr^2REc_xK_r\gamma + \frac{1}{40}Pr^2R^3\xi + \frac{3}{5}Pr^2REc_x\beta\alpha \\ &-\frac{2}{15}(PrEcR\alpha + PrEcK_r + PrEcM\alpha) - \frac{12}{5}PrEc\beta\alpha + \frac{1}{10}PrEK_r\gamma \\ &+\frac{2}{15}PrEc_xM\alpha\gamma - \frac{1}{30}(PrEc_xR\alpha\gamma + PrEc_xM\alpha\gamma^2) \\ &+\frac{1}{5}PrEc_x\beta\alpha\gamma - \frac{2}{5}PrEc_xK_r^2\alpha\end{aligned} \tag{6.83}$$

The above process is continuous. Substituting Eqs. (6.80)−(6.82) into the main equation based on DTM. It can be obtained that the closed form of the solutions is,

$$\begin{aligned}F(\eta) = &1 + \alpha\eta^2 + \beta\eta^3 + \left(\frac{1}{12}R\alpha + \frac{1}{12}K_r + \frac{1}{12}M\alpha\right)\eta^4 \\ &+\left(\frac{1}{20}M\beta + \frac{1}{30}R\alpha^2 + \frac{1}{30}K_r\alpha\right)\eta^5 \\ &+\left(\frac{1}{360}MK_r + \frac{1}{360}M^2\alpha + \frac{1}{30}R\beta\alpha - \frac{1}{360}R^2\alpha - \frac{1}{300}R\alpha + \frac{1}{60}\alpha\beta\right)\eta^6 + \ldots\end{aligned} \tag{6.84}$$

$$G(\eta) = \gamma\eta - K_r\eta^2 + \left(-\frac{2}{3}K_r\alpha + \frac{1}{6}M\gamma\right)\eta^3$$
$$+\left(-\frac{1}{12}MK_r + \frac{1}{12}RK_r + \frac{1}{12}R\alpha\gamma - \frac{1}{2}\beta K_r\right)\eta^4$$
$$+\left(-\frac{1}{15}M\alpha K_r + \frac{1}{120}M^2\gamma + \frac{1}{30}R\alpha K_r - \frac{1}{60}RM\gamma + \frac{1}{10}R\beta K_r - \frac{1}{30}K_r^2\right)$$
$$+\left(-\frac{1}{360}M^2K_r + \frac{1}{90}M\alpha K_r + \frac{1}{180}MR\alpha\gamma - \frac{1}{30}MK_r\beta - \frac{1}{120}R^2K_r\right.$$
$$\left.+\frac{1}{60}RK_r\beta + \frac{1}{90}RK_r\alpha^2 + \frac{1}{120}RK_r\alpha - \frac{1}{90}K_r^2\alpha\right)\eta^6 + \ldots \tag{6.85}$$

$$\vartheta(\eta) = 1 + \xi\eta + \left(-2PrEc - 2PrEc_x\alpha^2 - \frac{1}{2}PrEc_x\gamma^2\right)\eta^2$$
$$+\left(\frac{2}{3}PrEc_x\gamma K_r - \frac{1}{6}PrR\xi - \frac{8}{3}\alpha PrEc - 4PrEc_x\beta\alpha\right)\eta^3$$
$$+\left(-\frac{1}{3}PrEc_xR\alpha^2 - \frac{1}{3}PrEc_xK_r\alpha - \frac{1}{3}PrEc_xM\alpha^2 - 3PrEc_x\beta^2\right.$$
$$-\frac{1}{12}PrR\alpha\xi + \frac{1}{3}Pr^2REc + \frac{1}{3}Pr^2Ec_x\alpha^2 + \frac{1}{12}PrEc_xM\gamma^2 - \frac{4}{3}PrEc\alpha^2$$
$$\left.-2PrEc\beta - \frac{1}{12}PrEc\gamma^2 + \frac{1}{3}PrEc_x\gamma\alpha K_r - \frac{1}{12}PrEc_xM\gamma^2 - \frac{1}{3}PrEc_xK_r^2\right)\eta^4 \tag{6.86}$$

To obtain the values of α, β, γ *and* ξ we substitute the boundary conditions into Eqs. (6.83)–(6.85) in point $\eta = 1$. So we have,

$$F(1) = 1 + \alpha + \beta + \left(\frac{1}{12}R\alpha + \frac{1}{12}K_r + \frac{1}{12}M\alpha\right)$$
$$+\left(\frac{1}{20}M\beta + \frac{1}{30}R\alpha^2 + \frac{1}{30}K_r\alpha\right)$$
$$+\left(\frac{1}{360}MK_r + \frac{1}{360}M^2\alpha + \frac{1}{30}R\beta\alpha - \frac{1}{360}R^2\alpha - \frac{1}{300}R\alpha + \frac{1}{60}\alpha\beta\right)$$
$$+\ldots = \lambda \tag{6.87}$$

$$
\begin{aligned}
F'(1) = {} & 1 + 2\alpha + 3\beta + 4\left(\frac{1}{12}R\alpha + \frac{1}{12}K_r + \frac{1}{12}M\alpha\right) \\
& + 5\left(\frac{1}{20}M\beta + \frac{1}{30}R\alpha^2 + \frac{1}{30}K_r\alpha\right) \\
& + 6\left(\frac{1}{360}MK_r + \frac{1}{360}M^2\alpha + \frac{1}{30}R\beta\alpha - \frac{1}{360}R^2\alpha - \frac{1}{300}R\alpha + \frac{1}{60}\alpha\beta\right) \\
& + \ldots = 0
\end{aligned} \tag{6.88}
$$

$$
\begin{aligned}
G(1) = {} & \gamma - K_r + \left(-\frac{2}{3}K_r\alpha + \frac{1}{6}M\gamma\right) \\
& + \left(-\frac{1}{12}MK_r + \frac{1}{12}RK_r + \frac{1}{12}R\alpha\gamma - \frac{1}{2}\beta K_r\right) \\
& + \left(-\frac{1}{15}M\alpha K_r + \frac{1}{120}M^2\gamma + \frac{1}{30}R\alpha K_r - \frac{1}{60}RM\gamma + \frac{1}{10}R\beta K_r - \frac{1}{30}K_r^2\right) \\
& + \left(-\frac{1}{360}M^2K_r + \frac{1}{90}M\alpha K_r + \frac{1}{180}MR\alpha\gamma - \frac{1}{30}MK_r\beta - \frac{1}{120}R^2K_r\right. \\
& \left. + \frac{1}{60}RK_r\beta + \frac{1}{90}RK_r\alpha^2 + \frac{1}{120}RK_r\alpha - \frac{1}{90}K_r^2\alpha\right) + \ldots = 0
\end{aligned} \tag{6.89}
$$

$$
\begin{aligned}
\vartheta(1) = {} & 1 + \xi + \left(-2PrEc - 2PrEc_x\alpha^2 - \frac{1}{2}PrEc_x\gamma^2\right) \\
& + \left(\frac{2}{3}PrEc_x\gamma K_r - \frac{1}{6}PrR\xi - \frac{8}{3}\alpha PrEc - 4PrEc_x\beta\alpha\right) \\
& + \left(-\frac{1}{3}PrEc_xR\alpha^2 - \frac{1}{3}PrEc_xK_r\alpha - \frac{1}{3}PrEc_xM\alpha^2 - 3PrEc_x\beta^2\right. \\
& - \frac{1}{12}PrR\alpha\xi + \frac{1}{3}Pr^2REc + \frac{1}{3}Pr^2Ec_x\alpha^2 + \frac{1}{12}PrEc_xM\gamma^2 - \frac{4}{3}PrEc\alpha^2 \\
& \left. - 2PrEc\beta - \frac{1}{12}PrEc\gamma^2 + \frac{1}{3}PrEc_x\gamma\alpha K_r - \frac{1}{12}PrEc_xM\gamma^2 - \frac{1}{3}PrEc_xK_r^2\right) \\
& + \ldots = 0
\end{aligned} \tag{6.90}
$$

Solving Eqs. (6.86)–(6.89) gives the values of α, β, γ *and* ξ. By substituting obtained α, β, γ *and* ξ into Eqs. (6.83–6.85), we can find the expressions of $F(\eta), G(\eta), \vartheta(\eta)$.

The effects of acting parameters on velocity profile and temperature distribution are discussed. Constant values with different nondimensional parameters are shown in Table 6.3. $\theta(\eta)$ values in different steps of DTM at $R=2,\ K_r=0.5,\ M=1,\ Pr=1,\ \lambda=0.5,\ Ec=Ec_x=0.5$ are shown in Table 6.4. Also it shows that this method is converged in step 16 and error has been minimized. There is an acceptable agreement between the numerical solution obtained by four-order Runge–kutta method and DTM as shown in Table 6.5. In these tables, error is introduced as followed:

$$\text{Error} = \left| f(\eta)_{\text{NM}} - f(\eta)_{\text{DTM}} \right| \tag{6.91}$$

Figs. 6.18 and 6.19 show the magnetic field effect on nondimensional velocity component (f and f'). The decrease of f curve is observed by applying higher magnetic field intensity. Also f' values increase near stretching sheet and decrease under porous sheet while at the middle point these values are constant.

Blowing velocity parameter (λ) has noticeable effect on nondimensional velocity component, which by increasing λ profile of f and f' become nonlinear and the maximum amount of f and f' increase. Also velocity component in x-direction increase severely as shown in Figs. 6.20 and 6.21.

Fig. 6.22 shows by increasing rotating parameter (K_r), values of transverse velocity component (g) between two sheets increase and the location of maximum amount of g approaching stretching sheet. Lorentz force has an inverse effect on g in comparison with Coriolis force which means that with increasing magnetic field (M), transverse velocity component between two plates decrease as shown in Fig. 6.23. The viscosity parameter (R) effects on g profile similar to magnetic field, however, with less intensity changes. Also with increasing R the location of maximum amount of g approaching stretching sheet indicates decreasing boundary layer thickness near stretching plate as shown in Fig. 6.24. Increasing the blowing velocity parameter leads to g increase, which shows blowing velocity parameter and magnetic field effects on g are in opposite as shown in Fig. 6.25.

Fig. 6.26 shows that increasing viscosity parameter leads to increasing the curve of temperature profile (θ) and the decreasing of θ values. With increasing rotating parameter near stretching plate a small amount of θ increasing has been observed. This effect decreases while we are

Table 6.3 Constant Values With Different Nondimensional Parameters

R	K	M	Pr	λ	Ec	Ec_x	$\alpha = f''(0)$	$\beta = f'''(0)$	$\gamma = g'(0)$	$\xi = \theta'(0)$
2	0.5	0.5	1	0.5	0.5	0.5	−1.137160	0.938104	0.297786	−0.510060
0.5	2	0.5	1	0.5	0.5	0.5	−1.028720	−0.296040	1.260540	−0.214870
0.5	0.5	1	1	0.5	0.5	0.5	−1.104230	0.637858	0.305824	−0.313001
0.5	0.5	0.5	7	0.5	0.5	0.5	−1.063670	0.383391	0.315570	3.755608
0.5	0.5	0.5	1	1	0.5	0.5	2.095631	−6.765270	0.536101	0.845358
0.5	0.5	0.5	1	0.5	2	0.5	−1.063670	0.383391	0.315570	1.175612
0.5	0.5	0.5	1	0.5	0.5	2	−1.063670	0.383391	0.315570	0.460925

Table 6.4 $\theta(\eta)$ Values in Different Steps of Differential Transformation Method at: $R = 2$, $K_r = 0.5$, $M = 1$, $Pr = 1$, $\lambda = 0.5$, $Ec = Ec_x = 0.5$

η	NM	$n = 4$	Error	$n = 8$	Error	$n = 12$	Error	$N = 16$	Error
0	1	1	0	1	0	1	0	1	0
0.1	0.936372	0.927427	0.008944	0.937194	0.000822	0.936681	0.000311	0.936681	0.000309
0.2	0.852373	0.834493	0.017880	0.854101	0.001728	0.853094	0.000721	0.853093	0.000720
0.3	0.754207	0.727977	0.026229	0.756699	0.002493	0.755213	0.001006	0.755211	0.001005
0.4	0.647177	0.613845	0.033332	0.650215	0.003038	0.648265	0.001089	0.648263	0.001086
0.5	0.535656	0.497241	0.038415	0.539025	0.003370	0.536636	0.000980	0.536633	0.000977
0.6	0.423083	0.382494	0.040590	0.426604	0.003521	0.423818	0.000735	0.423816	0.000732
0.7	0.311987	0.273113	0.038875	0.315487	0.003499	0.312415	0.000428	0.312412	0.000425
0.8	0.204023	0.171790	0.032232	0.207229	0.003207	0.204161	0.000139	0.204159	0.000136
0.9	0.100014	0.080401	0.019613	0.102321	0.002307	0.099973	0.000040	0.099972	0.000041
1	0	0	0	0	0	0	0	0	0

Table 6.5 Comparison Between Numerical Results and Differential Transformation Method Solution at: $R = 2$, $K_r = 0.5$, $M = 1$, $Pr = 1$, $\lambda = 0.5$, $Ec = Ec_x = 0.5$

	f			*g*			θ		
η	NM	DTM	Error	NM	DTM	Error	NM	DTM	Error
0	0	0	0	0	0	0	1	1	0
0.1	0.094302	0.094486	0.000184	0.024198	0.024248	0.000050	0.936372	0.936681	0.000309
0.2	0.177851	0.178417	0.000566	0.039813	0.039882	0.000069	0.852373	0.853093	0.000720
0.3	0.251398	0.252343	0.000945	0.048179	0.048233	0.000054	0.754207	0.755211	0.001005
0.4	0.315423	0.316618	0.001195	0.05055	0.050562	0.000011	0.647177	0.648263	0.001086
0.5	0.370172	0.371425	0.001254	0.048099	0.048057	0.000041	0.535656	0.536633	0.000977
0.6	0.415676	0.416791	0.001115	0.041924	0.041837	0.000087	0.423083	0.423816	0.000732
0.7	0.451786	0.452607	0.000821	0.033062	0.032952	0.000109	0.311987	0.312412	0.000425
0.8	0.478190	0.478647	0.000458	0.022494	0.022394	0.000099	0.204023	0.204159	0.000136
0.9	0.494445	0.494583	0.000139	0.011166	0.011107	0.000059	0.100014	0.099972	0.000041
1	0.5	0.5	0	0	0	0	0	0	0

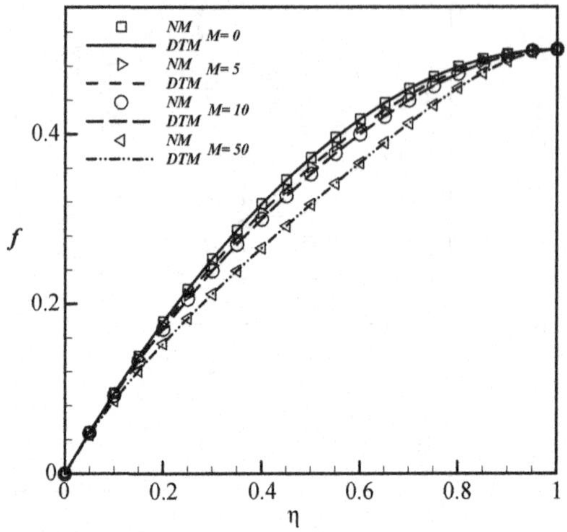

Figure 6.18 Velocity component profile (f) for variable M at $R=2$, $K_r=0.5$, $Pr=1$, $\lambda=0.5$, $Ec=Ec_x=0.5$.

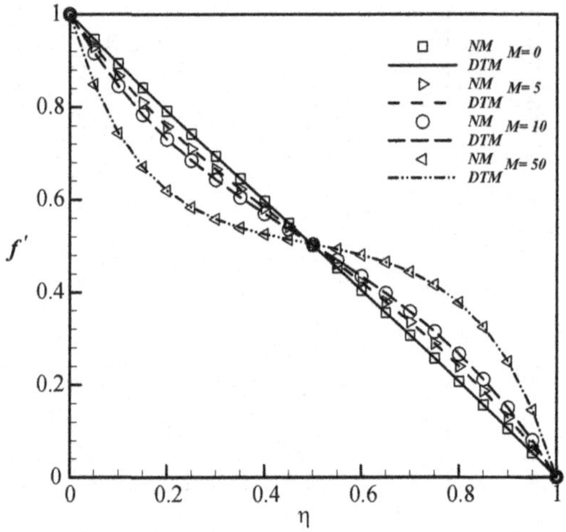

Figure 6.19 Velocity component profile (f') for variable M at $R=2$, $K_r=0.5$, $Pr=1$, $\lambda=0.5$, $Ec=Ec_x=0.5$.

approaching to porous plate as shown in Fig. 6.27 .The magnetic field does not have a noticeable effect on (θ) as shown in Fig. 6.28.

Different effects of prandtl number on temperature profile in the presence and absence of viscous dissipation are discussed in Figs. 6.29 and 6.30 as follows. Increasing prandtl, in the presence of viscous dissipation, leads to increasing temperature between two plates, while in absence of

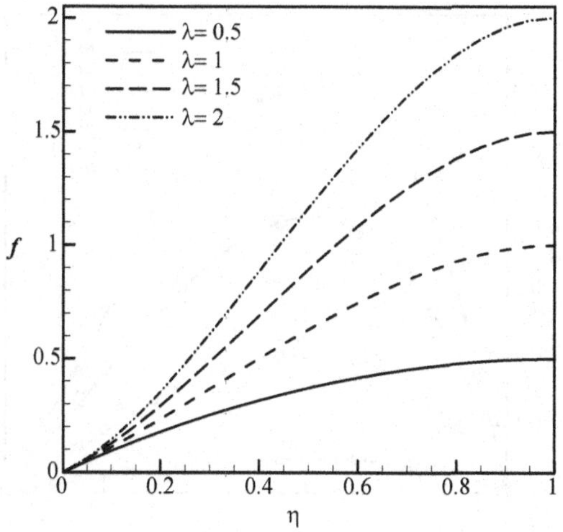

Figure 6.20 Velocity component profile (f) for variable λ at $R=2$, $K_r=0.5$, $M=1$, $Pr=1$, $Ec=Ec_x=0.5$.

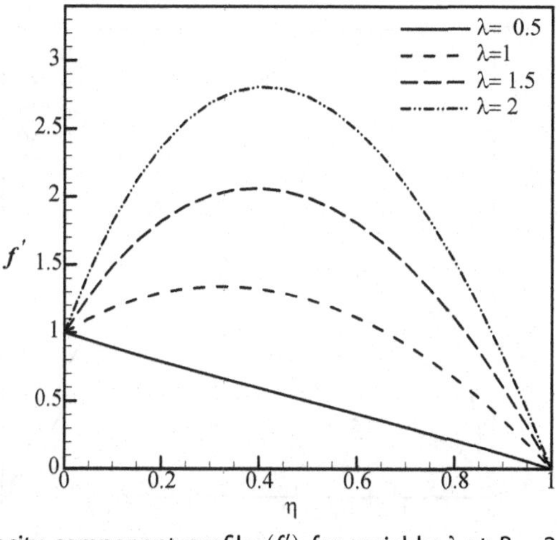

Figure 6.21 Velocity component profile (f') for variable λ at $R=2$, $K_r=0.5$, $M=1$, $Pr=1$, $Ec=Ec_x=0.5$.

viscous dissipation the changes are inverse. Increasing temperature between two plates observed, which caused by increasing this effect, is more sensible near stretching plate. (It can be seen in Fig. 6.31). The effects of viscous dissipation for which the Eckert number (Ec) and the local Eckert number (Ec_x) are responsible as shown in Figs. 6.32 and 6.33. It is obvious from the graphs that by increasing Ec and Ec_x the temperature increase near the

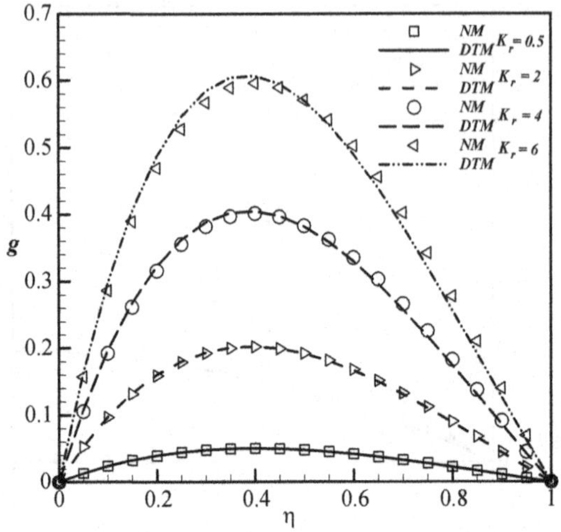

Figure 6.22 Velocity component profile (g) for variable K_r at $R=2$, $M=1$, $Pr=1$, $\lambda=0.5$, $Ec=Ec_x=0.5$.

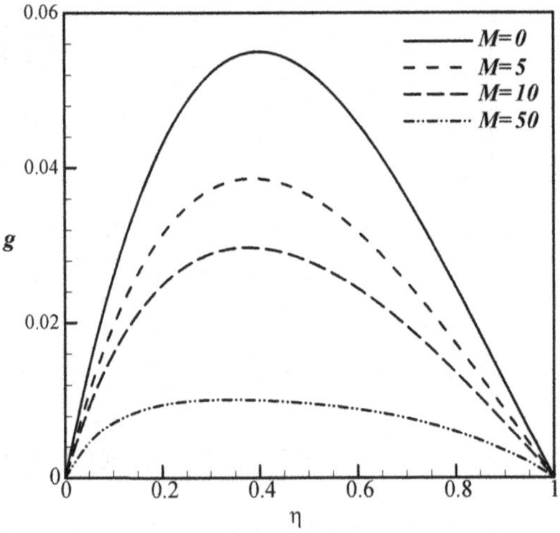

Figure 6.23 Velocity component profile (g) for variable M at $R=2$, $K_r=0.5$, $Pr=1$, λ=0.5, $Ec=Ec_x=0.5$.

stretching wall. This is due to the fact that heat energy is stored in the fluid due to the frictional heating. This phenomenon is more sensible on Ec than Ec_x. In Figs. 6.18, 6.19, 6.22, and 6.26 these results are compared with numerical method. It can be seen that there is good agreement between numerical solution and DTM.

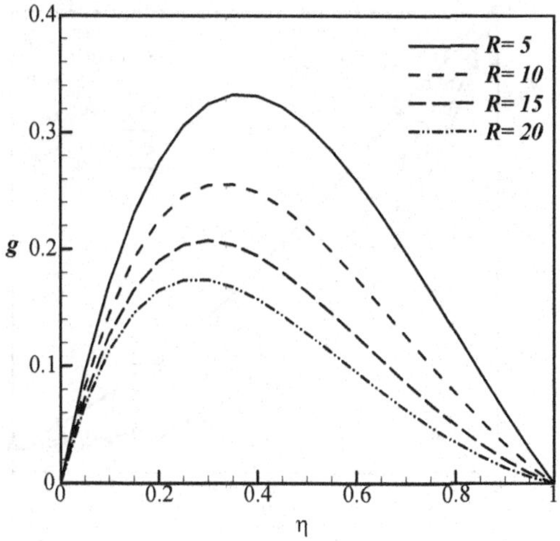

Figure 6.24 Velocity component profile (g) for variable R at $K_r = 4$, $M = 1$, $Pr = 1$, $\lambda = 0.5$, $Ec = Ec_x = 0.5$.

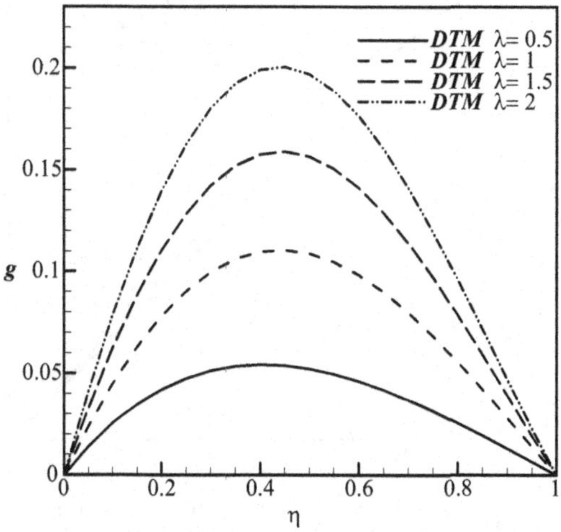

Figure 6.25 Velocity component profile (g) for variable λ at $R = 2$, $K_r = 0.5$, $M = 1$, $Pr = 1$, $Ec = Ec_x = 0.5$.

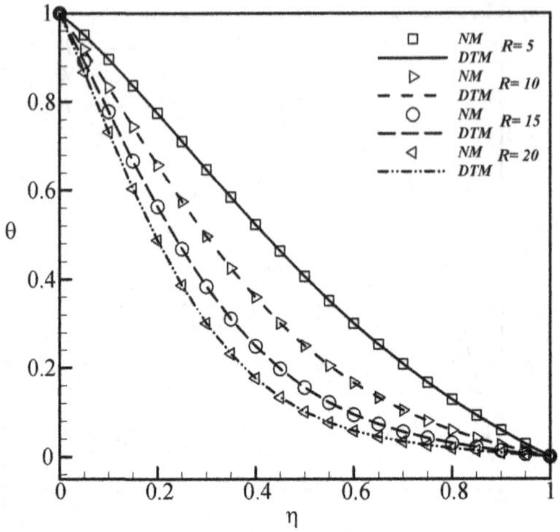

Figure 6.26 Temperature profile (θ) for variable R at $K_r = 0.5$, $M = 1$, $Pr = 1$, $\lambda = 0.5$, $Ec = Ec_x = 0.5$.

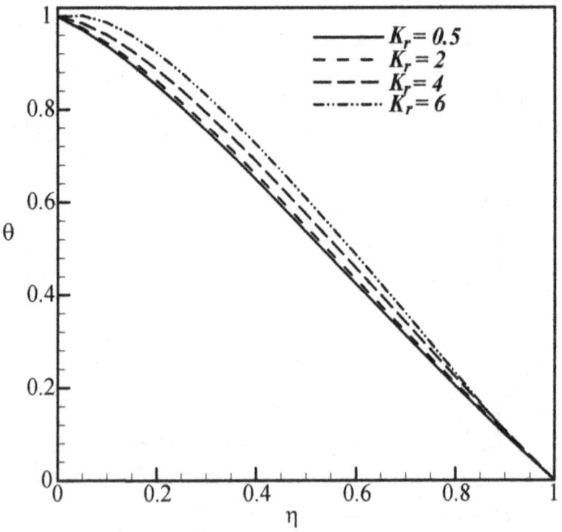

Figure 6.27 Temperature profile (θ) for variable K_r at $R = 2$, $M = 1$, $Pr = 1$, $\lambda = 0.5$, $Ec = Ec_x = 0.5$.

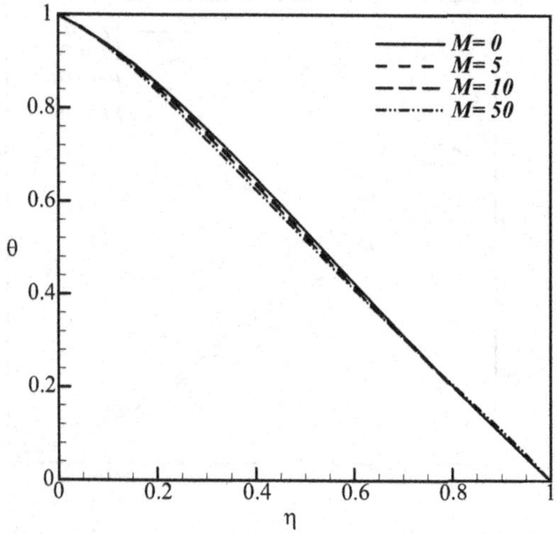

Figure 6.28 Temperature profile (θ) for variable M at $R = 2$, $K_r = 0.5$, $Pr = 1$, $\lambda = 0.5$, $Ec = Ec_x = 0.5$.

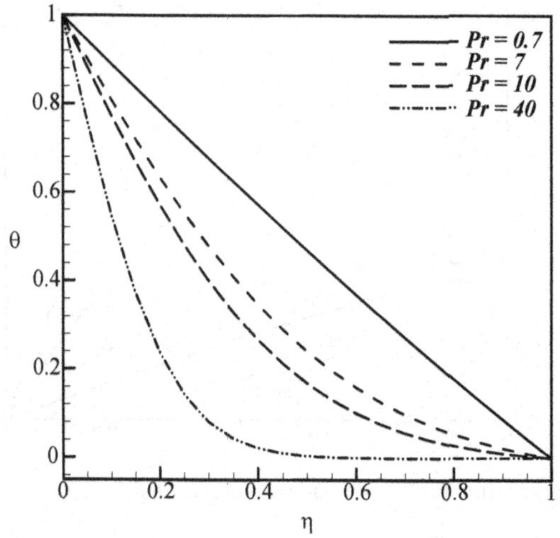

Figure 6.29 Temperature profile (θ) for variable Pr at $R = 2$, $K_r = 0.5$, $M = 1$, $\lambda = 0.5$, $Ec = Ec_x = 0$.

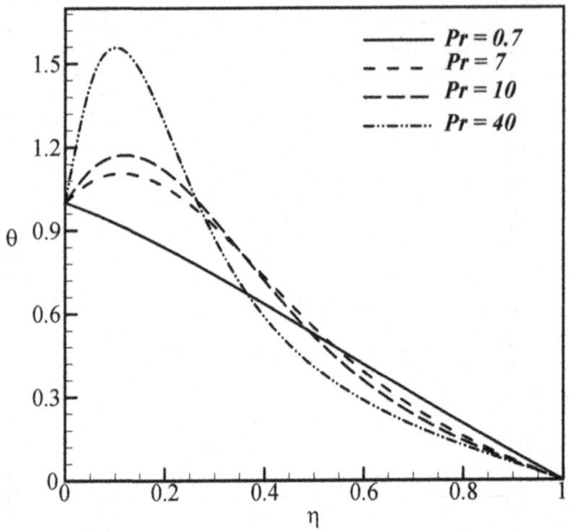

Figure 6.30 Temperature profile (θ) for variable Pr at $R = 2$, $K_r = 0.5$, $M = 1$, $\lambda = 0.5$, $Ec = Ec_x = 0.5$.

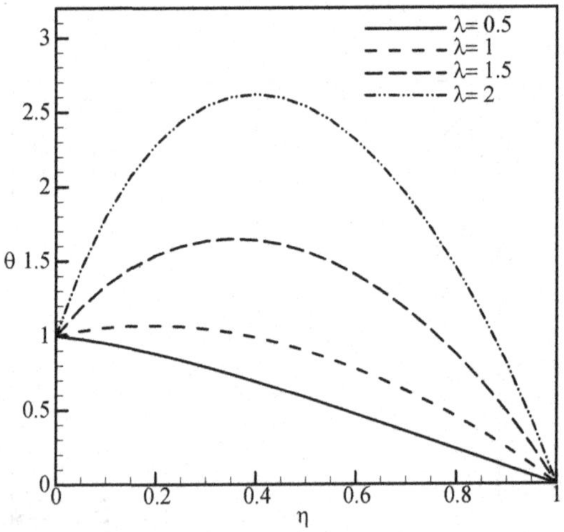

Figure 6.31 Temperature profile (θ) for variable λ at $R = 2$, $K_r = 0.5$, $M = 1$, $Pr = 1$, $Ec = Ec_x = 0.5$.

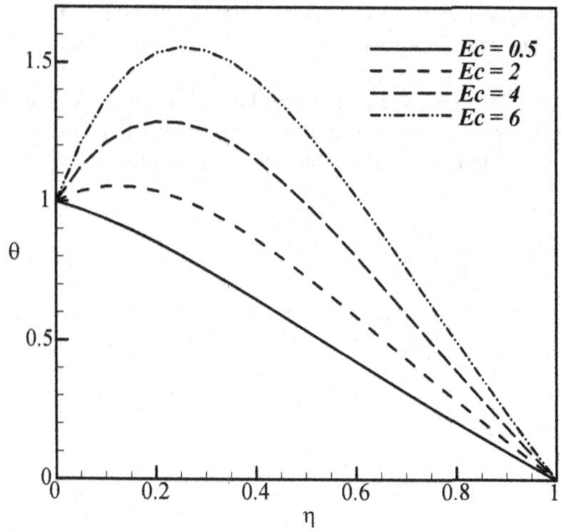

Figure 6.32 Temperature profile (θ) for variable Ec at $R = 2$, $K_r = 0.5$, $M = 1$, $Pr = 1$, $\lambda = 0.5$, $Ec_x = 0.5$.

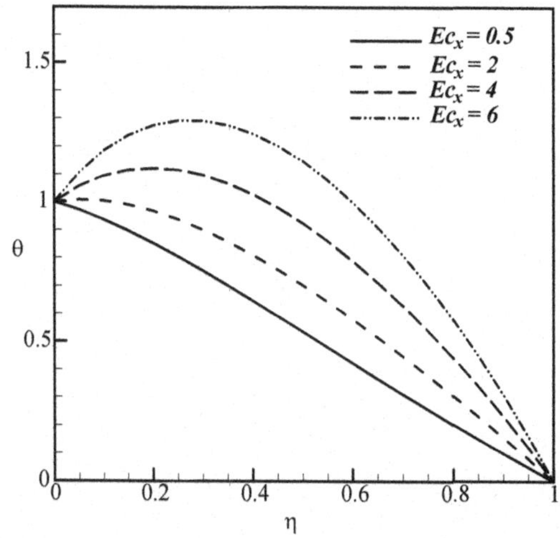

Figure 6.33 Temperature profile (θ) for variable Ec_x at $R = 2$, $K_r = 0.5$, $M = 1$, $Pr = 1$, $\lambda = 0.5$, $Ec = 0.5$.

REFERENCES

[1] Ellahi R. The effects of MHD and temperature dependent viscosity on the flow of non-Newtonian nanofluid in a pipe: analytical solutions. Applied Mathematical Modelling 2013;37(3):1451–67.

[2] Mosayebidorcheh S, Sheikholeslami M, Hatami M, Ganji DD. Analysis of turbulent MHD Couette nanofluid flow and heat transfer using hybrid DTM–FDM. Particuology 2016;26:95–101.

[3] Sheikholeslami M, Ashorynejad HR, Ganji DD, Rashidi MM. Heat and mass transfer of a micropolar fluid in a porous channel. Communications in Numerical Analysis 2014;2014.
[4] Sheikholeslami M, Ashorynejad HR, Ganji DD, Kolahdooz A. Investigation of rotating MHD viscous flow and heat transfer between stretching and porous surfaces using analytical method. Mathematical Problems in Engineering 2011;2011.

CHAPTER 7

DTM for Particles Motion, Sedimentation, and Combustion

7.1 INTRODUCTION

Many phenomena are existed in the environment in which particle's motion can be observed on them, such as centrifugation, centrifugal filters, industrial hopper, etc. Surface of the motion has different shapes especially for rotating application; it can be a circular, parabolic, or conical surface. It's necessary for scientists to analyze the motion of the particles on these surfaces, so an analytical solution is usually the more preferred and convenient method in engineering area because of less computational work as well as high accuracy which is widely used for predicting the motion of particles. Some of them are introduced in these sections:

7.2 MOTION OF A SPHERICAL PARTICLE ON A ROTATING PARABOLA

Consider a particle slides along a surface that has the shape of a parabola $z = cr^2$ (see Fig. 7.1). Following assumptions are considered for particles motion modeling [1]:

- Particle is at equilibrium.
- The particle rotates in a circle of radius R.
- The surface is rotating about its vertical symmetry axis with angular velocity ω.

Differential Transformation Method for Mechanical Engineering Problems
ISBN 978-0-12-805190-0
http://dx.doi.org/10.1016/B978-0-12-805190-0.00007-3

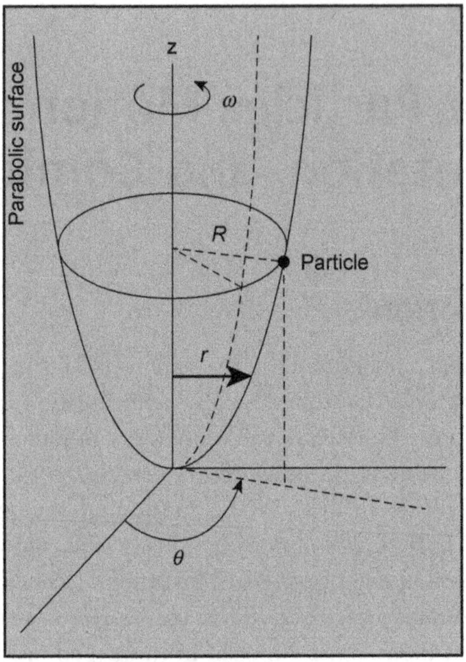

Figure 7.1 Schematic view of a spherical particle on a rotating parabolic surface.

By choosing the cylindrical coordinates r, θ, and z as generalized coordinates, the kinetic and potential energies can be given as,

$$T = \frac{1}{2}m\left(\dot{r}^2 + r^2\dot{\theta}^2 + \dot{z}^2\right)$$
$$U = mgz \tag{7.1}$$

We have in this case some equations of constraints that we must take into account, namely

$$z = cr^2$$
$$\dot{z} = 2c\dot{r}r \tag{7.2}$$

and

$$\theta = \omega t$$
$$\dot{\theta} = \omega \tag{7.3}$$

Inserting Eqs. (7.3) and (7.2) in Eq. (7.1), we can calculate the Lagrangian for the problem

$$L = T - U = \frac{1}{2}m(\dot{r}^2 + 4c^2r^2\dot{r}^2 + r^2\omega^2) - mgcr^2 \tag{7.4}$$

It is important to note that the inclusion of the equations of constraints in the Lagrangian has reduced the number of degrees of freedom to only one, i.e., $\dot{r}$. We now calculate the equation of motion using Lagrange's equation

$$\begin{aligned}\frac{\partial L}{\partial r} &= m(4c^2 r\dot{r}^2 + r\omega^2 - 2gcr) \\ \frac{d}{dt}\frac{\partial L}{\partial \dot{r}} &= m(\ddot{r} + 4c^2 r^2 \ddot{r} + 8c^2 r\dot{r}^2)\end{aligned} \tag{7.5}$$

and

$$\ddot{r}(1 + 4c^2 r^2) + \dot{r}^2(4c^2 r) + r(2gc - \omega^2) = 0 \tag{7.6}$$

By considering $2gc - \omega^2 = \varepsilon^2$;

$$\ddot{r} + 4c^2 \ddot{r} r^2 + 4c^2 r\dot{r}^2 + \varepsilon^2 r = 0 \tag{7.7}$$

It's considered that initial particle position is in radius A, and its initial velocity is zero. So, its initial conditions are

$$r(0) = A, \quad \dot{r}(0) = 0 \tag{7.8}$$

For solving the particle motion on a rotating parabola by an efficient, fast, and high accurate method, Eq. (7.7) is solved by multi-step differential transformation method (Ms-DTM),

$$\begin{aligned}&(k+2)(k+1)R_j(k+2) \\ &-4c^2 \sum_{k1}^{k} \sum_{l=0}^{k1} R_j(l)R_j(k1-l)(k-k1+1)(k-k1+2)R_j(k-k1+2) + \varepsilon^2 R_j(k) \\ &+4c^2 \sum_{k1=0}^{k} \sum_{l=0}^{k1} R_j(l)(k1-l+1)R_j(k1-l+1)R_j(k-k1+1)(k-k1+1) = 0\end{aligned} \tag{7.9}$$

With initial condition as,

$$\begin{aligned}&R_0(0) = A, & &R_0(1) = 0 \\ &R_i(0) = r_{i-1}(t_i), & &R_i(1) = r'_{i-1}(t_i), \quad i = 1, 2, \ldots, K \le M\end{aligned} \tag{7.10}$$

Since the procedure of solving Eq. (7.9) is autonomous of constants A, ε, and c, for generalization and simplification of problem for future cases with different physical conditions the constants that represent physical properties are assumed to be:

$$A = c = \varepsilon = 1, \tag{7.11}$$

After solving Eq. (7.9) and using initial condition Eqs. (7.10) and (7.11), position of the particle, $r(t)$, will be appeared as following equation for each 0.25-s time step,

$$
r(t)=\begin{cases}
r_0(t)=1-\dfrac{1}{10}t^2-\dfrac{11}{3000}t^4-\dfrac{553}{2250000}t^6-\dfrac{59363}{3150000000}t^8-\dfrac{10256209}{7087500000000}t^{10} & t\in[0,0.25]\\
r_{0.25}(t)=1.006293-0.050230616t-0.1013895(t-0.25)^2-0.003744513(t-0.25)^3-0.00390231(t-0.25)^4-0.38551e-3(t-0.25)^5 & \\
\quad -0.279966e-3(t-0.25)^6-0.40481e-4(t-0.25)^7-0.23109e-4(t-0.25)^8-0.396e-5(t-0.25)^9-0.18719e-5(t-0.25)^{10} & t\in[0.25,0.5]\\
r_{0.50}(t)=1.025707-0.10188t-0.105738(t-0.5)^2-0.0079821(t-0.5)^3-0.0046757(t-0.5)^4-0.88127e-3(t-0.5)^5 & \\
\quad -0.398166e-3(t-0.5)^6-0.99646e-4(t-0.5)^7-0.383726e-4(t-0.5)^8-0.10192e-4(t-0.5)^9-0.32931e-5(t-0.5)^{10} & t\in[0.5,0.75]\\
r_{0.75}(t)=1.05996-0.1565587t-0.113642(t-0.75)^2-0.013349(t-0.75)^3-0.00621702(t-0.75)^4-0.0016488(t-0.75)^5 & \\
\quad -0.65612e-3(t-0.75)^6-0.206483e-3(t-0.75)^7-0.7254e-4(t-0.75)^8-0.2103e-4(t-0.75)^9-0.5254e-5(t-0.75)^{10} & t\in[0.75,1.0]\\
\cdots & \\
r_{9.0}(t)=-2.63269+0.374569t-0.1812(t-9)^2+0.053389(t-9)^3-0.024782(t-9)^4+0.0101267(t-9)^5 & \\
\quad -0.00325035(t-9)^6-0.1527698e-3(t-9)^7+0.001640656(t-9)^8-0.0020994(t-9)^9+0.0020165285(t-9)^{10} & t\in[9.0,9.25]\\
r_{9.25}(t)=-1.88514+.29261t-0.1490558(t-9.25)^2+0.033957(t-9.25)^3-0.014987(t-9.25)^4+0.00580704(t-9.25)^5 & \\
\quad -0.00229197(t-9.25)^6+0.72248e-3(t-9.25)^7-0.8865e-4(t-9.25)^8-0.14027e-3(t-9.25)^9+0.19074e-3(t-9.25)^{10} & t\in[9.25,9.5]\\
r_{9.5}(t)=-1.2385+0.223613t-0.1284215(t-9.5)^2+0.0219744(t-9.5)^3-0.009517(t-9.5)^4+0.0032032(t-9.5)^5 & \\
\quad -0.001265(t-9.5)^6+0.43899e-3(t-9.5)^7-0.1384e-3(t-9.5)^8+0.2617e-4(t-9.5)^9+0.85178e-5(t-9.5)^{10} & t\in[9.5,9.75]\\
r_{9.75}(t)=-0.6551+0.162983t-0.11507(t-9.75)^2+0.0141168(t-9.75)^3-0.0064938(t-9.75)^4+0.001773667(t-9.75)^5 & \\
\quad -0.70295e-3(t-9.75)^6+0.22445e-3(t-9.75)^7-0.78165e-4(t-9.75)^8+0.22326e-4(t-9.75)^9-0.5106e-5(t-9.75)^{10} & t\in[9.75,10]
\end{cases}
\tag{7.12}
$$

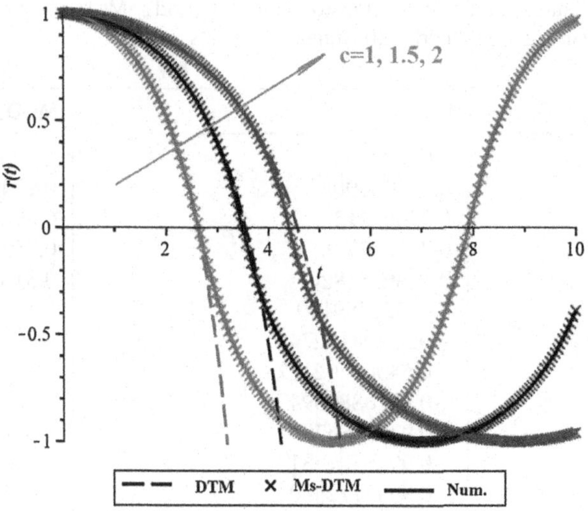

Figure 7.2 Multi-step differential transformation method (Ms-DTM) efficiency in particle motion analysis, compared with DTM and numerical solution. *DTM*, differential transformation method.

Eq. (7.12) is depicted in Fig. 7.2 and is compared with obtained result from differential transformation method (DTM). The values are presented in Table 7.1. Fig. 7.2 shows the particle's position for three different c constants when $A = \varepsilon = 1$. Influence of constant parameter (c) on stability and phase plane is investigated through Fig. 7.3.

7.3 MOTION OF A SPHERICAL PARTICLE IN PLANE COUETTE FLUID FLOW

The Vander Werff model for particle motion in Couette flow is adopted in this modeling while the positive direction rotation of particle is clockwise, and combined effects of inertia, gravity, and buoyancy forces are assumed to be negligible [2]. So, the inertia force in left-hand side of force balance equation is the product of spherical particle mass by its acceleration [3]

$$T = \frac{4\pi a^3}{3}\rho_s \dot{V} = \frac{4\pi a^3}{3}\rho_s(\ddot{x}, \ddot{y}, 0) \tag{7.13}$$

where a, ρ, and V denote the radius, density, and velocity of spherical particle, respectively. $\dot{V}$ is the first derivative of particle's velocity, and $\ddot{x}$ and $\ddot{y}$

Table 7.1 Multi-step Differential Transformation Method's (Ms-DTM) Values for Position of Particle, Compared With Numerical Results

	$r(t)$	
t	**Num.**	**Ms-DTM**
0	1.0000000	1.0000000
0.5	0.974766908	0.9747669183
1.0	0.896067130	0.8960671597
1.5	0.753057776	0.7530578586
2.0	0.5189251820	0.518925278
2.5	0.13321789149	0.133218454
3.0	−0.33404132749	−0.333782926
3.5	−0.64064202117	−0.639815924
4.0	−0.82788893947	−0.8266833025
4.5	−0.9393086789	−0.9377677280
5.0	−0.9926709981	−0.9907876234
5.5	−0.9947172580	−0.9924488554
6.0	−0.94568648462	−0.9429458988
6.5	−0.83945793817	−0.836078069
7.0	−0.65951111277	−0.655144210
7.5	0.36531058744	0.359190250
8.0	0.09376192157	0.100601846
8.5	0.49425716809	0.497276130
9.0	0.73793715747	0.738436224
9.5	0.88704545678	0.885813784
10	0.97044161417	0.9677503517

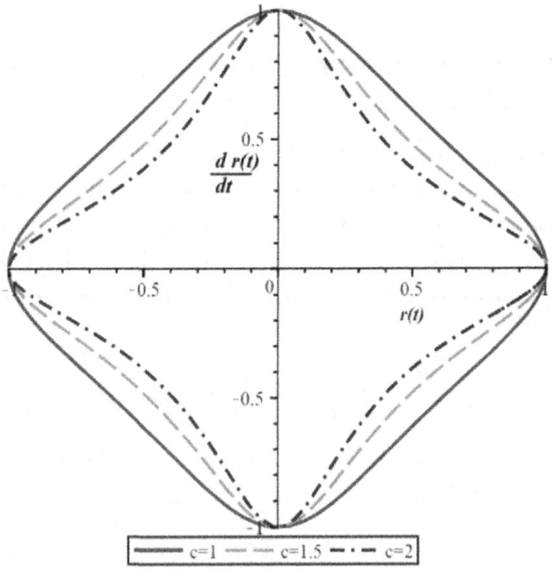

Figure 7.3 Phase plane for particle in different constant using multi-step differential transformation method (Ms-DTM) when $A = \varepsilon = 1$.

are the second derivatives of particle motion in horizontal and vertical directions respect to time. To calculate the drag force, the velocities of the spherical particle are considered small adequately so that the Stokes law can be governed:

$$T_{Dx} = 6\pi\mu a V_{rx} = 6\pi\mu a(\dot{x} - \alpha y) \tag{7.14a}$$

$$T_{Dy} = 6\pi\mu a V_{ry} = 6\pi\mu a\dot{y} \tag{7.14b}$$

where μ signifies the viscosity of fluid. The rotation and shear portion of the lift force is obtained as:

$$T_{Rx} = \frac{1}{2}\pi a^3 \rho\alpha\dot{y} \tag{7.15a}$$

$$T_{Ry} = \frac{1}{2}\pi a^3 \rho\alpha(\alpha y - \dot{x}) \tag{7.15b}$$

$$T_{Sx} = 0 \tag{7.16a}$$

$$T_{Sy} = 6.46a^2 \rho\alpha^{1/2}\nu^{1/2}(\alpha y - \dot{x}) \tag{7.16b}$$

where ν is the dynamic viscosity, $(\frac{\mu}{\rho})$, and α is defined as positive proportionality constant.

An illustration of the spherical particle in plane Couette fluid flow and applied forces on particle are shown in Fig. 7.4. The mass of particle is assumed in the center of sphere, and the forces caused from the rotation and shear fields and their interactions on drag and lift forces of particle are illustrated in Fig. 7.4(a) and (b), respectively. By forming the force balance equation of the inertia force to the drag and lift forces; the equations of motion for the particle are driven as:

$$T = T_R + T_S - T_D \tag{7.17}$$

Eventually, by substituting Eqs. (7.14) and (7.16) into Eq. (7.17) the system of equation of motion for spherical particle in plane Couette flow yields:

$$\begin{cases} \dfrac{4\pi a^3}{3}\rho_s\ddot{x} = \dfrac{1}{2}\pi a^3\rho\alpha\dot{y} - 6\pi\mu a(\dot{x} - \alpha y) \\ \dfrac{4\pi a^3}{3}\rho_s\ddot{y} = \left(\dfrac{1}{2}\pi a^3\rho\alpha + 6.46a^2\rho\alpha^{1/2}\nu^{1/2}\right)(\alpha y - \dot{x}) - 6\pi\mu a\dot{y} \end{cases} \tag{7.18}$$

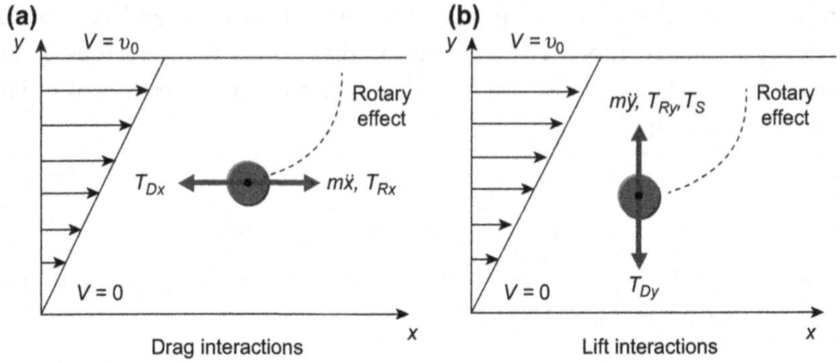

Figure 7.4 Schematic view of exerted forces on a spherical particle in Couette fluid flow. (a) Drag interactions; (b) lift interactions.

For simplicity the governing equations have been expressed as:

$$\begin{cases} \ddot{x} = A\dot{y} - B(\dot{x} - \alpha y) \\ \ddot{y} = B\dot{y} - (A + C)(\dot{x} - \alpha y) \end{cases} \tag{7.19}$$

where the coefficients $A-C$ are defined as follows:

$$\begin{aligned} A &= \left(\frac{3\alpha}{8}\right)\left(\frac{\rho}{\rho_S}\right) \\ B &= \left(\frac{9v}{2a^2}\right)\left(\frac{\rho}{\rho_S}\right) \\ C &= 4.845\left(\frac{\sqrt{\alpha v}}{a}\right)\left(\frac{\rho}{\rho_S}\right) \end{aligned} \tag{7.20}$$

An appropriate initial condition is required to avoid trapping the procedure in nontrivial solution:

$$\begin{aligned} x(t=0) &= 0, \ \dot{x}(t=0) = u_0 \\ y(t=0) &= 0, \ \dot{y}(t=0) = v_0 \end{aligned} \tag{7.21}$$

For applying Ms-DTM to the present problem, in first step traditional DTM using Table 7.1 should be applied to Eq. (7.19),

$$\begin{aligned}&(k+2)(k+1)X_j(k+2)-A(k+1)Y_j(k+1)\\&+B((k+1)X_j(k+1)-\alpha Y_j(k))=0\\&(k+2)(k+1)Y_j(k+2)+B(k+1)Y_j(k+1)\\&+(A+C)((k+1)X_j(k+1)-\alpha Y_j(k))=0\end{aligned} \tag{7.22}$$

With initial condition as,

$$\begin{aligned}&X_0(0)=0, && X_0(1)=1\\&X_i(0)=x_{i-1}(t_i), && X_i(1)=x'_{i-1}(t_i), \quad i=1,2,\ldots,K\le M\\&Y_0(0)=0, && Y_0(1)=1\\&Y_i(0)=y_{i-1}(t_i), && Y_i(1)=y'_{i-1}(t_i), \quad i=1,2,\ldots,K\le M\end{aligned} \tag{7.23}$$

For example, if we divide the domain $t \in [0, 8]$ to eight subintervals (each 1 s) for the first subinterval $t \in [0, 1]$, we should obtain the DTM solution of Eq. (7.22) using initial condition in Eq. (7.21) and for next subinterval, $t \in [1, 2]$, DTM solution of Eq. (7.22) should be obtained as the first subinterval but with a new initial condition. This new initial condition, as previously mentioned, can be calculated from the obtained equation from the first subinterval by substituting $t=1$ in it. And for the third subinterval, initial condition can be determined by substituting $t=2$ in the DTM solution obtained in the second step. This process should be repeated, which generates a sequence of approximate solutions for $x_i(t)$ and $y_i(t)$, $i=1, 2,\ldots,8$ for the solution $x(t)$ and $y(t)$ after derivative with respect to time, $u(t)$ and $v(t)$ will be obtained.

Since the procedure of solving Eq. (7.22) is autonomous of constants A, B, and C, for generalization and simplification of problem for future cases with different physical conditions the constants that represent physical properties are assumed to be as following:

$$A=B=\alpha=u_0=v_0=1,\ C=0 \tag{7.24}$$

After solving Eq. (7.22) using initial condition Eqs. (7.23) and (7.21), position of the particle, $x(t)$ and $y(t)$, will be appeared, and after derivative, velocity profiles will be calculated as following for each 1-s time step,

$$
u(t)=\begin{cases}
u_0(t)=1-\dfrac{1}{2}t^2+\dfrac{1}{3}t^3-\dfrac{1}{8}t^4+\dfrac{1}{30}t^5-\dfrac{1}{144}t^6+\dfrac{1}{840}t^7 & t\in[0,1]\\
u_1(t)=1.104167-0.368254t+0.992063e-4(t-1)^2+0.0613095(t-1)^3-0.030663(t-1)^4+0.0091997(t-1)^5- & \\
\quad 0.002044477(t-1)^6+0.36509e-3(t-1)^7 & t\in[1,2]\\
u_2(t)=0.947912-0.270994t+0.0677689(t-2)^2-0.13686e-4(t-2)^3-0.005641(t-2)^4+0.0022569(t-2)^5- & \\
\quad 0.56428e-3(t-2)^6+0.1074883e-3(t-2)^7 & t\in[2,3]\\
u_3(t)=0.647533-0.149562t+0.0498569(t-3)^2-0.0083108(t-3)^3+6.9893e-7(t-3)^4+0.415265e-3(t-3)^5- & \\
\quad 0.138445e-3(t-3)^6+0.2966841e-4(t-3)^7 & t\in[3,4]\\
u_4(t)=0.384619-0.073370t+0.0275137(t-4)^2-0.0061140(t-4)^3+0.0007642(t-4)^4+2.0687e-8(t-4)^5- & \\
\quad 0.00002548(t-4)^6+0.000007279586(t-4)^7 & t\in[4,5]\\
u_5(t)=0.20862-0.033743t+0.0134968(t-5)^2-0.00337401(t-5)^3+0.56227e-3(t-5)^4-0.562082e-4(t-5)^5- & \\
\quad 6.32323e-9(t-5)^6+0.134009e-5(t-5)^7 & t\in[5,6]\\
u_6(t)=0.106183-0.014897t+0.0062069(t-6)^2-0.0016551(t-6)^3+0.3103e-3(t-6)^4-0.413657e-4(t-6)^5+ & \\
\quad 0.34452e-5(t-6)^6+5.6481e-10(t-6)^7 & t\in[6,7]\\
u_7(t)=0.051485-0.0063942t+0.00274(t-7)^2-0.00076115(t-7)^3+0.000152(t-7)^4-0.228e-4(t-7)^5+ & \\
\quad 0.254e-5(t-7)^6-1.81e-7(t-7)^7 & t\in[7,8]
\end{cases}
\qquad (7.25)
$$

and velocity in y direction, $v(t)$, will be:

$$
v(t)=\begin{cases}
v_0(t)=1-2t+\frac{3}{2}t^2-\frac{2}{3}t^3+\frac{5}{24}t^4-\frac{1}{20}t^5+\frac{7}{720}t^6-\frac{1}{630}t^7 & t\in[0,1]\\
v_1(t)=0.36766-0.36785t+0.367956(t-1)^2-0.18399(t-1)^3+0.061334(t-1)^4- & \\
\quad 0.015334(t-1)^5+0.00306685(t-1)^6-0.511148e-3(t-1)^7 & t\in[1,2]\\
v_2(t)=-0.135702+0.8212e-4t+0.067687(t-2)^2-0.045138(t-2)^3+0.0169285(t-2)^4- & \\
\quad 0.004514507(t-2)^5+0.940551e-3(t-2)^6-0.161241e-3(t-2)^7 & t\in[2,3]\\
v_3(t)=-0.24931+0.049865t-0.83872e-5(t-3)^2-0.0083053(t-3)^3+0.00415334(t-3)^4 & \\
\quad -0.00124607(t-3)^5+0.276913e-3(t-3)^6-0.49449e-4(t-3)^7 & t\in[3,4]\\
v_4(t)=-0.201764+0.0366842t-0.0091704(t-4)^2-4.1374e-7(t-4)^3+0.000764(t-4)^4- & \\
\quad 0.000306(t-4)^5+0.0000764(t-4)^6-0.000014559(t-4)^7 & t\in[4,5]\\
v_5(t)=-0.128214+0.020244t-0.006747(t-5)^2+0.00112416(t-5)^3+1.897e-7(t-5)^4- & \\
\quad 0.56284e-4(t-5)^5+0.18755e-4(t-5)^6-0.40185e-5(t-5)^7 & t\in[5,6]\\
v_6(t)=-0.071998+0.009931t-0.003724(t-6)^2+0.000827(t-6)^3-0.00010336(t-6)^4- & \\
\quad 2.3722e-8(t-6)^5+0.0000034531(t-6)^6-9.86028e-7(t-6)^7 & t\in[6,7]\\
v_7(t)=-0.037449+0.0045669t-0.001826(t-7)^2+0.000457(t-7)^3-0.76083e-4(t-7)^4 & \\
\quad +0.76029e-5(t-7)^5+1.7969e-9(t-7)^6-1.81536e-7(t-7)^7 & t\in[7,8]
\end{cases}
\tag{7.26}
$$

Eqs. (7.25) and (7.26) are depicted in Fig. 7.5 and are compared with those obtained by homotopy perturbation method (HPM)—Pade, also their values are presented in Tables 7.2 and 7.3 for $u(t)$ and $v(t)$, respectively. Fig. 7.6 shows the acceleration of the particle obtained from Fig. 7.5. As seen in these figures and tables,

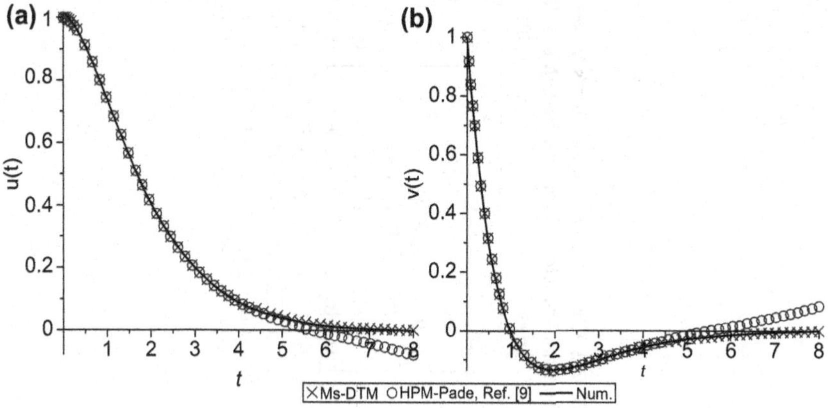

Figure 7.5 Multi-step differential transformation method (Ms-DTM) efficiency in particle velocity estimating, compared with homotopy perturbation method—Pade [2]. (a) velocity in x direction, $u(t)$; (b) velocity in y direction, $v(t)$.

Table 7.2 Comparison Between $u(t)$'s Values and Errors Obtained From Homotopy Perturbation Method (HPM)—Pade and Multi-step Differential Transformation Method (Ms-DTM) When $A = B = \alpha = 1$, $C = 0$

t	Num.	HPM—Pade	Ms-DTM	% Error HPM—Pade	% Error Ms-DTM
0	1.0	1.0	1.0	0	0
1	0.735758911174129	0.7357586420	0.7359126984	3.66E-07	0.00021
2	0.406005878786318	0.4059621586	0.4059247848	0.000108	0.0002
3	0.199148297016095	0.1985113312	0.1988459460	0.003198	0.001518
4	0.0915782073185634	0.08815295207	0.09113681554	0.037403	0.00482
5	0.0404276801389359	0.02950000148	0.03991176120	0.270302	0.012762
6	0.0173512487117868	−0.00813908291	0.01679880445	1.469078	0.031839
7	0.0072950308693911	−0.04120493255	0.006725715286	6.648356	0.078042
8	0.0030191406936549	−0.07720068776	0.002442357340	26.57042	0.191042

Table 7.3 Comparison Between $v(t)$'s Values and Errors Obtained From Homotopy Perturbation Method (HPM)—Pade and Multi-step Differential Transformation Method (Ms-DTM) When $A = B = \alpha = 1$, $C = 0$

t	Num.	HPM—Pade	Ms-DTM	% Error HPM—Pade	% Error Ms-DTM
0	1.0	1.0	1.0	0	0
1	−4.38556843 E−8	2.36772637 E−7	−1.984126984E-7	1.838492	4.514
2	−0.1353353333998	−0.1352915191	−0.1355378794	0.000324	0.0015
3	−0.0995741854390	−0.09892589377	−0.09971382172	0.006511	0.0014
4	−0.0549469585114	−0.05140825685	−0.05502746765	0.064402	0.00147
5	−0.0269518219371	−0.01550417243	−0.02699360654	0.424745	0.00155
6	−0.0123937803694	0.01465314367	−0.01241399659	2.182298	0.00163
7	−0.0054712929795	0.04659627467	−0.005480594942	9.516501	0.0017
8	−0.0023482314014	0.08469760161	−0.002352349259	37.06868	0.00175

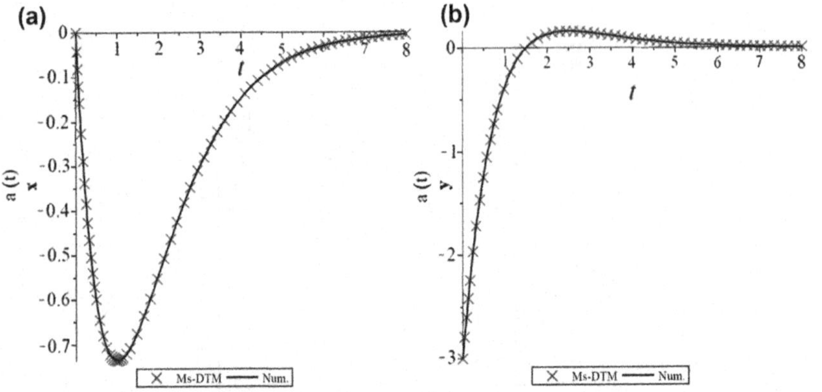

Figure 7.6 Multi-step differential transformation method (Ms-DTM) efficiency in particle acceleration estimating, $A = B = \alpha = 1$, $C = 0$. (a) Acceleration in x direction, $a_x(t)$; (b) acceleration in y direction, $a_x(t)$.

Ms-DTM is more accurate than HPM—Pade, especially when time increased. This may be due to perturbation, small parameter, p, and linearization in HPM—Pade method. Also Ms-DTM with eight time steps has excellent agreement with numerical solution, but it is easily capable to increase the time steps.

Figs. 7.7 and 7.8 demonstrate the velocity variation of the particle for other values of the A, B, C, and α. These two figures, with mathematical purpose, are presented to show that Ms-DTM is completely equal to

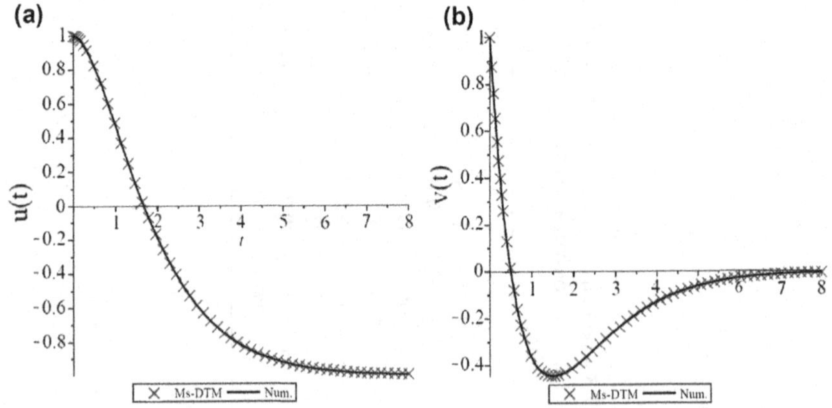

Figure 7.7 Particle velocity when $A = B = C = \alpha = 1$. (a) Velocity in x direction, $u(t)$; (b) velocity in y direction, $v(t)$.

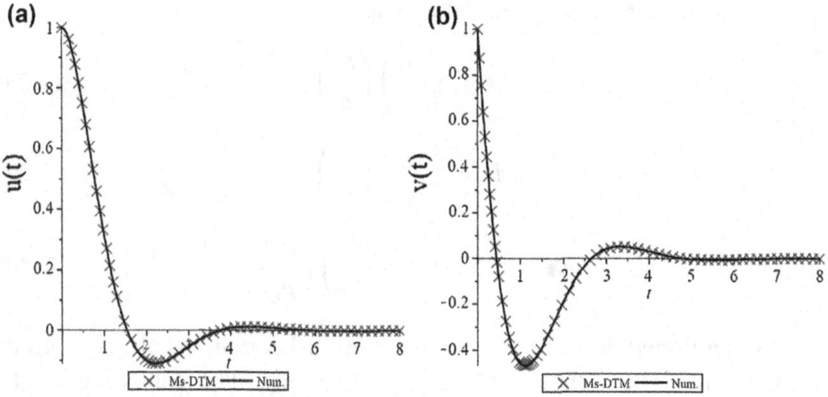

Figure 7.8 Particle velocity when $A = B = C = 1$, $\alpha = 0$. (a) Velocity in x direction, $u(t)$; (b) velocity in y direction, $v(t)$.

numerical solution for other values of constants appeared in the mathematical formulation. The particle positions in a 2D plane couette fluid flow when $A = B = C = 1$ and $A = B = 1$, $C = 0$ are presented graphically in Fig. 7.9(a) and (b), respectively, for each 1-s time step. As these figures reveal, when $C = 0$, due to nonshear portion effect in y direction or T_{sy}, the rotation of particle in y direction is lower than that in the case when $C = 1$.

Eq. (7.19) can be also solved by differential transformation method with Padé approximation (DTM−Pade), for this aim it can be rewritten as;

$$\begin{cases} \ddot{x} - A\dot{y} + B(\dot{x} - \alpha y) = 0 \\ \ddot{y} + B\dot{y} + (A + C)(\dot{x} - \alpha y) = 0 \end{cases} \tag{7.27}$$

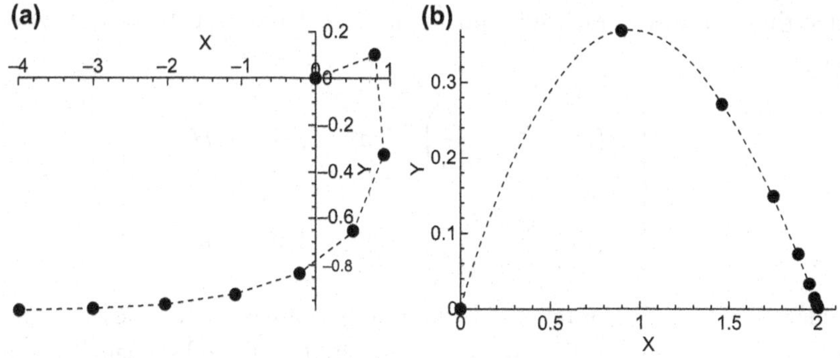

Figure 7.9 Particle position in each 1-s time step using multi-step differential transformation method when (a) $A = B = C = 1$, (b) $A = B = 1$, $C = 0$.

where coefficients A to C are defined as:

$$A = \left(\frac{3\alpha}{8}\right)\left(\frac{\rho_f}{\rho_s}\right) \tag{7.28}$$

$$B = \left(\frac{9\nu}{2r^2}\right)\left(\frac{\rho_f}{\rho_s}\right) \tag{7.29}$$

$$C = 1.542\left(\frac{\alpha^{1/2}\nu^{1/2}}{r}\right)\left(\frac{\rho_f}{\rho_s}\right) \tag{7.30}$$

As mentioned in the text, the nonzero initial conditions of equations of motion could be different for unlike situations. The following might represent either injection of the particle into the fluid or statistical fluctuations:

$$\begin{cases} x = 0\,, \quad \dot{x} = u_0 \quad \text{at} \quad t = 0 \\ y = 0\,, \quad \dot{y} = v_0 \quad \text{at} \quad t = 0 \end{cases} \tag{7.31}$$

To simplify the solution, constants dependent on physical properties of solid–fluid combination are considered to be $A = B = C = \alpha = u_0 = v_0 = 1$.

Now we apply the differential transformation method for Eq. (7.27), and taking the differential transform of Eq. (7.27) with respect to t, gives:

$$\begin{cases} (k+1)(k+2)X(k+2) + (k+1)X(k+1) - (k+1)Y(k+1) - Y(k) = 0 \\ \\ (k+1)(k+2)Y(k+2) + (k+1)Y(k+1) - (k+1)X(k+1) - Y(k) = 0 \end{cases} \tag{7.32}$$

From a process of inverse differential transformation, it can be shown that the solutions of each subdomain take $n+1$ term for the power series, i.e.,

$$\begin{cases} x_i(t) = \sum\limits_{k=0}^{n}\left(\frac{t}{H_i}\right)^k X_i(k), \quad 0 \le t \le H_i \\ \\ y_i(t) = \sum\limits_{k=0}^{n}\left(\frac{t}{H_i}\right)^k Y_i(k), \quad 0 \le t \le H_i \end{cases} \tag{7.33}$$

where $k = 0, 1, 2,\ldots,n$ represents the number of term of the power series, $i = 0, 1, 2,\ldots$ expresses the ith subdomain, and H_i is the subdomain interval.

From initial condition in Eq. (7.31), we have it in point $t = 0$ and $t = 1$, and exerting transformation

$$\begin{cases} X(0) = 0 \\ Y(0) = 0 \end{cases} \tag{7.34}$$

$$\begin{cases} X(1) = 1 \\ Y(1) = 1 \end{cases} \tag{7.35}$$

Accordingly, from a process of inverse differential transformation, in this problem we calculated $X(k+2)$ and $Y(k+2)$ from Eq. (7.32) as following:

$$\begin{cases} X(2) = 0 \\ Y(2) = -1 \end{cases} \tag{7.36}$$

$$\begin{cases} X(3) = -\dfrac{1}{6} \\ Y(3) = \dfrac{1}{2} \end{cases} \tag{7.37}$$

$$\begin{cases} X(4) = \dfrac{1}{12} \\ Y(4) = -\dfrac{1}{6} \end{cases} \tag{7.38}$$

$$\begin{cases} X(5) = -\dfrac{1}{40} \\ Y(5) = \dfrac{1}{24} \end{cases} \tag{7.39}$$

$$\begin{cases} X(6) = \dfrac{1}{180} \\ Y(6) = -\dfrac{1}{120} \end{cases} \tag{7.40}$$

The above process is continuous. Substituting Eqs. (7.36)–(7.40) into the main equation based on DTM, the closed form of the solutions can be obtained as:

$$\begin{aligned}
x(t) &= \sum_{K=0}^{22} X(K)t^K \\
&= t - \frac{1}{6}t^3 + \frac{1}{12}t^4 - \frac{1}{40}t^5 + \frac{1}{180}t^6 - \frac{1}{1008}t^7 + \frac{1}{6720}t^8 - \frac{1}{51840}t^9 \\
&+ \frac{1}{453600}t^{10} - \frac{1}{4435200}t^{11} + \frac{1}{47900160}t^{12} - \frac{1}{566092800}t^{13} \\
&+ \frac{1}{7264857600}t^{14} - \frac{1}{100590336000}t^{15} + \frac{1}{1494484992000}t^{16} \\
&- \frac{1}{23712495206400}t^{17} + \frac{1}{400148356608000}t^{18} \\
&- \frac{1}{7155594141696000}t^{19} + \frac{1}{135161222676480000}t^{20} \\
&- \frac{1}{2688996956405760000}t^{21} + \frac{1}{56200036388880384000}t^{22}
\end{aligned} \tag{7.41}$$

$$\begin{aligned}
y(t) &= \sum_{K=0}^{22} Y(K)t^K \\
&= t - t^2 + \frac{1}{2}t^3 - \frac{1}{6}t^4 + \frac{1}{24}t^5 - \frac{1}{120}t^6 + \frac{1}{720}t^7 - \frac{1}{5040}t^8 + \frac{1}{40320}t^9 \\
&- \frac{1}{362880}t^{10} + \frac{1}{3628800}t^{11} - \frac{1}{39916800}t^{12} + \frac{1}{479001600}t^{13} \\
&- \frac{1}{6227020800}t^{14} + \frac{1}{87178291200}t^{15} - \frac{1}{1307674368000}t^{16} \\
&+ \frac{1}{20922789888000}t^{17} - \frac{1}{355687428096000}t^{18} \\
&+ \frac{1}{6402373705728000}t^{19} - \frac{1}{121645100408832000}t^{20} \\
&+ \frac{1}{2432902008176640000}t^{21} - \frac{1}{51090942171709440000}t^{22}
\end{aligned} \tag{7.42}$$

As it is obvious, solution of terms vary periodically and in each step more duration of particle motion is covered. By increasing series terms, the accuracy of DTM solution is improved and a larger period of acceleration motion of the particle is covered. Basically, by estimating the constants $A - C$ for each selected combinations of solid–fluid, results can be derived easily.

After obtaining the result of 22nd iteration for DTM, it is seen from graphs and tables that the DTM for this problem do not have exact solution; therefore we will apply the Pade approximation for variations of velocities of the particle as follows

$$u(t)_{[4/4]} = \dot{x}(t)_{[4/4]}$$
$$= \frac{\frac{31}{58060800} - \left(\frac{1537}{8128512000}\right)t^2 + \left(\frac{47}{1354752000}\right)t^3 - \left(\frac{3539}{1463132160000}\right)t^4 + \left(\frac{67}{217728000}\right)t}{\frac{31}{58060800} + \left(\frac{67}{217728000}\right)t + \left(\frac{211}{2709504000}\right)t^2 + \left(\frac{43}{4064256000}\right)t^3 + \left(\frac{143}{209018880000}\right)t^4} \tag{7.43}$$

$$v(t)_{[4/4]} = \dot{y}(t)_{[4/4]}$$
$$= \frac{\frac{1457}{870912000} - \left(\frac{2089}{870912000}\right)t + \left(\frac{6899}{8128512000}\right)t^2 - \left(\frac{481}{3657830400}\right)t^3 + \left(\frac{1763}{209018880000}\right)t^4}{\frac{1457}{870912000} + \left(\frac{11}{11612160}\right)t + \left(\frac{1901}{8128512000}\right)t^2 + \left(\frac{1}{32659200}\right)t^3 + \left(\frac{907}{487710720000}\right)t^4} \tag{7.44}$$

The results are compared with DTM and numerical solution. Fig. 7.10 depicts the horizontal and vertical velocities versus time. It is observed that the DTM–Pade approximant solution is more accurate than DTM. Comparing DTM–Pade [4/4] and DTM–Pade [10/10], DTM–Pade [10/10] gives closer results to numerical solution. This fact is more pronounced for large values of time, i.e., $t = 10$.

7.4 NONSPHERICAL PARTICLES SEDIMENTATION

Consider a rigid, nonspherical particle with sphericity ϕ, equivalent volume diameter D, mass m, and density ρ_s is falling in an infinite extent of an incompressible Newtonian fluid of density ρ and viscosity μ (Fig. 7.11).

U represents the velocity of the particle at any instant time, t, and g is the acceleration due to gravity. The unsteady motion of the particle in a fluid can be described by the Basset–Boussinesq–Oseen (BBO) equation. For a dense particle falling in light fluids (assuming $\rho \ll \rho_s$), Basset History force is negligible. Thus, the equation of particle motion is given as [4]

$$m\frac{du}{dt} = mg\left(1 - \frac{\rho}{\rho_s}\right) - \frac{1}{8}\pi D^2 \rho C_D u^2 - \frac{1}{12}\pi D^3 \rho \frac{du}{dt} \tag{7.45}$$

where C_D is the drag coefficient. In the right-hand side of Eq. (7.45), the first term represents the buoyancy effect, the second term corresponds to

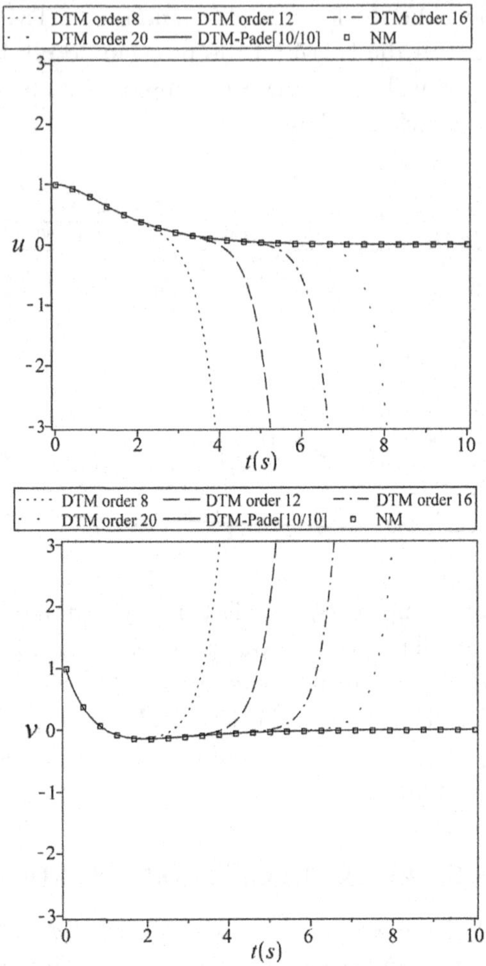

Figure 7.10 Comparison of horizontal velocity (u) and vertical velocity (v) obtained by differential transformation method (DTM) and DTM–Pade when $A = B = \alpha = u_0 = v_0 = 1$.

drag resistance, and the last term is associated with the added mass effect which is due to acceleration of fluid around the particle.

The nonlinear terms due to nonlinearity nature of the drag coefficient C_D is the main difficulty in solving Eq. (7.45). By substituting C_D in Eq. (7.45), following expression is gained (assuming $\mathrm{Re} = \frac{\rho u D}{\mu}$):

$$\left(m + \frac{1}{12}\pi D^3 \rho\right)\frac{du}{dt} + \frac{1}{8}\pi D^2 \rho\left(\frac{30\mu}{\rho u D} + 67.289e^{-5.03\phi}\right)u^2 - mg\left(1 - \frac{\rho}{\rho_s}\right) = 0 \tag{7.46}$$

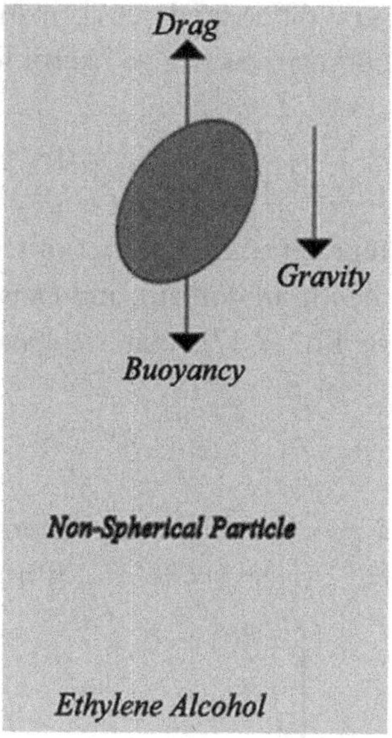

Figure 7.11 Schematic view of vertically falling particle in ethylene alcohol.

by rearranging parameters, Eq. (7.46) could be rewritten as follows:

$$a\frac{du}{dt} + bu + cu^2 - d = 0, \quad u(0) = 0 \tag{7.47}$$

where

$$a = \left(m + \frac{1}{12}\pi D^3 \rho\right) \tag{7.48}$$

$$b = 3.75\pi D\mu \tag{7.49}$$

$$c = \frac{67.289e^{-5.03\phi}}{8}\pi D^2 \rho \tag{7.50}$$

$$d = mg\left(1 - \frac{\rho}{\rho_s}\right) \tag{7.51}$$

By the transforming of Eq. (7.47) considered by using the related definitions of DTM, and considering $a = b = c = d = 1$, we have the following:

$$(k+1)U(k+1) + U(k) + \sum_{l=0}^{k} U(l)U(k-l) - \delta(k) = 0 \tag{7.52}$$

From a process of inverse differential transformation, it can be shown that the solutions of each subdomain takes $n+1$ term for the power series, i.e.,

$$u_i(t)=\sum_{k=0}^{n}\left(\frac{t}{H_i}\right)^k U_i(k), \quad 0\le t\le H_i \tag{7.53}$$

where $k=0, 1, 2,\ldots,n$ represents the number of terms of the power series, $i=0, 1, 2,\ldots$ expresses the ith subdomain, and H_i is the subdomain interval. From initial condition in Eq. (7.47), that we have it in point $t=0$, and exerting transformation,

$$U(0)=0 \tag{7.54}$$

Accordingly, from a process of inverse differential transformation, in this problem we calculated $U(K+1)$ $U(k+1)$ from Eq. (7.52) as following:

$$\begin{cases} U(0)=0, \\ U(1)=1, \\ U(2)=-1/2, \\ U(3)=-1/6, \\ \vdots \end{cases} \tag{7.55}$$

The above process is continuous. Substituting Eq. (7.55) into the main equation based on DTM, the closed form of the solutions can be obtained as:

$$\begin{aligned} u(t)&=\sum_{K=0}^{20} U(K)t^K \\ &=t-\frac{1}{2}t^2-\frac{1}{6}t^3+\frac{7}{24}t^4-\frac{1}{24}t^5-\frac{17}{144}t^6+\frac{67}{1008}t^7+\frac{227}{8064}t^8+\ldots \end{aligned} \tag{7.56}$$

In a similar manner, we will obtain another subdomain's series solution, and we can present the solution of Eq. (7.47) accurately.

The calculations presented in this chapter adopt a value of $n=20$. Having determined the various values of $U(K+1)$ from Eq. (7.52) with the transformed initial condition of Eq. (7.54), the first subdomain solutions of Eq. (7.47) can be obtained by means of the inverse transformed equations of Eq. (7.53). The final values of the first subdomain, i.e., the solutions of the previous calculation, are then taken as the initial condition of the second subdomain, which is subsequently calculated using the same procedure as described above. By repeatedly adopting the final values of one subdomain as the initial condition of the following subdomain, the differential equation can be solved from its first subdomain to its final subdomain. Therefore, the proposed method enables the solutions of Eq. (7.47) to be solved over the entire time domain, but not exactly, therefore after obtaining the results, the Pade approximation is applied as follows:

$$u(t)_{[4/4]} = \frac{\frac{125}{6967296}t + \frac{625}{292626432}t^3}{\frac{125}{6967296} + \frac{125}{13934592}t + \frac{625}{65028098}t^2 + \frac{625}{585252864}t^3 + \frac{625}{2341011456}t^4} \tag{7.57}$$

An analytical solution for velocity and acceleration of the nonspherical particle during the unsteady motion by DTM—Pade approximant is obtained. The results are compared with DTM and numerical method (NM). Fig. 7.12 depicts the velocity versus time for the three methods. It is observed that the DTM—Pade approximate solution is more accurate than DTM. DTM—Pade [4/4] gives closer results to numerical solution. This fact is more pronounced for larger values of time, i.e., $t = 2$. Moreover, this interesting agreement between DTM—Pade approximation and numerical solution is shown in Table 7.4.

As noted previously, DTM—Pade [4/4] gives closer results to numerical solution; therefore, it will produce more acceptable results regarding

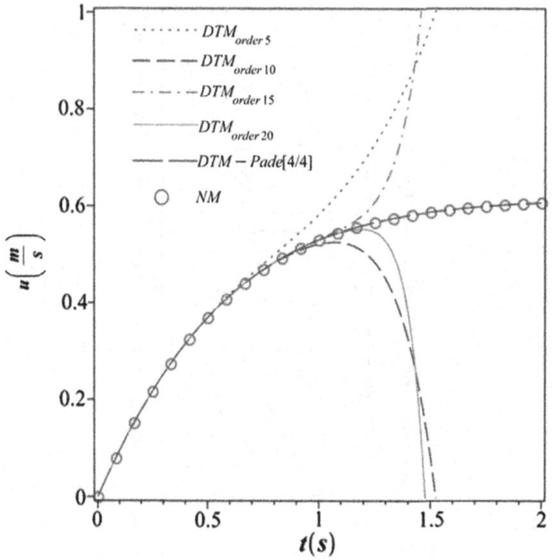

Figure 7.12 Velocity profiles of the particle ($a = b = c = d = 1$). *DTM*, differential transformation method.

Table 7.4 The Results of the Differential Transformation Method (DTM) and DTM–Pade and NM at $a = b = c = d = 1$

t	$u(t)_{\text{DTM–Padé[4/4]}}$	$u(t)_{\text{DTM order 20}}$	$u(t)_{\text{NM}}$	$\text{Error}_{\text{DTM–Padé[4/4]}}$	$\text{Error}_{\text{DTM order 20}}$
0	0	0	0	0	0
0.4	0.316007088	0.316007092	0.316007131	0.0000000434	0.0000000394
0.8	0.483836437	0.483835084	0.483837560	0.0000011223	0.0000024755
1.2	0.561131387	0.552128895	0.561150879	0.0000194924	0.0090219842
1.6	0.59429563	−2.319449392	0.594422825	0.0001271951	2.9138722166
2	0.607843137	−246.4477393	0.608320171	0.0004770338	247.0560594709

DTM–Pade, differential transformation method with Padé approximation; *NM*, numerical method.

accelerations of the particle during the falling process. This is confirmed by the curves in Fig. 7.13.

Figs. 7.14 and 7.15 show the position of the particle in the sedimentation process for each time step equal to 0.02 s. The effect of particle sphericity on settling position is shown in Fig. 7.14. From this figure, it can be concluded that the particle with larger sphericity moves faster than smaller ones. Finally, Fig. 7.15 explains the material of particle's effect on settling position.

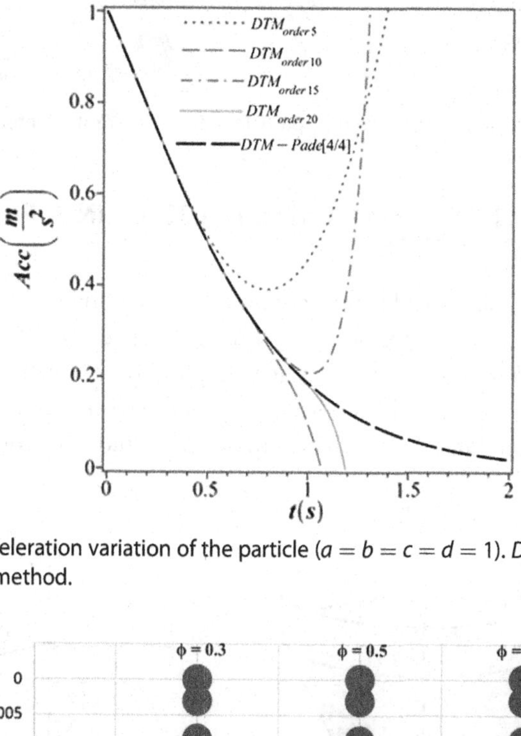

Figure 7.13 Acceleration variation of the particle ($a = b = c = d = 1$). *DTM*, differential transformation method.

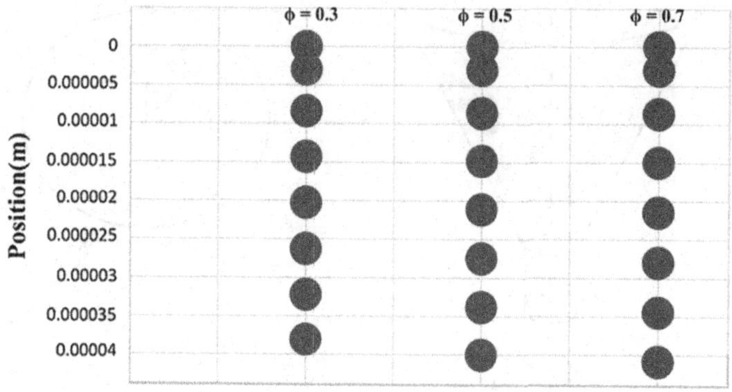

Figure 7.14 Positions of falling particles for different sphericity of copper particles with time interval = 0.02 s.

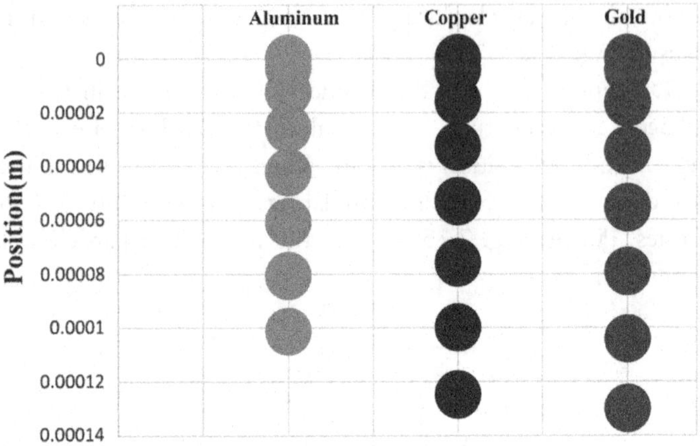

Figure 7.15 Positions of falling particles for different materials with time interval = 0.02 s.

7.5 MOTION OF A SPHERICAL PARTICLE IN A FLUID FORCED VORTEX

Consider the polar coordinates r-θ with pole at the origin $r = 0$. Considering the 2D model for this problem, when rotating speed is approximately constant, is an acceptable assumption and simplification which is considered in the literature [5]. So, in this study, motion of the particle in a plane is studied according to Fig. 7.16. The mass of fluid is rotating in the

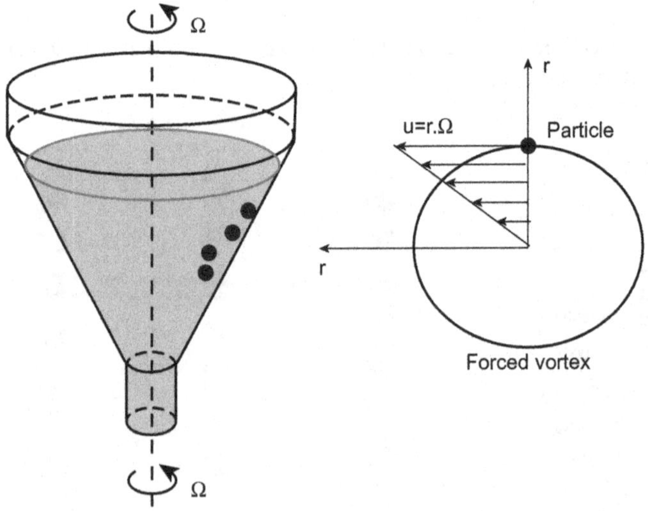

Figure 7.16 Schematic view of spherical particles in a sample fluid forced vortex (industrial hopper).

counterclockwise direction around the origin (see Fig. 7.16). The streamlines are concentric circles with common center at the origin. For forced vortex, the tangential velocity is given by $u_\theta = \Omega \cdot r$ whereas for free vortex $u_\theta = c/r$, Ω and c are constants. In both cases the radial components of velocity of fluid are zero. At the instant $t = 0$, a small spherical particle is released in the flow field with initial radius r_0; zero angular velocity; and u_0 as radial component of velocity. The fluid will drive the particle to rotation, exerting drag force [5],

$$F_D = c_D A \frac{1}{2} \rho u^2 \tag{7.58}$$

where u is the relative velocity between the particle and the flow; ρ is the mass density of fluid; A is the projected area of sphere; and c_D is the drag coefficient and is a function of the Reynolds number. The equations of motion of the particle can be written as $m \cdot a = F$; m is the mass of the particle; a is the acceleration; and F is the exerted external force on particle. The equations in the radial and tangential directions are given respectively [5],

$$m\left(\ddot{r} - r\dot{\theta}^2\right) = -c_D A \frac{1}{2} \rho \dot{r}^2 \tag{7.59}$$

$$m\left(r\ddot{\theta} + 2\dot{r}\dot{\theta}\right) = c_D A \frac{1}{2} \rho \left(r\dot{\theta} - u_{\theta fluid}\right)^2 \tag{7.60}$$

where $u_{\theta fluid} = \Omega \cdot r$ for the forced vortex and $u_{\theta fluid} = c/r$ for free vortex. The quantities to the left side are the radial and tangential components of acceleration in polar coordinates. Now, by using $\Omega = \frac{\omega}{\omega_0}$, $U = \frac{u}{u_0}$, $R = \frac{r}{r_0}$ nondimensionalized equations will be:

$$\begin{cases} \dot{R}(\tau) - \dfrac{u_0}{r_0\omega_0} U(\tau) = 0 \\ \dfrac{u_0}{r_0\omega_0} \dot{U}(\tau) - R(\tau)\Omega(\tau)^2 + \alpha^2 \left(\dfrac{u_0}{r_0\omega_0}\right)^2 U(\tau)^2 = 0 \\ R(\tau)\dot{\Omega}(\tau) + \dfrac{u_0}{r_0\omega_0} 2U(\tau)\Omega(\tau) - \alpha^2 R(\tau)^2 \begin{cases} (\Omega(\tau) - 1)^2 & \text{Forced Vortex} \\ \left(\Omega(\tau) - \dfrac{1}{R(\tau)^2}\right)^2 & \text{Free Vortex} \end{cases} = 0 \end{cases} \tag{7.61}$$

The nondimensional parameter $\alpha^2 = \dfrac{c_D A \frac{1}{2} \rho r_0}{m}$ represents the drag to inertia ratio. Due to importance of particle motion analysis which was introduced in Section 7.1, this system of equation should be solved, and according to its rational function nonlinearity form, analytical and numerical methods should be applied. By obtaining the particle position and velocities, its motion and treatment in vortices are completely understandable. For solving the particle motion in a forced vortex by an efficient, fast, and high accurate method, system of Eq. (7.61) is solved by DTM–Padé,

$$\begin{cases} (k+1)\overline{R}(k+1) - \dfrac{u_0}{r_0\omega_0}\overline{U}(k) = 0 \\ \dfrac{u_0}{r_0\omega_0}(k+1)\overline{U}(k+1) - \sum\limits_{k1}^{k}\sum\limits_{l=0}^{k1}\overline{R}(l)\overline{\Omega}(k1-l)\overline{\Omega}(k-k1) \\ +\alpha^2\left(\dfrac{u_0}{r_0\omega_0}\right)^2\sum\limits_{l=0}^{k}\overline{U}(l)\overline{U}(k-l) = 0 \\ \sum\limits_{l=0}^{k}\overline{R}(l)(k+1-l)\overline{\Omega}(k-l+1) + \dfrac{2u_0}{r_0\omega_0}\sum\limits_{l=0}^{k}\overline{U}(l)\overline{\Omega}(k-l) \\ -\alpha^2\left(\begin{array}{l} \sum\limits_{k2=0}^{k}\sum\limits_{k1=0}^{k2}\sum\limits_{l=0}^{k1}\overline{R}(l)\overline{R}(k1-l)\overline{\Omega}(k2-k1)\overline{\Omega}(k-k2) - \\ 2\sum\limits_{k1=0}^{k}\sum\limits_{l=0}^{k1}\overline{R}(l)\overline{R}(k1-l)\overline{\Omega}(k-k1) + \sum\limits_{l=0}^{k}\overline{R}(l)\overline{R}(k-l) \end{array}\right) = 0 \end{cases} \tag{7.62}$$

where $\overline{U}(k)$, $\overline{R}(k)$, $\overline{\Omega}(k)$ are DTM transformed forms of $U(k)$, $R(k)$, $\Omega(k)$ in Eq. (7.61). With initial condition,

$$\overline{U}(0) = 1,\ \overline{R}(0) = 1,\ \overline{\Omega}(0) = 0 \tag{7.63}$$

Since the procedure of solving Eq. (7.62) is autonomous of constants u_0, r_0, ω_0, and α^2, for generalization and simplification of the problem for future cases with different physical conditions the constants that represent physical properties are assumed to be as following:

$$u_0 = r_0 = \omega_0 = \alpha^2 = 1 \tag{7.64}$$

After solving Eq. (7.62) by using initial condition Eq. (7.63), other terms of particle position $R(\tau)$, radial velocity $U(\tau)$, and angular velocity $\Omega(\tau)$ will be appeared as follows,

$$\begin{aligned}
&\overline{R}(1) = \frac{u_0}{r_0\omega_0}, \overline{U}(1) = -\frac{\alpha^2 u_0}{r_0\omega}, \overline{\Omega}(1) = \alpha^2 \\
&\overline{R}(2) = -\frac{1}{2}\frac{u_0^2\alpha^2}{r_0^2\omega_0^2}, \overline{U}(2) = \frac{u_0^2\alpha^4}{r_0^2\omega_0^2}, \overline{\Omega}(2) = -\frac{1}{2}\frac{\alpha^2(u_0 + 2\alpha^2 r_0\omega_0)}{r_0\omega_0} \\
&\overline{R}(3) = -\frac{1}{3}\frac{u_0^3\alpha^4}{r_0^3\omega_0^3}, \overline{U}(3) = \frac{1}{3}\frac{\alpha^4\left(r_0^4\omega_0^4 - 3\alpha^2 u_0^4\right)}{r_0^3\omega_0^3 u_0}, \overline{\Omega}(3) \\
&= \frac{1}{6}\frac{\alpha^2\left(6\alpha^4 r_0^2\omega_0^2 + 6u_0^2 + 2\alpha^2 u_0 r_0\omega_0 + 3u_0^2\alpha^2\right)}{r_0^2\omega_0^2} \\
&\ldots
\end{aligned} \tag{7.65}$$

For example, when using constant values in Eq. (7.64), final functions will be calculated as:

$$\begin{cases}
R(t) = 1 + t - \frac{1}{2}t^2 + \frac{1}{3}t^3 - \frac{1}{6}t^4 + \frac{1}{15}t^5 + \frac{17}{360}t^6 - \frac{499}{2520}t^7 + \frac{8147}{20160}t^8 - \frac{21079}{30240}t^9 \\
+\frac{509813}{453600}t^{10} - \frac{17505161}{9979200}t^{11} + \frac{80536271}{29937600}t^{12} - \frac{6355323463}{1556755200}t^{13} + \frac{10313621603}{1676505600}t^{14} \\
U(t) = 1 - t + t^2 - \frac{2}{3}t^3 + \frac{1}{3}t^4 + \frac{17}{60}t^5 - \frac{499}{360}t^6 + \frac{8147}{2520}t^7 - \frac{21079}{3360}t^8 + \frac{509813}{45360}t^9 \\
-\frac{17505161}{907200}t^{10} + \frac{80536271}{2494800}t^{11} - \frac{6355323463}{119750400}t^{12} + \frac{10313621603}{119750400}t^{13} - \frac{5520919291}{39916800}t^{14} \\
\Omega(t) = t - \frac{3}{2}t^2 + \frac{17}{6}t^3 - \frac{11}{2}t^4 + \frac{41}{4}t^5 - \frac{6641}{360}t^6 + \frac{3398}{105}t^7 - \frac{1122019}{20160}t^8 + \frac{2439641}{25920}t^9 \\
-\frac{28445111}{181440}t^{10} + \frac{45874001}{178200}t^{11} - \frac{6241044329}{14968800}t^{12} + \frac{1037085707459}{1556755200}t^{13} - \frac{22883640602789}{21794572800}t^{14}
\end{cases} \tag{7.66}$$

Eq. (7.66) is the DTM solution of Eq. (7.61), results are depicted in Figs. 7.17–7.19 and presented in Tables 7.5 and 7.6. Also the outcomes were compared with numerical fourth order Runge–Kutta and differential quadrature method (DQM). As seen in this figure DTM solution after a short agreement with numerical result, suddenly reaches to infinity (for all functions, $R(t)$, $U(t)$, and $\Omega(t)$). To overcome this shortcoming and increasing the convergence of this method, Padé approximation is applied.

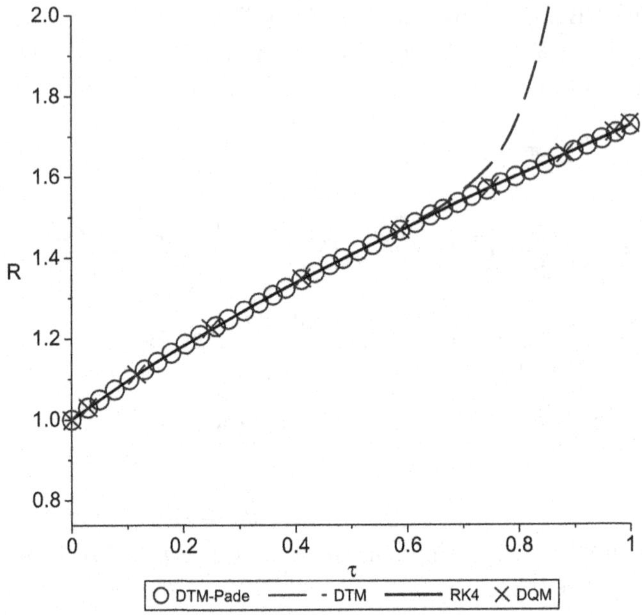

Figure 7.17 Position trajectory for the particle (*R*) obtained by different methods. *DQM*, differential quadrature method; *DTM*, differential transformation method; *DTM—Pade*, differential transformation method with Padé approximation.

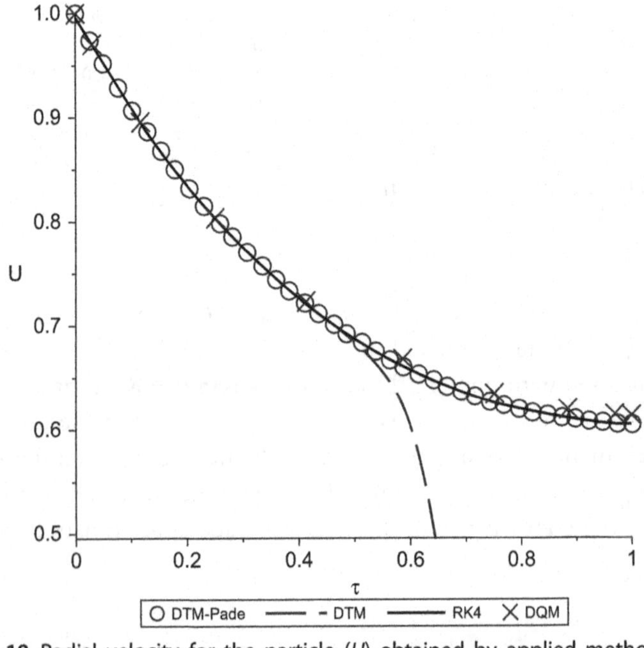

Figure 7.18 Radial velocity for the particle (*U*) obtained by applied methods. *DQM*, differential quadrature method; *DTM*, differential transformation method; *DTM—Pade*, differential transformation method with Padé approximation.

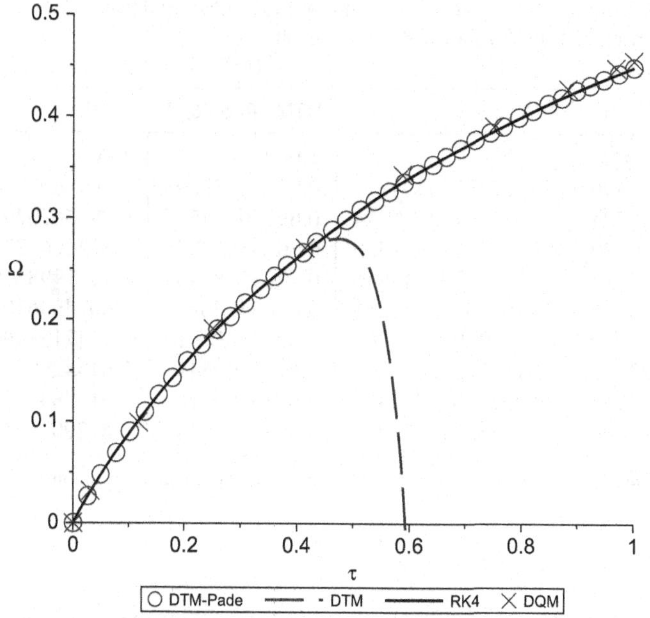

Figure 7.19 Angular velocity for the particle (Ω) obtained by applied methods. *DQM*, differential quadrature method; *DTM*, differential transformation method; *DTM–Pade*, differential transformation method with Padé approximation.

Table 7.5 Comparison Between Applied Methods in Nonuniform and Nondimensional Time for Angular Velocity Results

τ	$\Omega(\tau)$		
	RK4	**DTM–Padé[6,6]**	**DQM**
0.0000000000	0.000	0.000	0.000
0.0301536896	0.0288632031758779	0.02886319921	0.0315590037870364
0.1169777784	0.100142863870931	0.1001428582	0.0995637064474157
0.2500000000	0.185921658986676	0.1859216559	0.191245742076593
0.4131759114	0.267081663618700	0.2670816367	0.270099418999533
0.5868240886	0.334822176009758	0.3348218891	0.342447069510179
0.7500000000	0.386168789277334	0.3861657749	0.390261173414334
0.8830222216	0.421185866197261	0.4211728745	0.426568351587405
0.9698463104	0.441257247764227	0.4412281585	0.446310565708062
1.0000000000	0.447764342038054	0.4477267112	0.453066160230719

DTM, differential transformation method; *DTM–Pade*, differential transformation method with Padé approximation.

Table 7.6 Comparison Between Applied Methods in Nonuniform and NonDimensional Time for Radial Velocity Results

	$U(\tau)$		
τ	**RK4**	**DTM–Padé[6,6]**	**DQM**
0.0000000000	1.000	1.000	1.000
0.0301536896	0.970737559394717	0.9707375589	0.969860793657630
0.1169777784	0.895704744070214	0.8957047452	0.895367579781897
0.2500000000	0.803455212921360	0.8034552207	0.803206925729227
0.4131759114	0.720357770547646	0.7203578045	0.723199889802020
0.5868240886	0.660545915322201	0.6605462648	0.667454630759341
0.7500000000	0.625835951196769	0.6258387861	0.634712849664366
0.8830222216	0.610250219603200	0.6102609542	0.619428876555633
0.9698463104	0.605365864743497	0.6053880716	0.613984475505496
1.0000000000	0.604553915451281	0.6045818949	0.612909563556282

DTM, differential transformation method; *DTM–Pade*, differential transformation method with Padé approximation.

Using Padé with accuracy of [6,6] results in Eq. (7.66) and will be transformed as the following form,

$$\begin{cases} R(t) = \dfrac{(1+5.5828532t+10.728458t^2+7.9406046t^3+1.7311221t^4+0.48787532t^5+0.27840808t^6)}{(1+4.5828532t+6.6456053t^2+3.2530926t^3+0.43988108t^4+0.15648097t^5+0.01235856t^6)} \\ U(t) = \dfrac{(1+3.6270954t+5.2449498t^2+3.6857199t^3+1.2403949t^4+1.4403629t^5+0.58851084t^6)}{(1+4.6270954t+8.8720452t^2+8.5973364t^3+3.7170831t^4+0.64910789t^5+0.3698456t^6)} \\ \Omega(t) = \dfrac{(0.091822563t^6+0.44660102t^5+1.2839632t^4+2.4523533t^3+2.4126985t^2+t)}{(1+3.9126985t+5.4880679t^2+3.93008583t^3+2.06204617t^4+0.576084t^5+0.2009531t^6)} \end{cases} \tag{7.67}$$

In this chapter, for the first time, analytical approaches called DTM–Padé and DQM have been successfully applied to find the most accurate analytical solution for the motion of a particle in a forced vortex. Radial position, angular velocity, and radial particle velocity were calculated and depicted. As a main outcome from the present study, it is observed that the results of DQM are in excellent agreement with numerical ones. Also this method is a simple, fast, powerful, and efficient technique for finding problems solution in science and engineering with coupled nonlinear differential equations. Also, it reduces the size of calculations. Results show that by passing time, particle recedes from the vortex center, and its radial velocity decreases while its angular velocity increases.

7.6 COMBUSTION OF MICROPARTICLES

Consider a spherical particle which due to high reaction with oxygen will be combusted. Since the thermal diffusivity of substance is large and the Biot number is small ($Bi_H << 0.1$), it is assumed that the particle is isothermal. In this state, a lumped system analysis is applicable. When this criterion is satisfied, the variation of temperature with location within the particle will be slight and can be approximated as being uniform, so particle has a spatially uniform temperature; therefore the temperature of particle is a function of time only, $T = T(t)$ and is not a function of radial coordinate, $T \neq T(r)$. The assumptions used in this modeling include the following [6].

1. The spherical particle burns in a quiescent, infinite ambient medium, and there are no interactions with other particles, also the effects of forced convection are ignored.
2. Thermophysical properties for the particle and ambient gaseous oxidizer are assumed to be constant.
3. The particle radiates as a gray body to the surroundings without contribution of the intervening medium.

By these assumptions and considering the particle as a thermodynamic system, and by using the principle of conservation of energy (first law of thermodynamics), the energy balance equation for this particle can be written as (Fig. 7.20);

$$\dot{E}_{in} - \dot{E}_{out} + \dot{E}_{gen} = \left(\frac{dE}{dt}\right)_p \tag{7.68}$$

where $\dot{E}_{in}$ is the rate of energy entering the system which is owing to absorption of total radiation incident on the particle surface from the

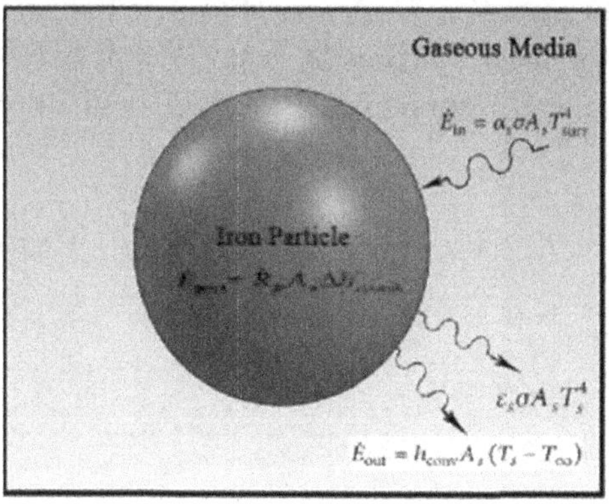

Figure 7.20 Iron microparticle and its energy balance.

surrounding, $\dot{E}_{out}$ is the rate of energy leaving the system by mechanisms of convection on the particle surface and thermal radiation that emits from the outer surface of particle, $\dot{E}_{gen}$ is the rate of generation of energy inside the particle due to the combustion process and equals to the heat released from the chemical reaction, and $(dE/dt)p$ is the rate of change in total energy of particle. These energy terms can be calculated by:

$$\dot{E}_{in} = \alpha_s \sigma A_s T_{surr}^4 \tag{7.69}$$

$$\dot{E}_{out} = h_{conv} A_s (T_s - T_\infty) + \varepsilon_s \sigma A_s T_s^4 \tag{7.70}$$

$$\dot{E}_{gen} = \dot{Q}_{comb} = \dot{R}_p A_s \Delta h_{comb}^{\circ} \tag{7.71}$$

$$\left(\frac{dE}{dt}\right)_p = \rho_p V_p c_p \frac{dT_s}{dt} \tag{7.72}$$

By substituting of Eqs. (7.69)−(7.72) in Eq. (7.68),

$$\alpha_s \sigma A_s T_{surr}^4 - \left(h_{conv} A_s (T_s - T_\infty) + \varepsilon_s \sigma A_s T_s^4\right) + \dot{R}_p A_s \Delta h_{comb}^{\circ} = \rho_p V_p c_p \frac{dT_s}{dt} \tag{7.73}$$

Three reasonable assumptions are used for improving Eq. (7.73):

1. Both absorptivity and emissivity of the surface depend on the temperature and the wavelength of radiation. Kirchhoff's law of radiation states that the absorptivity and the emissivity of a surface at a given temperature and wavelength are equal $(\varepsilon_s \simeq \alpha_s)$.
2. The initial temperature of the particle at the beginning of combustion can be regarded as the initial condition. This temperature is known as ignition temperature $(T(0) = T_{ig})$.
3. The density of particle is a function of particle temperature, so it can be considered as a linear function $(\rho_p = \rho_p(T) = \rho_{p,\infty}[1 + \beta(T - T_\infty)])$.

By applying these assumptions, Eq. (7.73) will be converted to the following:

$$\rho_{p,\infty}[1 + \beta(T - T_\infty)] V_p c_p \frac{dT_s}{dt} + h_{conv} A_s (T_s - T_\infty) + \varepsilon_s \sigma A_s \left(T_s^4 - T_{surr}^4\right) - \dot{R}_p A_s \Delta h_{comb}^{\circ} = 0 \tag{7.74}$$

For solving this nonlinear differential equation, it's more suitable that all the terms be converted to the dimensionless form. The following set of dimensionless variables is defined:

$$\begin{cases} \theta = \dfrac{T}{T_{ig}}, \quad \theta_{\infty} = \dfrac{T_{\infty}}{T_{ig}}, \quad \theta_{surr} = \dfrac{T_{surr}}{T_{ig}}, \quad \varepsilon_1 = \beta T_{ig} \\ \tau = \dfrac{t}{\left(\dfrac{\rho_{p,\infty} V_p c_p}{h_{conv} A_s}\right)}, \psi = \dfrac{\dot{Q}_{comb}}{h_{conv} A_s T_{ig}}, \varepsilon_2 = \dfrac{\varepsilon_s \sigma T_{ig}^3}{h_{conv}} \end{cases} \tag{7.75}$$

Consequently, the nonlinear differential equation and its initial condition can be expressed in the dimensionless form

$$\varepsilon_1 \theta \frac{d\theta}{d\tau} + (1 - \varepsilon_1 \theta_{\infty}) \frac{d\theta}{d\tau} + \varepsilon_2 \left(\theta^4 - \theta_{surr}^4\right) + \theta - \psi - \theta_{\infty} = 0 \tag{7.76}$$

$$\theta(0) = 1 \tag{7.77}$$

By applying DTM from Chapter 1 principle, transformed form of Eq. (7.76) will be,

$$\begin{aligned} &(k+1)\Theta(k+1) + \varepsilon_1 \sum_{l=0}^{k} \Theta(l) \cdot (k+1-l) \cdot \Theta(k+1-l) \\ &\quad -\varepsilon_1 \theta_{\infty} (k+1) \cdot \Theta(k+1) + \Theta(k) + \\ &\varepsilon_2 \sum_{k2=0}^{k} \sum_{k1=0}^{k2} \sum_{l=0}^{k1} \Theta(l) \cdot \Theta(k2-k1) \cdot \Theta(k-k2) \cdot \Theta(k1-l) \\ &\quad -\delta(k) \cdot \left(\theta_{\infty} + \varepsilon_2 \theta_{surr}^4 + \psi\right) = 0 \end{aligned} \tag{7.78}$$

where Θ is transformed form of θ and

$$\delta(k) = \begin{cases} 1 & k = 0 \\ 0 & k \neq 1 \end{cases} \tag{7.79}$$

Transformed form of initial condition (Eq. 7.77) will be,

$$\Theta(0) = 1 \tag{7.80}$$

For example, for an iron particle with 20 μm diameter (see Table 2 in Ref. [6]) solving Eq. (7.78) makes,

$$\begin{aligned} &\Theta(0) = 1, \Theta(1) = 1.170326597, \Theta(2) = -0.6324112173 \\ &\Theta(3) = 0.2462459086, \Theta(4) = -0.08656001439, \ldots \end{aligned} \tag{7.81}$$

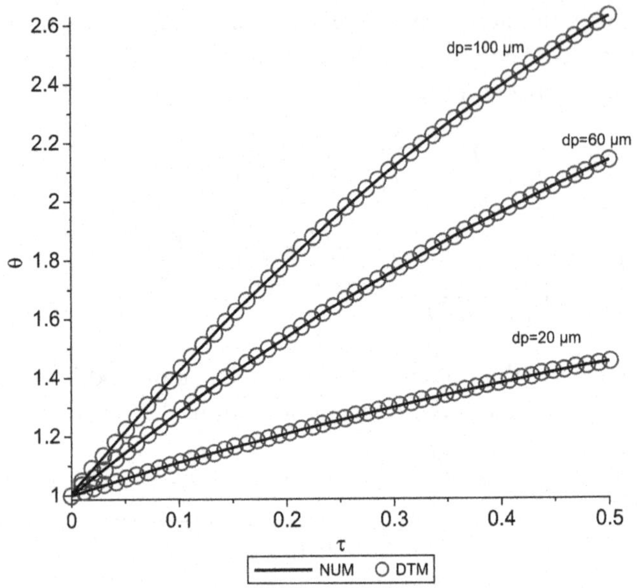

Figure 7.21 Comparison between DTM and numerical method in different particle diameters. *DTM*, differential transformation method.

By substituting DTM transformed terms of T (Eq. 7.81) into transformed DTM equation, $\theta(\tau)$ can be determined as (Figs. 7.21 and 7.22),

$$\begin{aligned}\theta(\tau) = {} & 1 + 1.17033\tau - 0.632411\tau^2 + 0.246245\tau^3 - 0.08656\tau^4 \\ & + 0.0324505\tau^5 - 0.0130388\tau^6 + 0.00504942\tau^7 \\ & - 0.0017\tau^8 + 0.00042358\tau^9 - 0.114879e - 4\tau^{10}\end{aligned} \tag{7.82}$$

7.7 UNSTEADY SEDIMENTATION OF SPHERICAL PARTICLES

For modeling the particle sediment phenomenon, consider a small, rigid particle with a spherical shape of diameter D and mass of m and density of ρ_s falling in infinite extent filled water as an incompressible Newtonian fluid. Density of water, ρ, and its viscosity, μ, are known. We considered the gravity, buoyancy, Drag forces, and added mass (virtual mass) effect on particle. According to the BBO equation for the unsteady motion of the particle in a fluid, for a dense particle falling in light fluids and by assuming

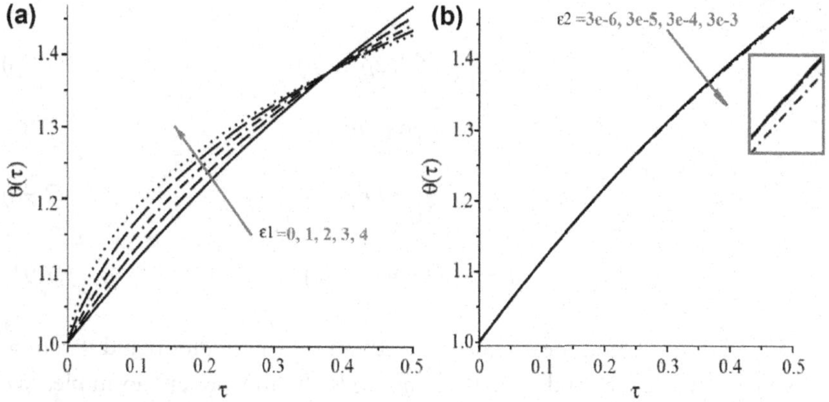

Figure 7.22 (a) Effect of $\varepsilon 1$ on nondimensional temperature profile for micro- and nanoparticles. (b) Effect of $\varepsilon 2$ on nondimensional temperature profile for micro- and nanoparticles.

$\rho << \rho_s$, Basset History force is negligible. So by rewriting force balance for the particle, the equation of motion is gained as follows [7],

$$m\frac{du}{dt} = mg\left(1 - \frac{\rho}{\rho_s}\right) - \frac{1}{8}\pi D^2 \rho C_D u^2 - \frac{1}{12}\pi D^3 \rho \frac{du}{dt} \tag{7.83}$$

where C_D is the drag coefficient. In the right-hand side of Eq. (7.83), the first term represents the buoyancy affect, the second term corresponds to drag resistance, and the last term is due to the added mass effect which is due to acceleration of fluid around the particle. The main difficulty of solving Eq. (7.83) is the nonlinear terms due to the nonlinearity nature of the drag coefficient C_D. By considering C_D in range of Reynolds number, $0 \leq \mathrm{Re} \leq 10^5$ as following equation

$$C_D = \frac{24}{\mathrm{Re}}\left(1 + \frac{1}{48}\mathrm{Re}\right) \tag{7.84}$$

Substituting Eq. (7.84) into Eq. (7.83), and mass of the spherical particle is

$$m = \frac{1}{6}\pi D^3 \rho_s \tag{7.85}$$

Eq. (7.83) can be rewritten as

$$a\frac{du}{dt} + bu + cu^2 - d = 0, \quad u(0) = 0 \tag{7.86}$$

where

$$a = \frac{1}{12}\pi D^3(2\rho_s + \rho) \tag{7.88}$$

$$b = 3\pi D\,\mu \tag{7.89}$$

$$c = \frac{1}{16}\pi D^2\rho \tag{7.90}$$

$$d = \frac{1}{6}\pi D^3 g(\rho_s - \rho) \tag{7.91}$$

Eq. (7.86) is a nonlinear equation with an initial condition and it can be solved by numerical and analytical methods. In the present example, we choose three different materials for solid particle, aluminum, copper, and lead and considered three different diameters (1, 3, and 5 mm) for them. A schematic of described problem is shown in Fig. 7.23. Physical properties of the selected material are shown in Table 7.7, and the resulted coefficients *a*, *b*, *c*, and *d* from Eqs. (7.88)–(7.91) are listed in Table 7.8, and Eq. (7.86) as a nonlinear equation, is solved by numerical method, DTM–Padé approximation compared to collocation method and Galerkin method. It is necessary to inform that professional version of this problem with more complexity can be solved by CFD methods that are available in the literature.

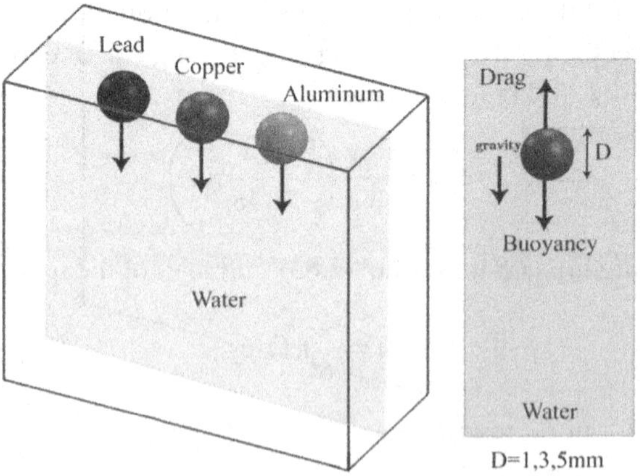

Figure 7.23 Schematic view of particles settling in water. It is assumed that particles are falling freely such that no particle–particle or particle–wall interaction exists.

Table 7.7 Properties of the Selected Materials

Material	Density (kg/m^3)	Viscosity (kg/m s)
Aluminum	2702	—
Copper	8940	—
Lead	11,340	—
Water	996.51	0.001

Applying the DTM transformation from Chapter 1 into Eq. (7.86) to find $u(t)$, we have:

$$a(k+1)U(k+1) + bU(k) + c\sum_{l=0}^{k} U(l)U(k-l) - d\delta(k) = 0, \quad U(0) = 0 \tag{7.92}$$

Rearranging Eq. (7.92), we have

$$U(k+1) = \frac{d\delta(k) - c\sum_{l=0}^{k}[U(l)U(k-l)] - bU(k)}{a(k+1)}, \quad U(0) = 0 \tag{7.93}$$

where

$$\delta(k) = \begin{cases} 1 & \text{if } k = 0 \\ 0 & \text{if } k \neq 0 \end{cases} \tag{7.94}$$

By solving Eq. (7.93) and using the initial condition, the DTM terms are obtained as

$$\begin{aligned} U(1) &= \frac{d}{a} \\ U(2) &= -\frac{1}{2}\frac{bd}{a^2} \\ U(3) &= \frac{1}{6}\frac{d(b^2 - 2cd)}{a^3} \\ U(4) &= -\frac{1}{24}\frac{bd(b^2 - 8cd)}{a^4} \\ &\vdots \end{aligned} \tag{7.95}$$

Now representing Eq. (7.95) in series form, $u(t)$ function will be obtained. After five iterations in DTM series for $a = b = c = d = 1$, the $u(t)$ function is obtained as

$$u(t) = t - \frac{1}{2}t^2 - \frac{1}{6}t^3 + \frac{7}{24}t^4 - \frac{1}{24}t^5 \tag{7.96}$$

Table 7.8 Coefficients in Eq. (7.86)

Liquid	Solid (Particle)	Diameter (mm)	*a*	*b*	*c*	*d*
Water	Aluminum	1	0.0000016756496	0.00000942477796	0.000195664281	0.000008760256
		3	0.0000452425392	0.00002827433389	0.001760978529	0.000236526917
		5	0.0002094562001	0.00004712388981	0.004891607024	0.001095032024
		1	0.0000049418587	0.00000942477796	0.000195664281	0.000040801768
	Copper	3	0.0001334301866	0.00002827433389	0.001760978529	0.001101647738
		5	0.0006177323456	0.00004712388981	0.004891607024	0.005100221010
		1	0.0000061984958	0.00000942477796	0.000195664281	0.000053129377
	Lead	3	0.0001673593872	0.00002827433389	0.001760978529	0.001434493196
		5	0.0007748119783	0.00004712388981	0.004891607024	0.006641172207

Subsequently, applying Padé approximation to Eq. (7.96) (for Padé [4,4] accuracy), we have,

$$Padé\,[4,4](u(t)) = \frac{\frac{3441}{13480}t^4 - \frac{224}{1011}t^3 - \frac{227}{674}t^2 + t}{1 + \frac{55}{337}t + \frac{9}{337}t^2 + \frac{7}{1685}t^3 + \frac{1}{1685}t^4} \qquad (7.97)$$

At first, a comparison between the three analytical methods and Runge–Kutta methods is provided to select the best and reliable analytical method for present problem. Eq. (7.86) is considered with all constants are equal to unity (i.e., $a = b = c = d = 1$). Results of the solutions are depicted in Fig. 7.24. From Fig. 7.24, it is revealed that GM and CM have a good agreement with numerical result; but in DTM when time tends to infinity, even in high iterations (20 iterates), it cannot estimate a constant velocity as "terminal velocity" and its value suddenly reaches to zero. For solving this problem, Padé approximation in different orders such as [2,2], [4,4], [8,8] (e.g., Eq. 7.97) is used, and convergence results are depicted in Fig. 7.25. As Fig. 7.25 shows, higher order of Padé approximation leads to obtain results closer to the numerical solution. Table 7.9 shows the velocity values versus time for applied analytical methods and are compared with numerical procedure. The errors (%) of these methods, with respect to the numerical method, were listed in Table 7.10. As seen in Table 7.10, DTM–Padé [8,8] is the best method for this equation. As GM also has a good agreement and

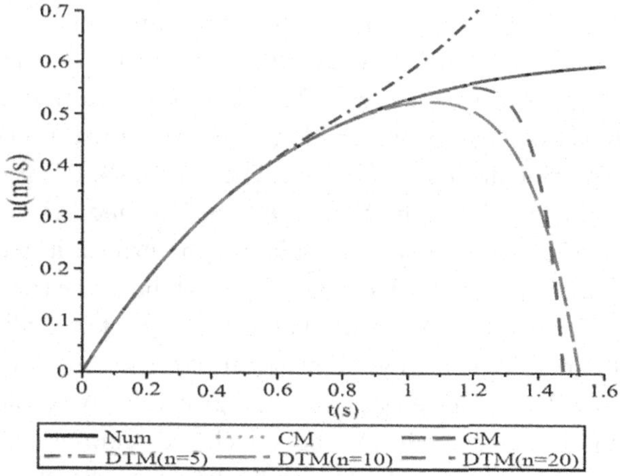

Figure 7.24 Comparison between numerical (NM), collocation (CM), Galerkin methods (GM) and DTM (in three different iterations) for Eq. (7.86) and $a = b = c = d = 1$. *DTM*, differential transformation method.

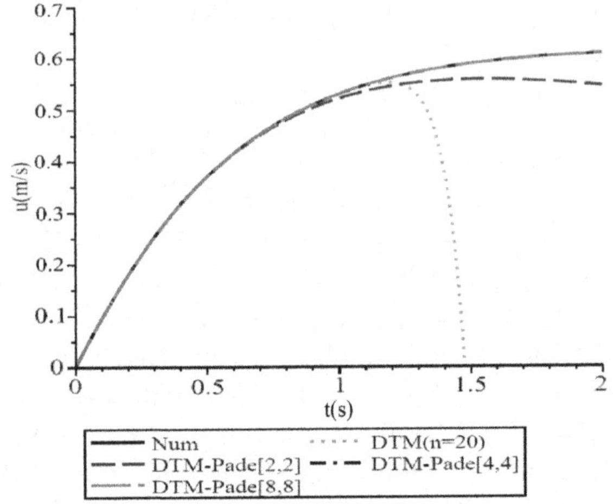

Figure 7.25 Convergence of the DTM−Padé regarding to the method order, [L, M], for Eq. (7.86) and $a = b = c = d = 1$. *DTM−Padé*, differential transformation method with Padé approximation.

acceptable accuracy, it can be considered as an efficient method. In the following section, these two methods are used for analysis of the practical settling motion of some spherical particles in water.

Afterward, to present some practical examples, aluminum, copper, and lead are selected in various sizes submerged in water. Physical properties of the materials and calculated coefficients for Eq. (7.86) in these practical examples are listed in Tables 7.7 and 7.8, respectively. Figs. 7.26 and 7.27 depict the variations of velocity and acceleration for aluminum in different diameters. Velocity variations for copper and lead are shown in Fig. 7.28 and 7.29. The effect of the particle material on velocity and acceleration for these three metals, aluminum, copper, and lead, are investigated in Figs. 7.30 and 7.31 for a constant diameter of 1 mm. These effects for larger diameters, $D = 3$ mm and $D = 5$ mm are shown in Figs. 7.32 and 7.33, respectively. Figs. 7.34 and 7.35 presented the position of the particle in sedimentation process for each time step equal to 0.02 s. Fig. 7.34 shows the effect of particle size, and Fig. 7.35 explains the material of particle's effect on settling position.

Generally speaking, the physical behavior of the settling particles is well captured. For all the selected particles, terminal velocity is calculated and presented in Table 7.11. It can be concluded from Table 7.11 that DTM−Pade [8,8] estimated the terminal velocity excellently, although its errors might increase compared with HPM when t tends to infinity. Outcomes reveal that the value of the terminal velocity increases

Table 7.9 Obtained Values for u(m/s) From Different Methods for Eq. (7.86) and $a = b = c = d = 1$

Time (s)	Numerical Method	Collocation Method	Galerkin Method	DTM ($n = 20$)	DTM–Pade [2,2]	DTM–Pade [4,4]	DTM–Pade [8,8]
0	0	0	0	0	0	0	0
0.2	0.17911335972	0.1792419	0.17916766	0.17911334	0.1791044	0.1791133	0.179113346
0.4	0.31600713125	0.3160935	0.31602345	0.31600709	0.3157894	0.3160070	0.316007092
0.6	0.41502836932	0.4150925	0.41506767	0.41502829	0.4137931	0.4150281	0.415028297
0.8	0.48383755960	0.4838791	0.48385852	0.48383508	0.4799999	0.4838364	0.483837459
1.0	0.53032985719	0.5303603	0.53032540	0.53010503	0.5217391	0.5303244	0.530329756
1.2	0.56115087915	0.5611641	0.56116612	0.55212889	0.5454545	0.5611313	0.561150740
1.4	0.58132582197	0.5813853	0.58132775	0.38037826	0.5562913	0.5812713	0.581325694
1.6	0.59442282455	0.5956222	0.59457135	—	0.5581395	0.5942956	0.594422665
1.8	0.60287916150	0.6103034	0.60512051	—	0.5538461	0.6026196	0.602879031
2	0.60832017092	0.6373060	0.62039362	—	0.5454545	0.6078431	0.608320058

DTM, differential transformation method; *DTM–Pade*, differential transformation method with Padé approximation.

Table 7.10 Calculated Errors (%) for Various Methods Solving Eq. (7.86) and $a = b = c = d = 1$

Time (s)	Collocation	Galerkin	DTM ($n = 20$)	DTM–Pade [2,2]	DTM–Pade [4,4]	DTM–Pade [8,8]
0	0	0	0	0	0	0
0.2	0.000303	0.000303	7.27454E-08	4.95894E-05	7.33037E-08	7.44203E-08
0.4	0.000274	5.17E-05	1.24524E-07	0.000688774	1.36866E-07	1.23258E-07
0.6	0.000155	9.47E-05	1.88481E-07	0.002976341	4.32802E-07	1.73302E-07
0.8	8.6E-05	4.33E-05	5.1164E-06	0.007931504	2.31877E-06	2.06488E-07
1.0	5.75E-05	8.39E-06	0.000423933	0.016198837	1.02894E-05	1.89122E-07
1.2	2.36E-05	2.72E-05	0.016077644	0.027971681	3.4736E-05	2.46742E-07
1.4	0.000102	3.32E-06	0.345671144	0.043064372	9.36325E-05	2.18938E-07
1.6	0.002018	0.00025	4.902019398	0.06103953	0.000213981	2.66898E-07
1.8	0.012315	0.003718	–	0.081331402	0.000430478	2.14978E-07
2	0.047649	0.019847	–	0.10334299	0.000784182	1.85468E-07

DTM, differential transformation method; *DTM–Pade*, differential transformation method with Padé approximation.

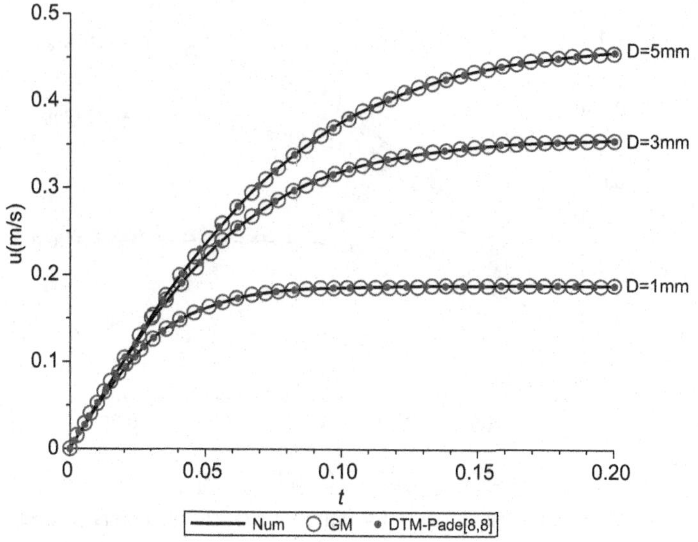

Figure 7.26 Velocity variation for different particle diameters (aluminum).

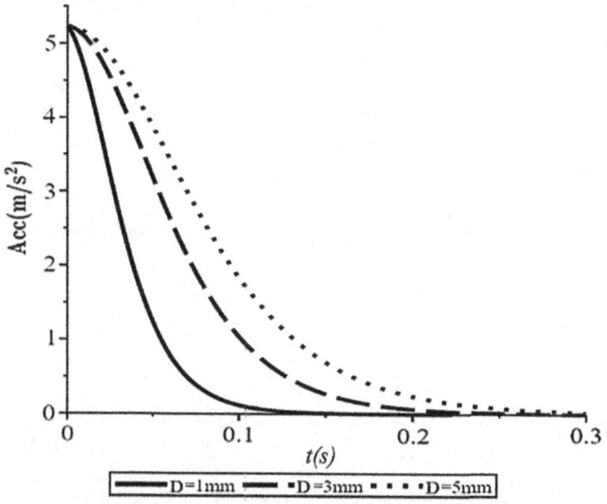

Figure 7.27 Acceleration variation for different particle diameters (aluminum).

significantly with particle diameter and density. Moreover, the smaller particle reaches to its terminal velocity earlier. Thus, the acceleration period is shorter for smaller and lighter particles. From a physical point of view, it can be concluded that larger particles reach zero acceleration (terminal velocity) more slowly. Results also explain that when particle is massive (because of greater density or diameter) it has lower position in the same time steps because of its greater terminal velocity.

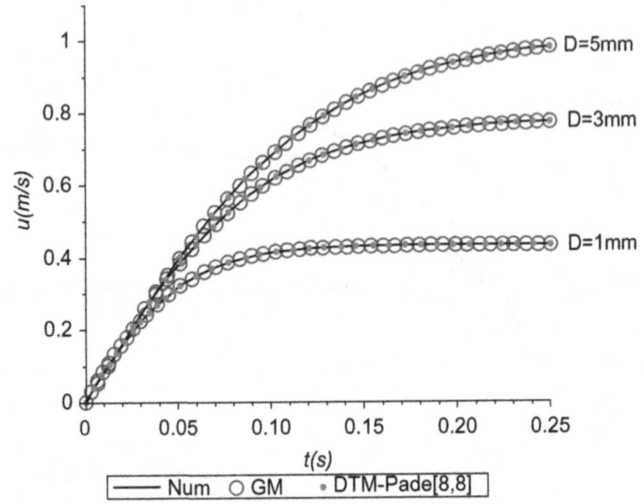

Figure 7.28 Velocity variation for different particle diameters (copper).

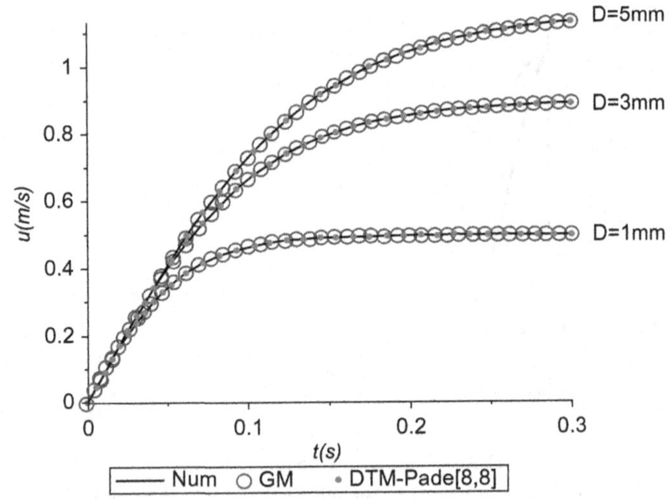

Figure 7.29 Velocity variation for different particle diameters (lead).

7.8 TRANSIENT VERTICALLY MOTION OF A SOLUBLE PARTICLE

For modeling the particle sediment phenomenon, consider a small particle with a spherical shape of variable diameter $D(t)$ and mass of $m(t)$ and density of ρ_s, falling in infinite extent filled by an incompressible Newtonian fluid.

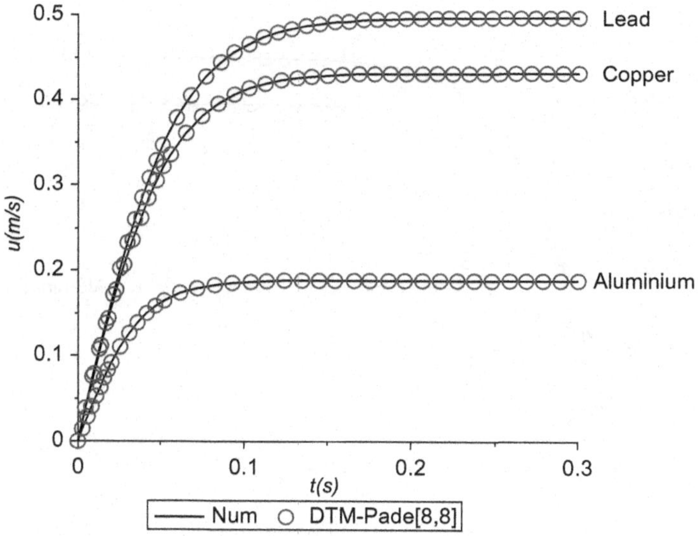

Figure 7.30 Comparison of velocity variation over time for different particle materials when $D = 1$ mm.

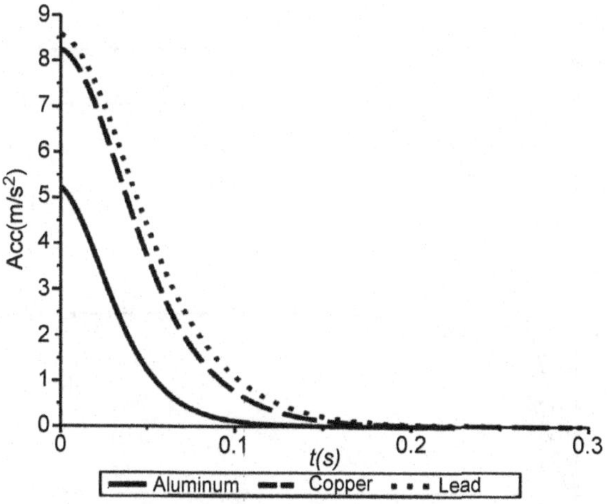

Figure 7.31 Comparison of acceleration variation for different particle materials ($D = 1$ mm). *DTM–Padé,* differential transformation method with Padé approximation.

Density of fluid, ρ, and its viscosity, μ, are known. We considered the gravity, buoyancy, drag forces, and added mass (virtual mass) effect on particle. According to the BBO equation for the unsteady motion of the particle in a fluid, for a dense particle falling in light fluids and by assuming

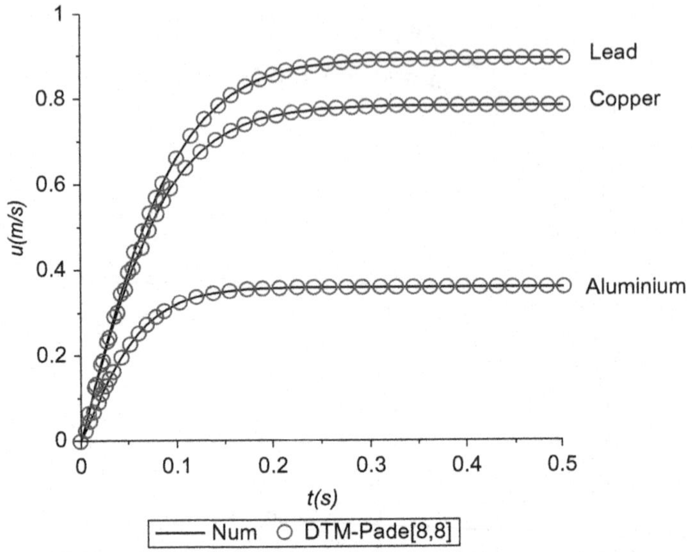

Figure 7.32 Velocity variation for different particle materials when $D = 3$ mm. *DTM–Padé*, differential transformation method with Padé approximation

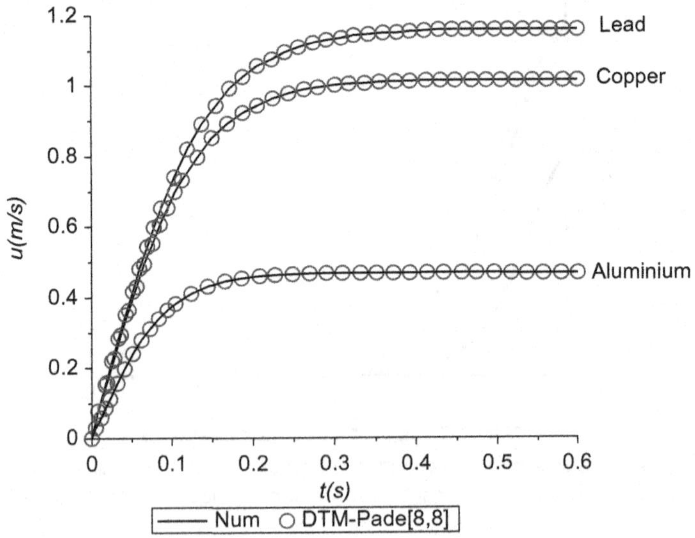

Figure 7.33 Velocity variation for different particle materials when $D = 5$ mm. *DTM–Padé*, differential transformation method with Padé approximation

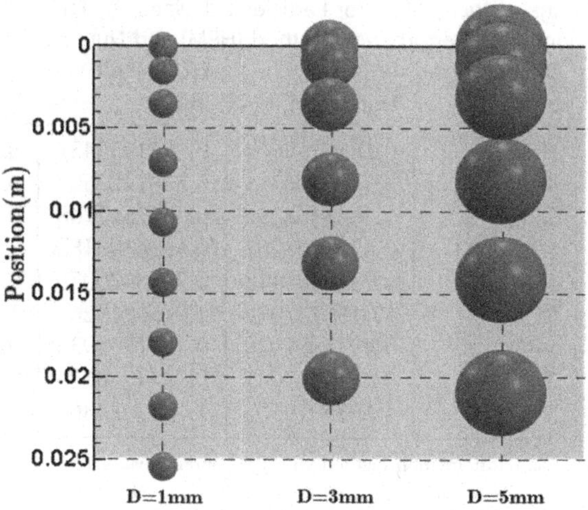

Figure 7.34 Positions of falling particles for different size of aluminum particles, time interval = 0.02 s.

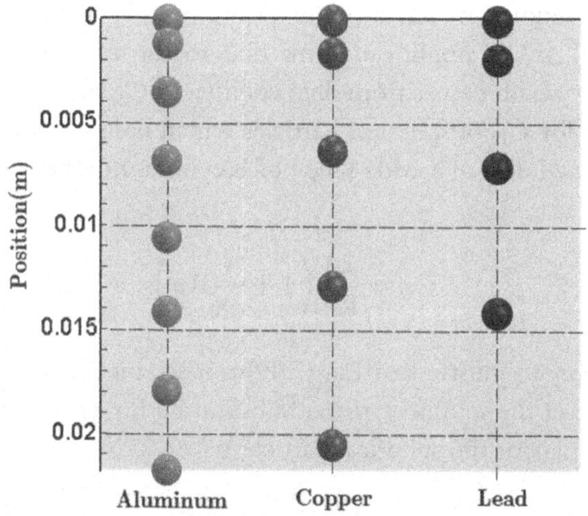

Figure 7.35 Positions of falling particle for different materials, $D = 1$ mm and time interval = 0.02 s.

$\rho << \rho_s$, Basset History force is negligible. So, by rewriting force balance for the particle, the equation of motion is gained as follows [8],

$$m(t)\frac{du(t)}{dt} = m(t)g\left(1 - \frac{\rho}{\rho_s}\right) - \frac{1}{8}\pi D(t)^2 \rho C_D u(t)^2 - \frac{1}{12}\pi D(t)^3 \rho \frac{du(t)}{dt} \tag{7.98}$$

Table 7.11 Terminal Velocity (m/s) for Particles Calculated by DTM–Padé and Compared by Homotopy Perturbation Method (HPM) and Numerical Method

Particle Material	Diameter (mm)	Numerical	DTM–Padé [8,8]	HPM
Aluminum	1	0.1888759074	0.1771953458	0.1888743212
	3	0.3585504311	0.3581483507	0.3585425621
	5	0.4683308257	0.4682662708	0.4683267958
	1	0.4332008767	0.4336398188	0.4332014752
Copper	3	0.7829089418	0.7828929076	0.7828921531
	5	1.0156731422	1.015681926	1.0156721043
	1	0.4975607949	0.4837738933	0.4965829147
Lead	3	0.8944267047	0.8944040156	0.8944159856
	5	1.1589079584	1.158902486	1.1589059863

DTM–Pade, differential transformation method with Padé approximation.

where C_D is the drag coefficient, in the right-hand side of Eq. (7.98), the first term represents the buoyancy effect, the second term corresponds to drag resistance, and the last term is due to the added mass effect which is due to acceleration of fluid around the particle. The main difficulty of solving Eq. (7.98) is nonlinear terms due to the nonlinearity nature of the drag force which comes from drag coefficient C_D, $u(t)^2$ and $D(t)^2$ terms. A correlation for C_D of spherical particles which has good agreement with the experimental data in a wide range of Reynolds number, $0 \leq \mathrm{Re} \leq 10^5$ is

$$C_D = \frac{24}{\mathrm{Re}}\left(1 + \frac{1}{48}\mathrm{Re}\right) \tag{7.99}$$

It's necessary to inform that Eq. (7.99) is based on the nonslip condition on the surface of the soluble particle. Substituting Eq. (7.99) into Eq. (7.98) and variable mass of the spherical particle is

$$m(t) = \frac{1}{6}\pi D(t)^3 \rho_s \tag{7.100}$$

Eq. (7.98) can be rewritten as

$$\begin{aligned}&\frac{1}{12}\pi D(t)^3(2\rho_s + \rho)\frac{du(t)}{dt} + 3\pi D(t)\mu u(t) + \frac{1}{16}\pi D(t)^2 \rho u(t)^2 \\ &- \frac{1}{6}\pi D(t)^3 g(\rho_s - \rho) = 0\end{aligned} \tag{7.101}$$

It is completely evident that in different industrial process and applications, diameter of the particle maybe varied by a known function which depends on its solubility. In this study, it is considered that diameter varies through a linear function, so for other functions it can be solved easily too by the same method.

$$D(t) = D_0 - \dot{D} \cdot t \tag{7.102}$$

where D_0 is the initial diameter, and $\dot{D}$ is the reduction rate of the diameter due to particle solubility. By substituting to Eq. (7.101),

$$\begin{aligned}&\frac{1}{12}\pi\left(D_0 - \dot{D}\cdot t\right)^3(2\rho_s + \rho)\frac{du(t)}{dt} + 3\pi\left(D_0 - \dot{D}\cdot t\right)\mu u(t)\\&+\frac{1}{16}\pi\left(D_0 - \dot{D}\cdot t\right)^2\rho u(t)^2\\&-\frac{1}{6}\pi\left(D_0 - \dot{D}\cdot t\right)^3 g(\rho_s - \rho) = 0\end{aligned} \tag{7.103}$$

Eq. (7.103) is a nonlinear equation with an initial condition ($u(0) = 0$) which can be solved by numerical and analytical methods. Here, DTM–Pade and Runge–Kutta methods are presented for solving the problem. A schematic of described problem is shown in Fig. 7.36.

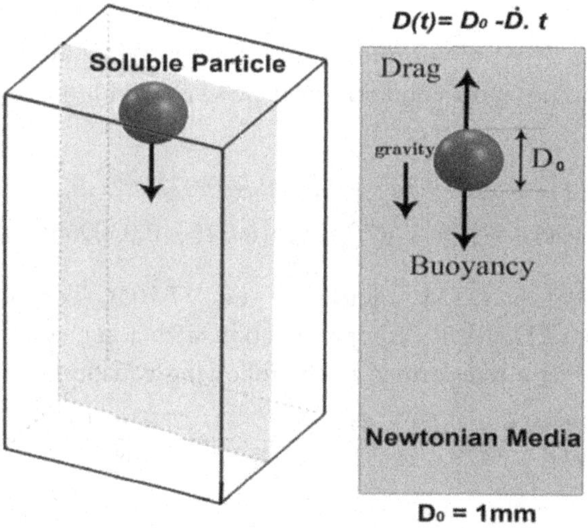

Figure 7.36 Schematic view of soluble particle settling in Newtonian media while its diameter reduces by a linear function due to dissolve in fluid.

Because the procedure of solving Eq. (7.103) is autonomous of constant coefficients $\frac{1}{12}\pi(2\rho_s+\rho)$, $3\pi\mu$, $\frac{1}{16}\pi\rho$, $\frac{1}{6}\pi g(\rho_s-\rho)$, and D_0, for generalization and simplification of the problem for future cases with different physical conditions, all the constants are assumed to be unity. The problem will be solved by this assumption and for $\dot{D}=0.4.0.2,\ 0.1\,\frac{mm}{s}$. For instance when $\dot{D}=0.2\,\frac{mm}{s}$, Eq. (7.103) will be rewritten as,

$$\begin{aligned}&\frac{du(t)}{dt}-0.6\left(\frac{du(t)}{dt}\right)t+0.12\left(\frac{du(t)}{dt}\right)t^2-0.008\left(\frac{du(t)}{dt}\right)t^3+\\&u(t)-0.2u(t)t+u(t)^2-0.4u(t)^2t+0.04u(t)^2t^2-1+0.6t\\&-0.12t^2+0.008t^3=0\end{aligned} \tag{7.104}$$

by applying DTM−Padé to Eq. (7.104),

$$\begin{aligned}&(k+1)U(k+1)+U(k)-0.6\sum_{l=0}^{k}(l+1)U(l+1)\delta_{k-l-1}\\&+0.12\sum_{l=0}^{k}(l+1)U(l+1)\delta_{k-l-2}\\&-0.008\sum_{l=0}^{k}(l+1)U(l+1)\delta_{k-l-3}-\delta_k-0.2\sum_{l=0}^{k}U(l)\delta_{k-l-1}+\sum_{l=0}^{k}U(l)U(k-l)\\&-0.4\sum_{k1=0}^{k}\sum_{l=0}^{k1}U(l)U(k1-l)\delta_{k-k1-l}+0.04\sum_{k1=0}^{k}\sum_{l=0}^{k1}U(l)U(k1-l)\delta_{k-k1-2}\\&+0.6\delta_{k-1}-0.12\delta_{k-2}+0.008\delta_{k-3}=0\end{aligned} \tag{7.105}$$

where $U(k)$ is DTM transformed forms of $u(t)$ in Eq. (7.103). After solving Eq. (7.105) using initial condition, particle's velocity, $u(t)$, will be appeared as fallows,

$$\begin{aligned}&u(t)=1\cdot t-0.5t^2-0.3t^3+0.295t^4+0.0726t^5-0.15577t^6\\&+0.00074381t^7+0.073487t^8-0.01683t^9-0.030966t^{10}\end{aligned} \tag{7.106}$$

Eq. (7.106) is DTM solution of Eq. (7.103). For increasing the convergency of DTM, Padé approximation with accuracy [6,6] is applied, and Eq. (7.106) is transformed to the following equation,

$$u(t)=\frac{(0.999999t-1.35275t^2+884.7181t^3-334.2273t^4+109.575t^5-21.7561t^6)}{(0.99999-0.85275t+884.5917t^2+107.52t^3+428.89t^4-35.793t^5+14.698t^6)} \tag{7.107}$$

Results of DTM−Padé [6,6] and numerical solution are presented in Fig. 7.37.

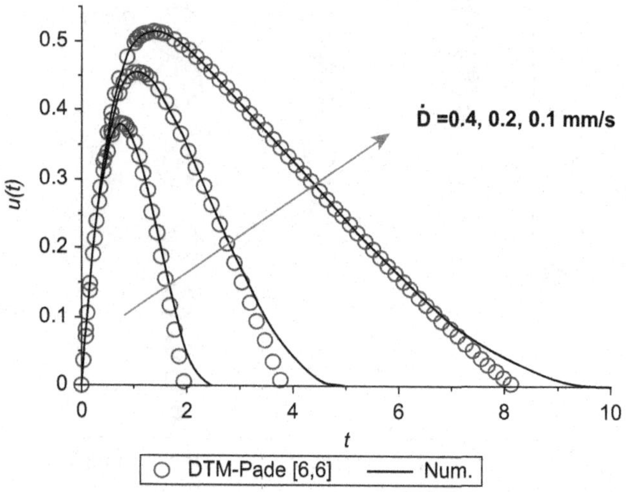

Figure 7.37 Velocity function for particle in three different reduction rate of diameter ($\dot{D}(mm/s)$) with numerical an DTM–Pade [6,6]. *DTM*, differential transformation method; *DTM–Padé*, differential transformation method with Padé approximation.

As seen DTM–Padé has an excellent agreement with numerical solution. Table 7.12 reveals that DTM–Padé results are completely acceptable; furthermore, it is possible to increase its accuracy by increasing the Padé order. As seen in Fig. 7.37 for soluble particle, against the insoluble and rigid particle, no terminal velocity is observed and velocity increases to a maximum value and then starts to reduce until it reaches zero. According to this figure, by increasing the solubility rate $\dot{D}$, maximum velocity increases,

Table 7.12 Comparison Between DTM–Padé [6,6] and Runge–Kutta Numerical Results for $u(t)$

t	Numerical Runge–Kutta Method	DTM–Padé [6,6]
0	0.000	0.000
0.5	0.356011026288972	0.3560109455
1.0	0.455673617777084	0.4556663757
1.5	0.424741800010445	0.4245943850
2.0	0.348494696237181	0.3474065553
2.5	0.260477668651496	0.2557216063
3.0	0.174182615316337	0.1590157555
3.5	0.099044601445866	0.06045697975
4.0	0.043185698190710	0.03799544552
4.5	0.010414005234102	0.01345128694

DTM–Padé, differential transformation method with Padé approximation.

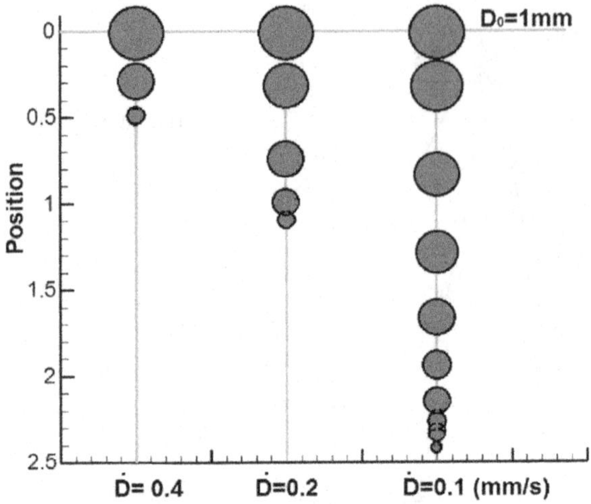

Figure 7.38 Positions of falling particle for different rates of solubility, $D_0 = 1$ mm, and time interval = 1 s.

but occurs some later. Positions and sizes of the particle are depicted graphically in Fig. 7.38 that demonstrates the falling process of the particle until it completely dissolves in Newtonian fluid media and disappears in the fluid.

REFERENCES

[1] Hatami M, Ganji DD. Motion of a spherical particle on a rotating parabola using Lagrangian and high accuracy Multi-step Differential Transformation Method. Powder Technology 2014;258:94–8.

[2] Hatami M, Sheikholeslami M, Domairry G. High accuracy analysis for motion of a spherical particle in plane Couette fluid flow by Multi-step Differential Transformation Method. Powder Technology 2014;260:59–67.

[3] Dogonchi AS, Hatami M, Domairry G. Motion analysis of a spherical solid particle in plane Couette Newtonian fluid flow. Powder Technology 2015;274:186–92.

[4] Dogonchi AS, Hatami M, Hosseinzadeh K, Domairry G. Non-spherical particles sedimentation in an incompressible Newtonian medium by Padé approximation. Powder Technology 2015;278:248–56.

[5] Hatami M, Ganji DD. Motion of a spherical particle in a fluid forced vortex by DQM and DTM. Particuology 2014;16:206–12.

[6] Hatami M, Ganji DD, Jafaryar M, Farkhadnia F. Transient combustion analysis for iron micro-particles in a gaseous media by weighted residual methods (WRMs). Case Studies in Thermal Engineering 2014;4:24–31.

[7] Nouri R, Ganji DD, Hatami M. Unsteady sedimentation analysis of spherical particles in Newtonian fluid media using analytical methods. Propulsion and Power Research 2014;3(2):96–105.

[8] Hatami M, Domairry G. Transient vertically motion of a soluble particle in a Newtonian fluid media. Powder Technology 2014;253:481–5.

CHAPTER 8

DTM for Solid Mechanics, Vibration, and Deflection

8.1 INTRODUCTION

Most scientific problems in solid mechanics are inherently nonlinear by nature, and, except for a limited number of cases, most of them do not have analytical solutions. Accordingly, the nonlinear equations are usually solved using other methods including numerical techniques or by using analytical methods such as differential transformation method (DTM). Therefore obtaining analytical limit state functions or using analytical techniques to obtain reliability index for nonlinear problems is almost impossible. Analytical methods which recently are widely used are one of the simple and reliable methods for solving system of coupled nonlinear differential equations. The most nonlinear problems in solid mechanics are vibration analysis, deflection and deformation of different beams, materials, or plates. In this chapter some of these problems are presented and solved by DTM, which are categorized in the following sections:

8.1 Introduction
8.2 Deflection Prediction of a Cantilever Beam
8.3 Vibration Analysis of Stepped FGM Beams
8.4 Piezoelectric Modal Sensors for Cantilever Beams
8.5 Damped System With High Nonlinearity
8.6 Free Vibration of a Centrifugally Stiffened Beam
8.7 Deflections of Orthotropic Rectangular Plate
8.8 Free Vibration of Circular Plates
8.9 Vibration of Pipes Conveying Fluid
8.10 Piezoelectric Modal Sensor for Nonuniform Euler–Bernoulli Beams With Rectangular Cross Section
8.11 Free Vibrations of Oscillators
8.12 Composite Sandwich Beams With Viscoelastic Core

Differential Transformation Method for Mechanical Engineering Problems
ISBN 978-0-12-805190-0
http://dx.doi.org/10.1016/B978-0-12-805190-0.00008-5

8.2 DEFLECTION PREDICTION OF A CANTILEVER BEAM

A cantilever beam OA is subjected to coplanar loading consisting of an axial compressive force F_A and of a transverse force Q_A (Fig. 8.1). F_A and Q_A are follower forces, i.e., they will rotate with the end section A of the beam during the deformation, and they will at all times remain tangential and perpendicular, respectively, to the deformed beam axis. It is assumed that the effect of the material nonlinearity is negligible in the mathematical derivation. Therefore at any point of coordinates $x(s)$ and $y(s)$ the external moment M is expressed as [1]:

$$M = (F_A \cos\theta_A + Q_A \sin\theta_A)(y_A \cdot y) + (F_A \sin\theta_A + Q_A \cos\theta_A)(x_A \cdot y) \tag{8.1}$$

where x and y are the longitudinal and transverse coordinates, respectively; q is the slope of the normal to the beam cross section; and x_A, y_A, and θ_A denote the coordinates and the normal slope at the end section. The classical Euler–Bernoulli hypothesis assumes that the bending moment M at any point of the beam is proportional to the corresponding curvature, i.e.,

$$M = EI\,\theta' \tag{8.2}$$

where E is Young's modulus and I is the area moment of inertia of the beam cross section about the x-axis. By using the following relations,

$$\frac{dx}{ds} = \cos\theta, \qquad \frac{dy}{ds} = \sin\theta \tag{8.3}$$

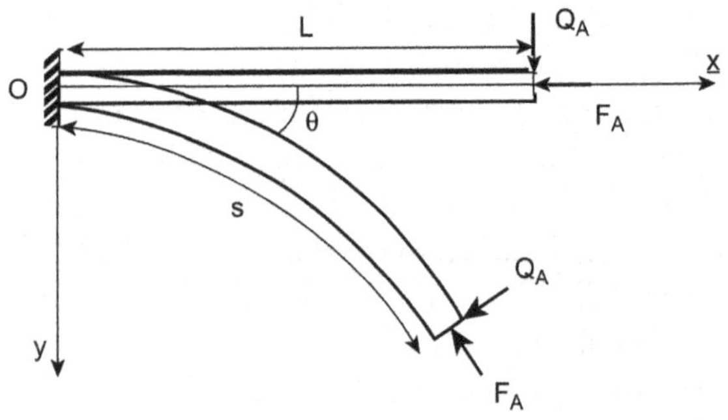

Figure 8.1 The geometry of a cantilever beam subjected to nonconservative external loading (follower forces).

and based on the trigonometric relations and by substituting Eqs. (8.2) and (8.3) into Eq. (8.1), the nonlinear differential equation governing the problem is obtained as follows,

$$\frac{\partial^2 \eta}{\partial s^2} + \frac{F_A}{EI}\sin(\eta(s)) + \frac{Q_A}{EI}\cos(\eta(s)) = 0 \tag{8.4}$$

The boundary conditions associated with the above equation are,

$$\eta(0) = 0, \quad \eta'(L) = 0 \tag{8.5}$$

where

$$\eta = \theta - \theta_A \tag{8.6}$$

Using only the two terms of a Taylor's series expansion for $\cos(\eta(s))$ and $\sin(\eta(s))$, and substituting in Eq. (8.4) yields,

$$\frac{\partial^2 \eta}{\partial s^2} + \frac{F_A}{EI}\left(\eta(s) - \frac{1}{6}\eta^3(s)\right) + \frac{Q_A}{EI}\left(1 - \frac{1}{2}\eta^2(s)\right) = 0 \tag{8.7}$$

Now DTM is applied from Chapter 1 into Eq. (8.7) to find $\eta(s)$. So,

$$\begin{aligned}
&(k+1)(k+2)H(k+2) + \frac{F_A}{EI}H(k) \\
&-\left(\frac{F_A}{EI}\right)\left(\frac{1}{6}\right)\sum_{k1=0}^{k}\sum_{l=0}^{k1} H(l)H(k1-l)H(k-k1) \\
&+\frac{Q_A}{EI}\delta[k] - \left(\frac{Q_A}{EI}\right)\left(\frac{1}{2}\right)\sum_{l=0}^{k} H(l)H(k-l) = 0
\end{aligned} \tag{8.8}$$

where

$$\delta[k] = \begin{cases} 1 & \text{if } k = 0 \\ 0 & \text{if } k \neq 0 \end{cases} \tag{8.9}$$

Similarly, the transformed form of boundary conditions (Eq. 8.5) can be written as,

$$H(0) = 0, \quad H(1) = a \tag{8.10}$$

in which the boundary condition of this case is the same as that of the previous case. By solving Eq. (8.8) and using boundary conditions Eq. (8.10), the DTM terms for this case for $F_A = 300$ KN, $Q_A = 350$ KN, $E = (40)$ GPa, and $I = (1.6)(10^{-5})$ m^4 can be,

$$\begin{aligned} H(2) &= -0.2734375000 \\ H(3) &= -0.7812500000a \\ H(4) &= 0.1068115234 + 0.02278645833a^2 \\ H(5) &= -0.005645751950a + 0.003906250000a^3 \end{aligned} \tag{8.11}$$

etc.

Now by applying Eq. (8.6) into Eq. (8.11) and using Eq. (8.10), the constant parameter "a" will be obtained so that the slope parameter equation will be estimated,

$$\begin{aligned} \eta(s) &= a\cdot s - 0.2734375000s^2 - 0.07812500000a\cdot s^3 \\ &+ (0.01068115234 + 0.02278645833a^2)s^4 \\ &+ (-0.005645751950a + 0.003906250000a^3)s^5 \\ &+ (0.0005145867667 - 0.003916422527a^2)s^6 \\ &+ (0.0008974756514a - 0.0001828632660a^3)s^7 \\ &+ (-0.00006135087461 + 0.0001701215902a^2 + 0.0001335144043a^4)s^8 \end{aligned} \tag{8.12}$$

By using boundary condition in $s = 1$, the "a" parameter will be determined as,

$$a = 0.6310718570 \tag{8.13}$$

and by substituting it into Eq. (8.12), $\eta(s)$ will be found as,

$$\begin{aligned} \eta(s) &= 0.6310718570s - 0.2734375000s^2 - 0.04930248883s^3 \\ &+ 0.01975589785s^4 - 0.002581135196s^5 - 0.001045135118s^6 \\ &+ 0.0005204134364s^7 + 0.00002757630897s^8 \end{aligned} \tag{8.14}$$

In this problem, deflection of a cantilever beam subjected to static coplanar loading is compared by two analytical methods called homotopy perturbation method (HPM) and DTM and fourth-order Runge–Kutta numerical method. For showing the efficiency of these analytical methods, Fig. 8.2 is presented.

As seen in most cases of these figures, HPM and DTM have an excellent agreement with numerical solution but in some cases (for example, when $F_A = 100$ KN, $Q_A = 100$ KN, $E = 10^9$ Pa, and $I = 10^{-4}$ m^4, see Fig. 8.2) the accuracy of the DTM is greater than the HPM. Table 8.1 compared the

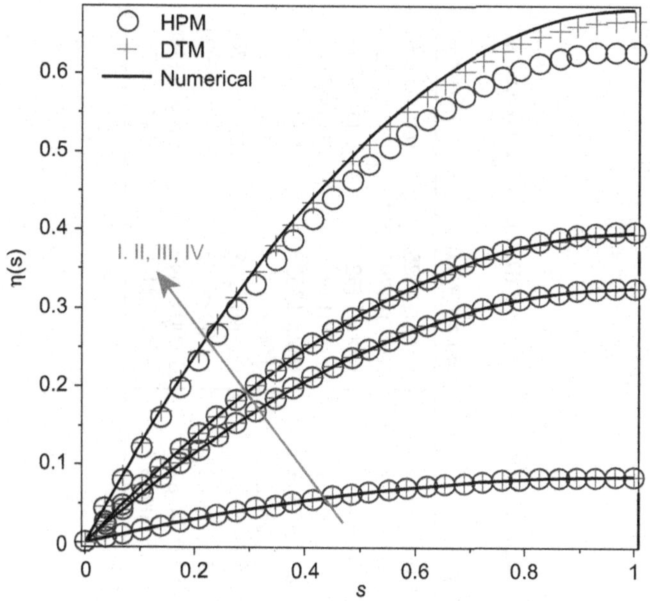

Figure 8.2 Comparison of slope parameter for homotopy perturbation method (HPM), differential transformation method (DTM), and numerical solution when

I: $F_A = 400$ KN, $Q_A = 450$ KN, $E = 40$ GPa and $I = 7 \times 10^{-5}$ m^4
II: $F_A = 300$ KN, $Q_A = 350$ KN, $E = 40$ GPa and $I = 1.6 \times 10^{-5}$ m^4
III: $F_A = 100$ KN, $Q_A = 50$ KN, $E = 10^9$ Pa and $I = 10^{-4}$ m^4
IV: $F_A = 100$ KN, $Q_A = 100$ KN, $E = 10^9$ Pa and $I = 10^{-4}$ m^4

results of the HPM and DTM with numerical procedure and homotopy analysis method which is presented in the literature when $F_A = 300$ KN, $Q_A = 350$ KN, $E = (40)$ GPa, and $I = (1.6)(10^{-5})$ m^4.

These data and calculated errors confirm that these two analytical methods are suitable and semi-exact for solving these kinds of problems. In the following step, effect of some parameters appeared in the mathematical formulation such as area moment of inertia (I), Young's modulus (E), transverse force (Q_A), and compressive force (F_A) on slope variation are investigated. Selected values for showing the variation of these parameters are considered according to the mean value of stress limit state concluded from [1]. These mean values are presented in Table 8.2.

Effect of compressive force (F_A) on slope parameter, $\eta(s)$, is presented in Fig. 8.3. As seen in this figure, slope parameter increases as well as compressive force increases. Variation of slope parameter based on

Table 8.1 Comparison Between Homotopy Perturbation Method (HPM), Differential Transformation Method (DTM), and Numerical Solution for $F_A = 300$ KN, $Q_A = 350$ KN, $E = (40)$ GPa, and $I = (1.6)(10^{-5})$ m^4

s	$\eta(s)_{Numerical}$	$\eta(s)_{HAM}$	$\eta(s)_{HPM}$	$\eta(s)_{DTM}$	(%) $Error_{HPM}$	(%) $Error_{DTM}$
0.1	0.06036251	0.06034326	0.06035369	0.06032545	0.00014625	0.00061397
0.2	0.11498708	0.11495123	0.11497007	0.11491317	0.00014793	0.00064274
0.3	0.16364459	0.16356398	0.16362082	0.16353411	0.00014527	0.00067508
0.4	0.20614594	0.20615968	0.20611719	0.20599929	0.00013944	0.00071139
0.5	0.24233792	0.24234267	0.24230515	0.24215566	0.00013523	0.00075207
0.6	0.27209923	0.27215896	0.27206024	0.27188220	0.00014331	0.00079763
0.7	0.29533656	0.29535986	0.29528238	0.29508623	0.00018348	0.00084760
0.8	0.31198116	0.31195631	0.31189042	0.31170062	0.00029085	0.00089921
0.9	0.32198607	0.32195632	0.32181657	0.32168184	0.00052643	0.00094486
1	0.32532409	0.32501035	0.32500059	0.32500948	0.00099440	0.00096706

Table 8.2 Mean Values for Parameters According to Stress Limit State Presented in Ref. [1]

Parameter	Mean Value
Shear force (Q_A)	350 (KN)
Normal force (F_A)	300 (KN)
Young's modulus (E)	40 (GPa)
Area moment of inertia (I)	2.6 E-5 (m^4)

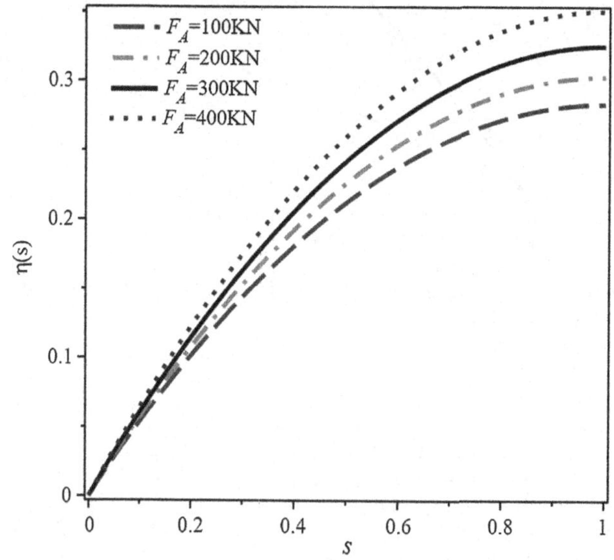

Figure 8.3 Effect of compressive force for $Q_A = 350$ KN, $E = (40)$ GPa and $I = (1.6)(10^{-5})$ m^4.

transverse force (Q_A) is depicted in Fig. 8.4. By increasing the Q_A, slope parameter is increased significantly. From these two figures, it is evident that the effect of Q_A for increasing the slope parameter is more than F_A. It is due to their direction applied to beam which as Fig. 8.1 shows, Q_A is a shear force and make more effects on beam slope parameter.

8.3 VIBRATION ANALYSIS OF STEPPED FGM BEAMS

Two types of stepped FGM beams made from ceramic–metal phases are chosen to investigate their vibration behavior. The geometries and descriptions of the beam types are shown in Fig. 8.5. The beams shown in this figure are supported by elastic conditions at both ends including

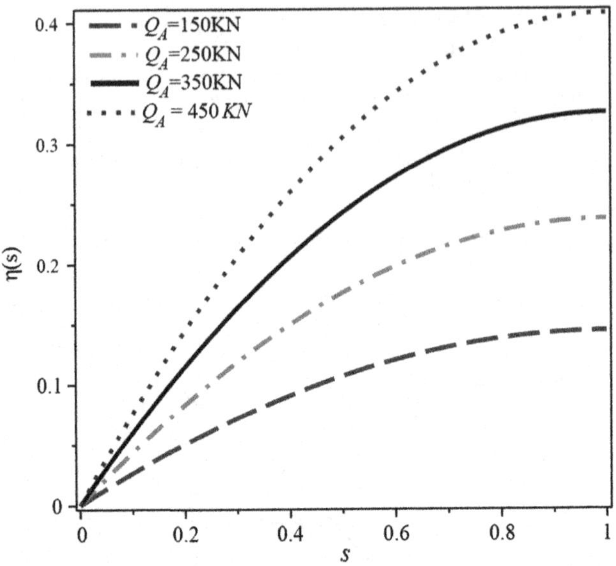

Figure 8.4 Effect of transverse force for $F_A = 300$ KN, $E = (40)$ GPa and $I = (1.6)(10^{-5})$ m^4.

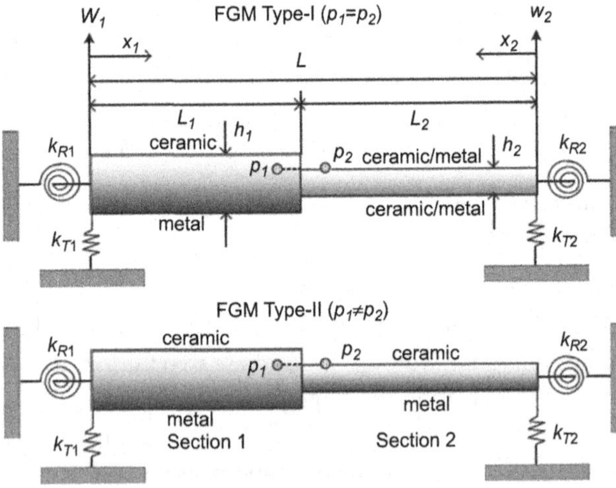

Figure 8.5 Two types of FGM beams under study.

translational and rotational springs which are defined as the E–E beams. It is known that FGMs are inhomogeneous spatial composite materials, typically composed of a ceramic–metal pair of materials. The material compositions are varied throughout the thickness direction from the top surface (ceramic 100%—metal 0%) to the bottom one (ceramic 0%—metal 100%). A schematic of these two beams is shown in Fig. 8.5 [2].

More assumptions of volume fractions of ceramic based on the power law distribution of the stepped FGM beams can be found in Ref. [2]. Consider classical beam theory: the partial differential equation used to describe the free vibration in each section of the stepped FGM beams can be expressed as:

$$\frac{\partial^4 w_j(x_j, t)}{\partial x_j^4} + \frac{I_{oj}}{\lambda_j} \frac{\partial^2 w_j(x_j, t)}{\partial t^2} - 0; \quad x_j \in [0, L] \ (j = 1, 2) \tag{8.15}$$

It is defined that I_{oj} is the moment of inertia, $\lambda_j = \left(D_{11} - \frac{B_{11}^2}{A_{11}} \right)$ is the material stiffness coefficient in each beam section. For harmonic vibration, $w_j(x_j, t) = W_j(x_j)e^{iwt}$ is substituted into Eq. (8.15) to obtain a time-independent governing equation as follows.

$$\frac{d^4 W_j(x_j)}{dx_j^4} - \frac{I_{oj}}{\lambda_j}\omega^2 W_j(x_j) - 0 \tag{8.16}$$

where ω is a natural frequency. On the principle of the DTM, the governing differential equation and boundary condition equations as well as the continuity conditions are transformed into a set of algebraic equations using transformation rule. The basic operations required in differential transformation for the governing differential equations, boundary conditions, and the continuity conditions are shown in Tables 2 and 3 in Ref. [2], respectively. The general function, $f_j(x_j)$ is considered as the transverse displacement $W_j(x_j)$:

$$W_j[r+4] = \frac{I_{oj}\omega^2}{\lambda_j(r+1)(r+2)(r+3)(r+4)} W_j[r]. \tag{8.17}$$

It can be seen that Eq. (8.17) is independent from boundary conditions. Therefore to obtain frequency results, the displacement function of Eq. (8.17) must be used to satisfy the corresponding boundary equations. To demonstrate the application of the DTM to vibration response of the stepped FGM beam, let us consider the *E*—*E* beams as shown in Fig. 8.5 which have elastic boundary conditions at both ends. The governing equation in the form of the recurrence equation for Section 1 of the beams can be expressed as:

$$W_1[r+4] = \frac{I_{01}\omega^2}{\lambda_1(r+1)(r+2)(r+3)(r+4)} W_1[r] \tag{8.18}$$

For this case, the boundary conditions at the left end can be expressed as:

$$\frac{d^3 W_1}{dx_1^3} + \frac{k_{T1}}{\lambda_1} W_1 = 0_1 \quad \frac{d^2 W_1}{dx_1^2} \frac{k_{R1}}{\lambda_1} \frac{dW_1}{dx_1} = 0_1 \tag{8.19}$$

where k_{T1} and k_{R1} are the translational spring and rotational spring constants at the left end, respectively. Let the nonzero values of deflection and slope at $x_1 = 0$ be C_0 and C_1, respectively. Applying the basic operations of DTM for these nonzero quantities at $x_1 = 0$, one obtains:

$$W_1[0] - W_1(x_1) - C_0 \quad W_1[1] - \frac{dW_1(x_1)}{dx_1} - C_1 \tag{8.20}$$

The expression for nonzero values of bending moment and shear force at $x_1 = 0$ can be written as:

$$W_1[2] - \frac{k_{R1}}{2\lambda_1} C_1, \quad W_1[3] - \frac{k_{R1}}{6\lambda_1} C_0 \tag{8.21}$$

To find $W_1(r)$ for all values of r, the components in Eqs. (8.20) and (8.21) are substituted into the recurrence equation in Eq. (8.18).

$$W_1[4r] - \frac{\omega^{2r} I_{01}^r}{\lambda_1^r (4r)!} C_0 \quad r = 0, 1, 2, 3, \ldots \tag{8.22a}$$

$$W_1[4r+1] - \frac{\omega^{2r} I_{01}^r}{\lambda_1^r (4r+1)!} C_1 \quad r = 0, 1, 2, 3, \ldots \tag{8.22b}$$

$$W_1[4r+2] - \frac{\omega^{2r} I_{01}^r k_{R1}}{\lambda_1^{(r+1)} (4r+2)!} C_1 \quad r = 0, 1, 2, 3, \ldots \tag{8.22c}$$

$$W_1[4r+3] - \frac{\omega^{2r} I_{01}^r k_{T1}}{\lambda_1^{(r+1)} (4r+3)!} C_0 \quad r = 0, 1, 2, 3, \ldots \tag{8.22d}$$

Next procedure is given for considering Section 2 of the beam, therefore, the governing recurrence equation of this section is

$$W_2[r+4] - \frac{I_{02}\omega^2}{\lambda_2 (r+1)(r+2)(r+3)(r+4)} W_2[r] \tag{8.23}$$

To consider the boundary conditions at the right end ($x_2 = 0$), the conditions can be expressed as:

$$\frac{d^3 W_2}{dx_2^3} + \frac{k_{T2}}{\lambda_2} W_2 - 0 \quad \frac{d^2 W_2}{dx_2^2} - \frac{k_{R2}}{\lambda_2} \frac{dW_2}{dx_2} - 0 \tag{8.24}$$

where k_{T2} and k_{R2} are the translational spring and rotational spring constants at the right end, respectively. The nonzero values of the deflection and slope account for C_2 and C_3, respectively. Again, applying the basic operations of DTM for the boundary condition at ($x_2 = 0$), one obtains:

$$W_2[0] = W_2(x_2) = C_2 \quad W_2[1] = \frac{dW_2(x_2)}{dx_2} = C_3 \tag{8.25}$$

The expression for nonzero values of bending moment and shear force at $x_2 = 0$ can be written as:

$$W_2[2] = \frac{k_{R2}}{2\lambda_2}C_3 \quad W_2[3] = \frac{k_{R1}}{6\lambda_1}C_2 \tag{8.26}$$

Again by using Eqs. (8.25) and (8.26) with Eq. (8.23), the expressions of $W_2(r)$ for all values of r can be written as follows.

$$W_2[4r] = \frac{\omega^{2r} I_{02}^r}{\lambda_2^r (4r)!}C_2 \quad r = 0, 1, 2, 3, \ldots \tag{8.27a}$$

$$W_2[4r+1] = \frac{\omega^{2r} I_{02}^r}{\lambda_2^r (4r+1)!}C_3 \quad r = 0, 1, 2, 3, \ldots \tag{8.27b}$$

$$W_2[4r+2] = \frac{\omega^{2r} I_{02}^r k_{R2}}{\lambda_2^{(r+1)} (4r+2)!}C_3 \quad r = 0, 1, 2, 3, \ldots \tag{8.27c}$$

It is assumed that, for an FGM beam having discontinuous cross section, stress concentration at the interchange or at the step location of such beam is neglected. Hence, the continuity conditions are

$$W_1(L_1) - W_2(L_2), \quad \frac{dW_1(L_1)}{dx_1} - \frac{dW_2(L_2)}{dx_2} \tag{8.28a}$$

$$\frac{d^2W_1(L_1)}{d^2x_1} - \frac{\lambda_2}{\lambda_1}\frac{d^2W_2(L_2)}{d^2x_2}, \quad \frac{d^3W_1(L_1)}{d^3x_1} - \frac{\lambda_2}{\lambda_1}\frac{d^3W_2(L_2)}{d^3x_2} \tag{8.28b}$$

According to the principle of the DTM, the continuity conditions are transformed into algebraic equations as well. The results of the transformation can be expressed as,

$$W_1[r]L_1^{(r)} - W_2[r]L_2^{(r)} = 0 \tag{8.29}$$

$$W_1[r]rL_1^{(r1)} + W_2[r]rL_2^{(r1)} = 0 \tag{8.30}$$

$$W_1[r]r(r-1)L_1^{(r2)} - \delta W_2[r]r(r-1)L_2^{(r2)} = 0 \tag{8.31}$$

$$W_1[r]r(r-1)L_1^{(r3)} + \delta W_2[r]r(r-1)L_2^{(r3)} = 0 \tag{8.32}$$

where $\delta = \frac{\lambda_2}{\lambda_1}$. The components of Eqs. (8.22) and (8.27) are substituted into the transformed continuity conditions in Eqs. (8.29)–(8.32). The results of this substitution can be arranged and presented in the matrix form as follows:

$$\begin{bmatrix} e_{11} & e_{12} & e_{13} & e_{14} \\ e_{21} & e_{22} & e_{23} & e_{24} \\ e_{31} & e_{32} & e_{33} & e_{34} \\ e_{41} & e_{42} & e_{43} & e_{44} \end{bmatrix} \begin{bmatrix} c_0 \\ c_1 \\ c_2 \\ c_3 \end{bmatrix} = 0 \tag{8.33}$$

where the elements in the matrix are

$$\begin{aligned}
e_{11} &= \sum_{r=0}^{\infty} \frac{I_{01}^r \omega^{2r} L_1^{(4r)}}{\lambda_1^r (4r)!} - k_{T1} \sum_{r=0}^{\infty} \frac{I_{01}^r \omega^{2r} L_1^{(4r+3)}}{\lambda_1^{(r+1)} (4r+3)!} \\
e_{12} &= \sum_{r=0}^{\infty} \frac{I_{01}^r \omega^{2r} L_1^{(4r+1)}}{\lambda_1^r (4r+1)!} + k_{R1} \sum_{r=0}^{\infty} \frac{I_{01}^r \omega^{2r} L_1^{(4r+2)}}{\lambda_1^{(r+1)} (4r+2)!} \\
e_{13} &= \sum_{r=0}^{\infty} \frac{I_{02}^r \omega^{2r} L_2^{(4r)}}{\lambda_2^r (4r)!} + k_{T1} \sum_{r=0}^{\infty} \frac{I_{02}^r \omega^{2r} L_1^{(4r+3)}}{\lambda_2^{(r+1)} (4r+3)!} \\
e_{14} &= \sum_{r=0}^{\infty} \frac{I_{02}^r \omega^{2r} L_2^{(4r+1)}}{\lambda_2^r (4r+1)!} - k_{R2} \sum_{r=0}^{\infty} \frac{I_{02}^r \omega^{2r} L_1^{(4r+2)}}{\lambda_2^{(r+1)} (4r+2)!} \\
e_{21} &= k_{T1} \sum_{r=0}^{\infty} \frac{I_{01}^r \omega^{2r} L_1^{(4r+2)}}{\lambda_1^r (4r+2)!} + \sum_{r=1}^{\infty} \frac{I_{01}^r \omega^{2r} L_1^{(4r-1)}}{\lambda_1^r (4r-1)!} \\
e_{22} &= \sum_{r=0}^{\infty} \frac{I_{01}^r \omega^{2r} L_1^{(4r)}}{\lambda_1^r (4r)!} + k_{R1} \sum_{r=0}^{\infty} \frac{I_{01}^r \omega^{2r} L_1^{(4r+1)}}{\lambda_1^r (4r+1)!} \\
e_{23} &= k_{T2} \sum_{r=0}^{\infty} \frac{I_{02}^r \omega^{2r} L_2^{(4r+2)}}{\lambda_2^{(r+1)} (4r+2)!} + \sum_{r=1}^{\infty} \frac{I_{02}^r \omega^{2r} L_2^{(4r-1)}}{\lambda_2^r (4r-1)!} \\
e_{24} &= \sum_{r=0}^{\infty} \frac{I_{02}^r \omega^{2r} L_2^{(4r)}}{\lambda_2^r (4r)!} + k_{R2} \sum_{r=0}^{\infty} \frac{I_{02}^r \omega^{2r} L_2^{(4r+1)}}{\lambda_2^{(r+1)} (4r+1)!} \\
e_{31} &= k_{T1} \sum_{r=0}^{\infty} \frac{I_{01}^r \omega^{2r} L_1^{(4r+1)}}{\lambda_1^{(r+1)} (4r+1)!} + \sum_{r=1}^{\infty} \frac{I_{01}^r \omega^{2r} L_1^{(4r-2)}}{\lambda_1^r (4r-2)!} \\
e_{32} &= k_{R1} \sum_{r=0}^{\infty} \frac{I_{01}^r \omega^{2r} L_1^{(4r)}}{\lambda_1^{(r+1)} (4r)!} + \sum_{r=1}^{\infty} \frac{I_{01}^r \omega^{2r} L_1^{(4r-1)}}{\lambda_1^r (4r-1)!}
\end{aligned} \tag{8.34}$$

$$\begin{aligned}
e_{33} &= \delta k_{T2}\sum_{r=0}^{\infty}\frac{I_{02}^{r}\omega^{2r}L_2^{(4r+1)}}{\lambda_2^{(r+1)}(4r+1)!}-\delta\sum_{r=1}^{\infty}\frac{I_{02}^{r}\omega^{2r}L_2^{(4r-2)}}{\lambda_2^{r}(4r-2)!}\\
e_{34} &= \delta k_{R2}\sum_{r=0}^{\infty}\frac{I_{02}^{r}\omega^{2r}L_2^{(4r)}}{\lambda_2^{(r+1)}(4r)!}-\delta\sum_{r=1}^{\infty}\frac{I_{02}^{r}\omega^{2r}L_2^{(4r-1)}}{\lambda_2^{r}(4r-1)!}\\
e_{41} &= k_{T1}\sum_{r=0}^{\infty}\frac{I_{01}^{r}\omega^{2r}L_1^{(4r)}}{\lambda_1^{(r+1)}(4r)!}+\sum_{r=1}^{\infty}\frac{I_{01}^{r}\omega^{2r}L_1^{(4r-3)}}{\lambda_1^{r}(4r-3)!}\\
e_{42} &= \sum_{r=1}^{\infty}\frac{I_{01}^{r}\omega^{2r}L_1^{(4r-2)}}{\lambda_1^{r}(4r-2)!}+k_{R1}\sum_{r=1}^{\infty}\frac{I_{01}^{r}\omega^{2r}L_1^{(4r-1)}}{\lambda_1^{(r+1)}(4r-1)!}\\
e_{43} &= \delta k_{T2}\sum_{r=0}^{\infty}\frac{I_{02}^{r}\omega^{2r}L_2^{(4r)}}{\lambda_2^{(r+1)}(4r)!}+\delta\sum_{r=1}^{\infty}\frac{I_{02}^{r}\omega^{2r}L_2^{(4r-3)}}{\lambda_2^{r}(4r-3)!}\\
e_{44} &= \delta\sum_{r=1}^{\infty}\frac{I_{02}^{r}\omega^{2r}L_2^{(4r-2)}}{\lambda_2^{r}(4r-2)!}+\delta k_{R2}\sum_{r=1}^{\infty}\frac{I_{02}^{r}\omega^{2r}L_2^{(4r-1)}}{\lambda_2^{(r+1)}(4r-1)!}
\end{aligned} \tag{8.35}$$

To obtain a nontrivial solution, the determinant of coefficient matrix in Eq. (8.33) could be set equal to zero. It is also noted that, for practical calculation, the finite number of r terms in each element of the matrix in Eq. (8.34) should be used. Therefore a convergence study will be performed to determine an appropriate maximum number of r in the upper limit of this equation. Mode shapes of the stepped FGM beams can be plotted by setting C_0 to unity in Eq. (8.33), so that the remaining nonzero constants (C_1, C_2, C_3) are solved. Thus, the mode shapes corresponding to any frequency can be expressed as the function of $W_j(x_j)=\sum_{r=0}^{\infty}x_j^r W_j[r]$. To plot the mode shapes with the whole length coordinate (x), for example, in the case of $E-E$ beam, its mode shape functions are

$$\begin{aligned}
W_1(x) &= \sum_{r=0}^{\infty}\frac{I_{01}^{r}\omega^{2r}}{\lambda_1^{r}(4r)!}x^{(4r)}+C_1\sum_{r=0}^{\infty}\frac{I_{01}^{r}\omega^{2r}}{\lambda_1^{r}(4r+1)!}x^{(4r+1)}\\
&+C_1k_{R1}\sum_{r=0}^{\infty}\frac{I_{01}^{r}\omega^{2r}}{\lambda_1^{(r+1)}(4r+2)!}x^{(4r+2)}-k_{T1}\sum_{r=0}^{\infty}\frac{I_{01}^{r}\omega^{2r}}{\lambda_1^{(4r+1)}(4r+3)!}x^{(4r+3)}x\in[0,L_1]\\
W_2(L-x) &= C_2\sum_{r=0}^{\infty}\frac{I_{02}^{r}\omega^{2r}}{\lambda_2^{r}(4r)!}(L-x)^{(4r)}+C_3\sum_{r=0}^{\infty}\frac{I_{02}^{r}\omega^{2r}}{\lambda_2^{r}(4r+1)!}(L-x)^{(4r+1)}\\
&+C_3k_{R2}\sum_{r=0}^{\infty}\frac{I_{02}^{r}\omega^{2r}}{\lambda_2^{(r+1)}(4r+2)!}(L-x)^{(4r+2)}\\
&-C_2k_{T2}\sum_{r=0}^{\infty}\frac{I_{02}^{r}\omega^{2r}}{\lambda_2^{(r+1)}(4r+3)!}(L-x)^{(4r+3)}x\in[L_1,L]
\end{aligned} \tag{8.36}$$

By following the same procedure, one can solve the vibration problem of the stepped FGM beams with other kinds of boundary conditions. The matrix elements for other boundary conditions such as clamped—elastic supported (*C—E*) and simply supported—elastic supported (*S—E*) beams are presented in Appendix A in Ref. [2]. In this section, the stepped FGM beams made from alumina (Al_2O_3) and aluminum (Al) are chosen to investigate their vibration behavior. The mechanical properties of the materials are:

For ceramic: (Al_2O_3), $Ec = 380$ GPa, $qc = 3960\ \text{kg/m}^3$, $m = 0.3$.

For metal: (Al), $Em = 70$ GPa, $qm = 2702\ \text{kg/m}^3$, $m = 0.3$.

The translational and rotational spring constants can be obtained by using the following forms $k_{Tj} = \frac{\beta_{Tj}\lambda_1}{L^3}$ and $k_{Rj} = \frac{\beta_{Rj}\lambda_1}{L}$ in which β_{Tj} and β_{Rj} are the translational and rotational spring factors. To clearly illustrate the changes of frequencies owing to the variations of spring constant factors, Fig. 8.6 plots the first to fourth frequency results of *E—E* beams against the spring constant factors ($\beta_{T1} = \beta_{R1} = \beta_{T2} = \beta_{R2}$). As can be observed, all frequencies remain constant when the beams are supported by soft springs, using small values of spring constant factors. The frequencies increase considerably as the increase of spring constant factors is in the range of moderate spring stiffness $\beta_{T1} = \beta_{R1} = \beta_{T2} = \beta_{R2} = 10^2 \rightarrow 10^4$.

Mode shape functions for *C—E*, *S—E*, and *E—E* beams provided in this problem are used to plot and illustrate the vibration mode shapes of stepped FGM beams in Fig. 8.4. The first to fourth mode shapes shown in this

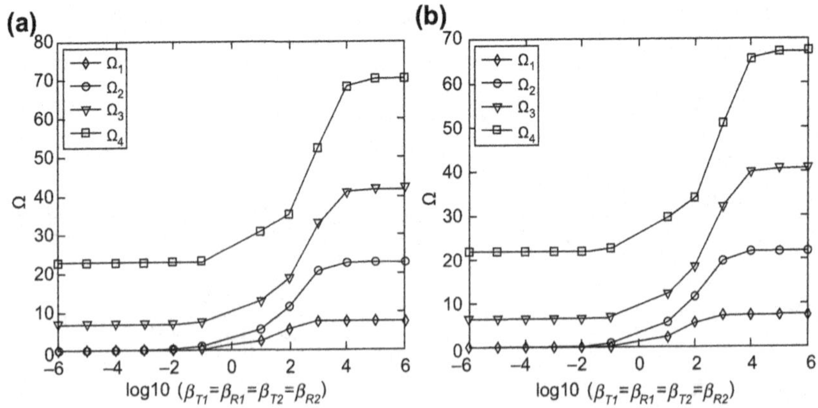

Figure 8.6 Dimensionless frequencies of *E—E* beams with the variations of spring constant factors. (a) FGM Type-I. (b) FGM Type-II ($L/h = 20$; $L1 = 0.5$ L; $n = 0.5$; $N = n = 0.5$).

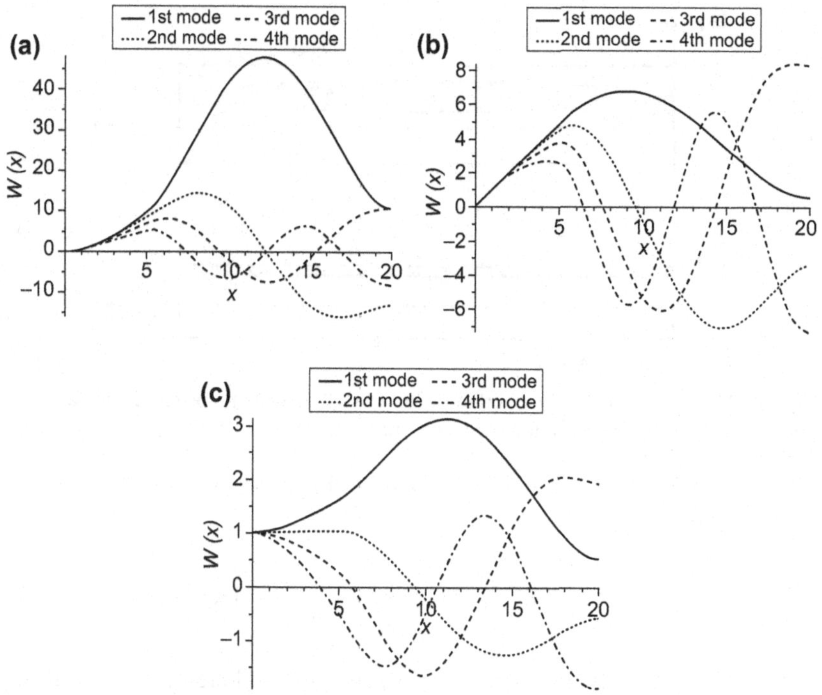

Figure 8.7 The first to fourth mode shapes of FGM Type-I beams. (a) Mode shapes of *C*–*E* beams, (b) *S*–*E* beams, and (c) *E*–*E* beams.

figure are obtained from FGM Type-I beams, using $\beta_{T1} = \beta_{R1} = \beta_{T2} = \beta_{R2} = 10$. Fig. 8.7(a) is for the mode shapes of *C*–*E* beams, (b) for *S*–*E* beams, and (c) for *E*–*E* beams.

8.4 PIEZOELECTRIC MODAL SENSORS FOR CANTILEVER BEAMS

In general, shaped piezoelectric sensors made of polyvinylidene fluoride (PVDF) are chosen since these add little loading to light structures and in addition are easy to cut into desired shapes. The PVDF sensors are designed by shaping the surface electrode, whereby the output of the sensor can be made sensitive to selected modal coordinates, other modal coordinates may be filtered out. Consider a beam with length *L*, width *b*, and thickness *h*. A shaped PVDF film of constant thickness is attached onto the top surface and spanned across the entire length of the beam, as shown in Fig. 8.8.

Assuming that the PVDF sensor thickness is much smaller than beam thickness *h*, the mass and stiffness of the sensor is then negligible compared

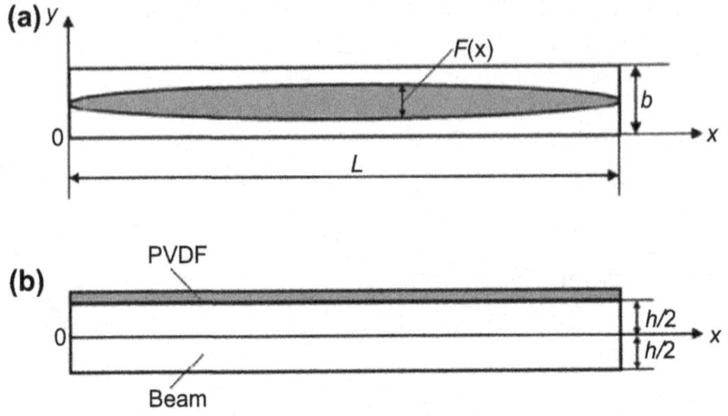

Figure 8.8 A shaped polyvinylidene fluoride (PVDF) film bonded on a beam (dimensions are not scaled). (a) Top view; (b) front view.

to the properties of the beam. The output charge $Q(t)$ of the PVDF sensor can be expressed as [3]:

$$Q(t) = \frac{h}{2}\int_0^L F(x)\cdot\left(e_{31}\frac{\partial^2 w(x,t)}{\partial x^2}\right)dx \tag{8.37}$$

where e_{31} is the PVDF sensor stress/charge coefficient, $w(x, t)$ is the displacement of the beam, and $F(x)$ is the PVDF film shape function. The displacement distribution of the vibrating beam may be represented by the form of a modal expansion:

$$w(x,t) = \sum_{n=1}^{N}\eta_n(t)\phi_n(x) \tag{8.38}$$

where $\eta_n(t)$ and $\phi_n(x)$ are the nth modal coordinates and structural mode shape function. N is the index for the highest order structural mode. Substituting Eq. (8.38) into Eq. (8.37), and expressing in dimensionless form, one obtains:

$$Q(t) = \frac{h}{2L}e_{31}\sum_{n=1}^{N}\eta_n(t)\int_0^L F(x)\cdot\frac{\partial^2\phi_n(x)}{\partial x^2}dx \tag{8.39}$$

where $X = \frac{x}{L}$, $\Phi_n(X) = \phi_n(x)$. We make an approximation by expanding $F(X)$ as a linear function of the second derivative of the mode shape function $\frac{d^2\phi_j(x)}{dx^2}$.

$$F(x) = \sum_{j=1}^{N}B_j\frac{\partial^2\phi_j(x)}{\partial X^2} \tag{8.40}$$

where B_j are the unknown shape coefficients for the PVDF sensor. Substituting Eq. (8.40) into Eq. (8.39), we obtain

$$Q(t) = \frac{h}{2L} e_{31} \sum_{n=1}^{N} \sum_{j=1}^{N} B_k \int_0^1 \frac{\partial^2 \phi_n(x)}{\partial X^2} \cdot \frac{\partial^2 \phi_j(x)}{\partial X^2} dX \cdot \eta_n(t) \tag{8.41}$$

Eq. (8.41) can be simplified as a matrix form:

$$Q(t) = B^T k \eta \tag{8.42}$$

where B and g are the vectors. K is an $N \times N$ matrix with element.

$$k(n,j) = \frac{h}{2L} e_{31} \int_0^1 \frac{\partial^2 \phi_n(x)}{\partial X^2} \cdot \frac{\partial^2 \phi_j(x)}{\partial X^2} dX \tag{8.43}$$

To construct a shaped PVDF sensor that accurately measures the Jth structural mode, the output signal of the sensor $Q(t)$ should be directly proportional to the Jth modal coordinate $\eta_j(t)$ and orthogonal to other modes. For simplicity, we set $Q(t) = \eta_j(t)$ to obtain

$$\eta_J(t) = B^T k \eta \tag{8.44}$$

From Eq. (8.44), the PVDF shape coefficients can be obtained:

$$B^T = k^{-1}(J, 1\!:\!N) \tag{8.45}$$

In accordance with the above analysis, to design a modal sensor, the second spatial derivative of the mode shapes must be obtained. This will be solved by using the DTM technique. For this purpose, one particular case, say, of a cantilever beam with intermediate support is considered, as shown in Fig. 8.9. The beam is divided into two sections with the two mirror systems of reference x_1 and x_2. The positive direction of the spatial coordinate x_1 is defined in the direction to the right for Section 1, and x_2 is defined in the direction to the left for Section 2. The ordinary differential equation describing the free vibration in each section is as follows:

$$\frac{d^4 \phi_j(x_j)}{dx_j^4} - \frac{m_s \omega^2}{EI} \phi_j(x_j) = 0 \quad x_j \in [0 \cdot L_j], \ (j = 1, 2) \tag{8.46}$$

where subscripts $j = 1$ and 2 denote Sections 1 and 2 of the beam, respectively. E is Young's modulus, I is the cross-sectional moment of inertia of

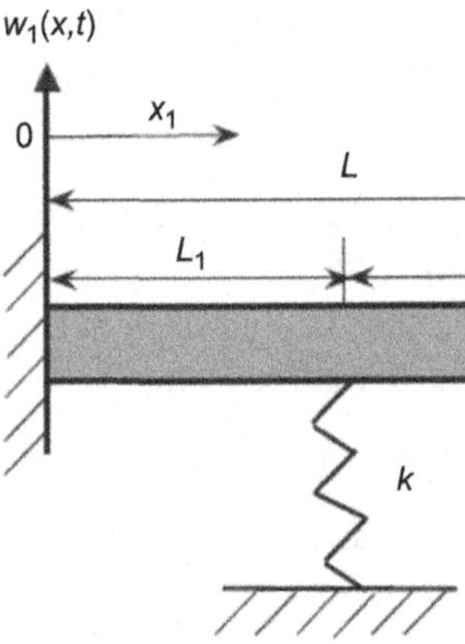

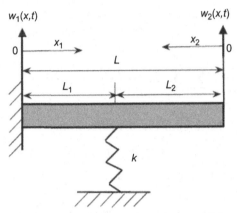

Figure 8.9 The coordinate system for a cantilever beam with intermediate support.

the beam $I = \frac{bh^3}{12}$, $m_s = \rho bh$ is the mass per unit length, and ρ is the density of the beam. Introduction of dimensionless variables,

$$X_j = \frac{x_j}{L}, \quad \Phi_j(X_j) = \frac{\phi_j(x_j)}{L} \tag{8.47}$$

can be rewritten in dimensionless form,

$$\frac{d^4\phi_j(x_j)}{dx_j^4} - \Omega^4\phi_j(x_j) = 0 \tag{8.48}$$

where $\Omega^4 = \frac{m_s\omega^2 L^4}{EI}$ Ω is the dimensionless natural frequency and the nth natural frequency is denoted as $\Omega(n)$. According to the DTM principle, $\phi_j(x_j)$ in Eq. (8.48) can be expressed in differential transformation formulation:

$$\phi_j(x_j) = \sum_{m=0}^{M} X_j^m \overline{\phi}_j(m) \tag{8.49}$$

And the second spatial derivative of the mode shapes can be expressed as

$$\frac{d^2\phi_j(x_j)}{dx_j^2} = \sum_{m=0}^{M} (m+1)(m+2)X_j^m \overline{\phi}_j(m+2) \tag{8.50}$$

By applying the DTM to Eq. (8.48), and using the transformation operations in Chapter 1 and some simplification, the following recurrence equation can be obtained:

$$\overline{\phi}_j(m+4) = \frac{\Omega^4}{(m+1)(m+2)(m+3)(m+4)}\overline{\phi}_j(m) \tag{8.51}$$

To calculate $\frac{d^2\phi_j(x_j)}{dx_j^2}$ in Eq. (16), the differential transformation $\overline{\phi_j}(m)$ should be solved. From Eq. (8.51), it can be seen $\overline{\phi_j}(m)$ $(m \geq 4)$ is a function of $\overline{\phi_j}(0), \overline{\phi_j}(1), \overline{\phi_j}(2), \overline{\phi_j}(3)$, and Ω. These nine unknown parameters, namely Ω and $\overline{\phi_j}(s)$ ($s = 0$, 1, 2, 3 and $j = 1$, 2), can be determined by the boundary condition equations of each section of the beam and the continuity conditions at intermediate support. By using boundary condition of the cantilever beam shown in Fig. 8.8, we get at left end (clamped):

$$\phi_1(x_1) = \frac{d\phi_1(x_1)}{dX_1} = 0, \quad (X_1 = 0) \tag{8.52}$$

At right end (free):

$$\frac{d^2\phi(x_2)}{dX_2^2} = \frac{d^3\phi(x_2)}{dX_2^3} = 0, \quad (X_2 = 0) \tag{8.53}$$

The differential transformations of Eq. (8.51) are obtained with the definition Eq. (8.49) as

$$\overline{\phi_1}(0) = 0, \quad \overline{\phi_1}(1) = 0 \tag{8.54}$$

Similarly, for differential transformations of Eq. (8.53) can be expressed as:

$$\overline{\phi_1}(2) = 0, \quad \overline{\phi_1}(3) = 0 \tag{8.55}$$

For a beam with intermediate support, the continuity conditions in dimensionless form are

$$\phi_1(R_1) = \phi_2(R_2), \qquad \frac{d\phi_1(R_1)}{dX_1} = -\frac{d\phi_2(R_2)}{dX_2} \tag{8.56}$$

$$\frac{d^2\phi_1(R_1)}{dX_1^2} = \frac{d^2\phi_2(R_2)}{dX_2^2}, \qquad \frac{d^3\phi_1(R_1)}{dX_1^3} - K_s\phi_1(R_1) = -\frac{d^3\phi_2(R_2)}{dX_2^3} \tag{8.57}$$

where $R_1 = \frac{L_1}{L}$ is denoted as dimensionless step location, $R_2 = \frac{L_2}{L}$ and $R_1 + R_2$, $K_s = \frac{K_s L^3}{EI}$ are the dimensionless stiffness, and K_s is the stiffness of the intermediate support. Substituting Eq. (8.49) into Eqs. (8.56) and (8.55), we obtain

$$\sum_{m=0}^{M} R_1^m \overline{\phi_1}(m) = \sum_{m=0}^{M} R_2^m \overline{\phi_2}(m) \tag{8.58}$$

$$\sum_{m=0}^{M}(m+1)R_1^m\overline{\phi_1}(m+1) = \sum_{m=0}^{M}(m+1)R_2^m\overline{\phi_2}(m+1) \tag{8.59}$$

$$\sum_{m=0}^{M}(m+1)(m+2)R_1^m\overline{\phi_1}(m+2) = \sum_{m=0}^{M}(m+1)(m+2)R_2^m\overline{\phi_2}(m+2) \tag{8.60}$$

$$\sum_{m=0}^{M}(m+1)(m+2)(m+3)R_2^m\overline{\phi_2}(m+3) - K_sR_1^m\overline{\phi_1}(m) = -\sum_{m=0}^{M}(m+1)(m+2)(m+3)R_2^m\overline{\phi_2}(m+3) \tag{8.61}$$

Substituting Eqs. (8.54) and (8.55) into Eqs. (8.58)–(8.61) and then rewriting in matrix form, we obtain:

$$\begin{bmatrix} f_{11}(\Omega) & f_{12}(\Omega) & f_{13}(\Omega) & f_{14}(\Omega) \\ f_{21}(\Omega) & f_{22}(\Omega) & f_{23}(\Omega) & f_{24}(\Omega) \\ f_{31}(\Omega) & f_{32}(\Omega) & f_{33}(\Omega) & f_{34}(\Omega) \\ f_{41}(\Omega) & f_{42}(\Omega) & f_{43}(\Omega) & f_{44}(\Omega) \end{bmatrix} \begin{bmatrix} \overline{\phi_1}(2) \\ \overline{\phi_1}(3) \\ \overline{\phi_2}(0) \\ \overline{\phi_2}(1) \end{bmatrix} = 0 \tag{8.62}$$

where $f_{ij}(\Omega)$ is a linear function of $\overline{\phi_2}(0)$, $\overline{\phi_1}(2)$, $\overline{\phi_1}(2)$ and $\overline{\phi_2}(1)$. From Eq. (8.62), the dimensionless natural frequencies Ω can be solved by

$$\det\begin{bmatrix} f_{11}(\Omega) & f_{12}(\Omega) & f_{13}(\Omega) & f_{14}(\Omega) \\ f_{21}(\Omega) & f_{22}(\Omega) & f_{23}(\Omega) & f_{24}(\Omega) \\ f_{31}(\Omega) & f_{32}(\Omega) & f_{33}(\Omega) & f_{34}(\Omega) \\ f_{41}(\Omega) & f_{42}(\Omega) & f_{43}(\Omega) & f_{44}(\Omega) \end{bmatrix} = 0 \tag{8.63}$$

Substituting the solved $\Omega(n)$ into Eq. (8.51) and using Eq. (8.50), the closed-form series solution for the second spatial derivative of the mode shapes $\frac{d^2\phi_1(x_1)}{dx_1^2}$ for Section 1 and $\frac{d^2\phi_2(x_2)}{dx_2^2}$ for Section 2 can be determined. Rewriting the second spatial derivative of the mode shape functions with a uniform coordinate X, we obtain:

$$\frac{\partial^2\phi(x)}{\partial X^2} = \begin{cases} \dfrac{\partial^2\phi_1(x)}{\partial X^2} \\ \dfrac{\partial^2\phi_2(1-x)}{\partial X^2} \end{cases} X\in[0,R], X\in[R_1,1] \tag{8.64}$$

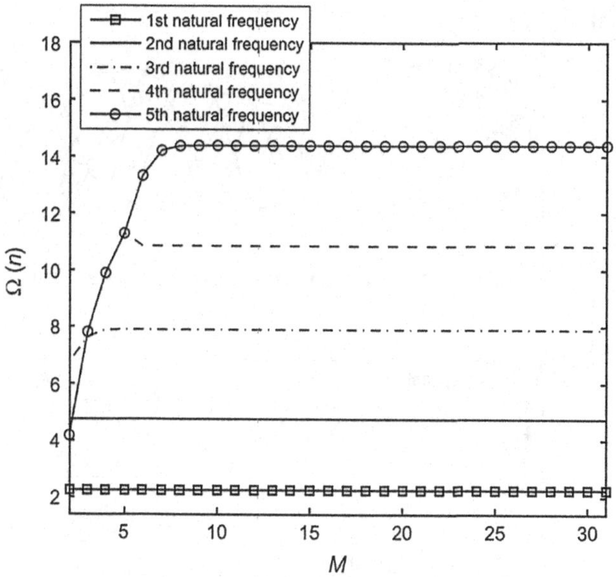

Figure 8.10 The first five dimensionless natural frequencies *X*(*n*) with different number of the series summation limit *M*.

It can be found that the solution of $\frac{d^2\phi(x)}{dx^2}$ using DTM is a continuous function and not discrete numerical values at knot point by finite element or finite difference methods.

To verify the proposed method to design the shaped PVDF modal sensor, an aluminum beam with dimensions of $L_x = 500$ mm, $h_o = 5$ mm, $b_o = 40$ mm is considered. The intermediate support is located at $R_1 = 0.85$ with dimensionless stiffness $k_s = 5$. Assume that modal damping is 0.01. A point force located at $x_0 = \frac{L_x}{20}$ 20 is used as the excitation. The excellent numerical stability of the solution can also be observed in Fig. 8.10.

For simplicity, the DTM solutions are truncated to $M = 20$ in all the subsequent calculations. Fig. 8.11 shows the first 10 mode shapes $\phi(x)$ and the corresponding second spatial derivative of the mode shapes $\frac{d^2\phi(x)}{dx^2}$. For the case of a uniform bending beam with classical boundary conditions (i.e., which are clamped, free, simply supported, or sliding), the second spatial derivative of the mode shapes form an orthogonal set.

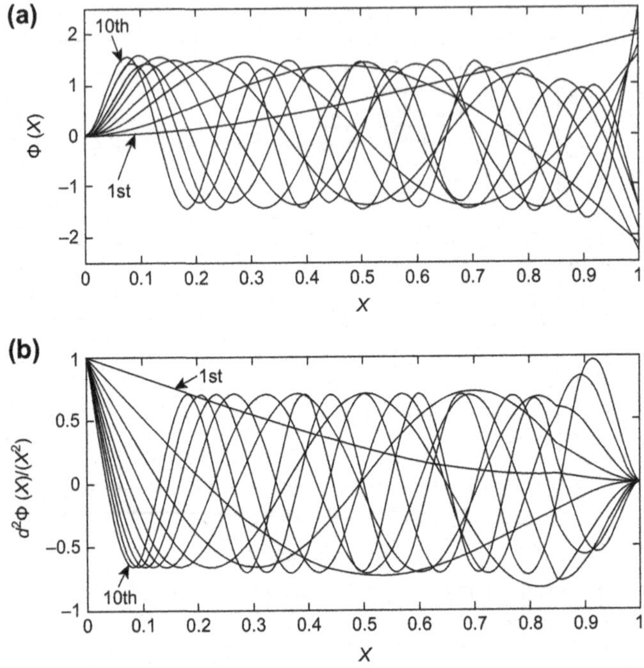

Figure 8.11 (a) The first 10 mode shapes and (b) the corresponding second spatial derivative of the mode shapes.

8.5 DAMPED SYSTEM WITH HIGH NONLINEARITY

The equation of motion of a damped vibration system with high nonlinearity can be expressed as follows [4]:

$$\ddot{x} + \zeta\dot{x} + x + cx^n = 0, \quad n = 2p + 1, \quad p = 0, 1, 2, \ldots \tag{8.65}$$

where the superposed dots (.) denote differentiation with respect to time, ζ is the damping coefficient, c is a constant parameter, and n is the degree of nonlinearity. The initial conditions of $x(t)$ are given by

$$x(0) = 1, \quad \dot{x}(0) = 0 \tag{8.66}$$

transformation domain.

$$\text{Let } y(t) = \dot{x}(t). \tag{8.67}$$

Substituting Eq. (8.67) into Eqs. (8.65) and (8.66) yields

$$\dot{y} + \zeta y + x + cx^n = 0. \tag{8.68}$$

The initial conditions become

$$x(0) = 1, \; y(0) = 0. \tag{8.69}$$

By a process of inverse differential transformation, the solutions of each subdomain take $m+1$ terms for the power series, i.e.,

$$x_i(t) = \sum_{k=0}^{m} \left(\frac{t}{H_i}\right)^k X_i(k), \quad 0 \le t \le H_i \tag{8.70}$$

$$Y_i(t) = \sum_{k=0}^{m} \left(\frac{t}{H_i}\right)^k Y_i(k), \quad 0 \le t \le H_i \tag{8.71}$$

where $i = 0, 1, 2,\ldots,n$ indicates the ith subdomain, $k = 0, 1,2,\ldots,m$ denotes the term of the power series, H_i represents the subdomain interval, and $X_i(k)$ and $Y_i(k)$ are the transformed functions of $x_i(t)$ and $y_i(t)$, respectively. From the initial conditions (Eq. 8.69), it can be seen that

$$X_0(0) = \delta(k), \quad \text{where } \delta(k) = \begin{cases} 1k = 0 \\ 0k \neq 0 \end{cases} \tag{8.72}$$

$$Y_0(0) = 0. \tag{8.73}$$

Eqs. (8.67) and (8.68) undergo differential transformation to yield the following

$$\frac{k+1}{H_i} X_i(k+1) = Y_i(k), \tag{8.74}$$

$$\begin{aligned} &\frac{k+1}{H_i} Y_i(k+1) + \zeta Y_i(k) + X_i(k) \\ &\quad + c \sum_{l=1}^{k} \left(\frac{n+1}{k} L - 1\right) \cdot \frac{X_i(l)}{X_i(0)} \cdot X_i^n(k-L) = 0. \end{aligned} \tag{8.75}$$

The present problem employs the DTM described above to generate a number of numerical results for the response of a damped system with high nonlinearity. The responses of $x(t)$ for different values of nonlinearity, n, and damping coefficient, ζ, are plotted in Fig. 8.12. It is noted that the present results are in excellent agreement with the numerical results obtained from the fourth-order Runge–Kutta approach. Fig. 8.13 presents the displacement and velocity time responses of a damped system with nonlinearity orders of $n = 3$. The results indicate that the amplitude decays are more rapidly for higher values of nonlinearity, particularly at higher values of the damping coefficient.

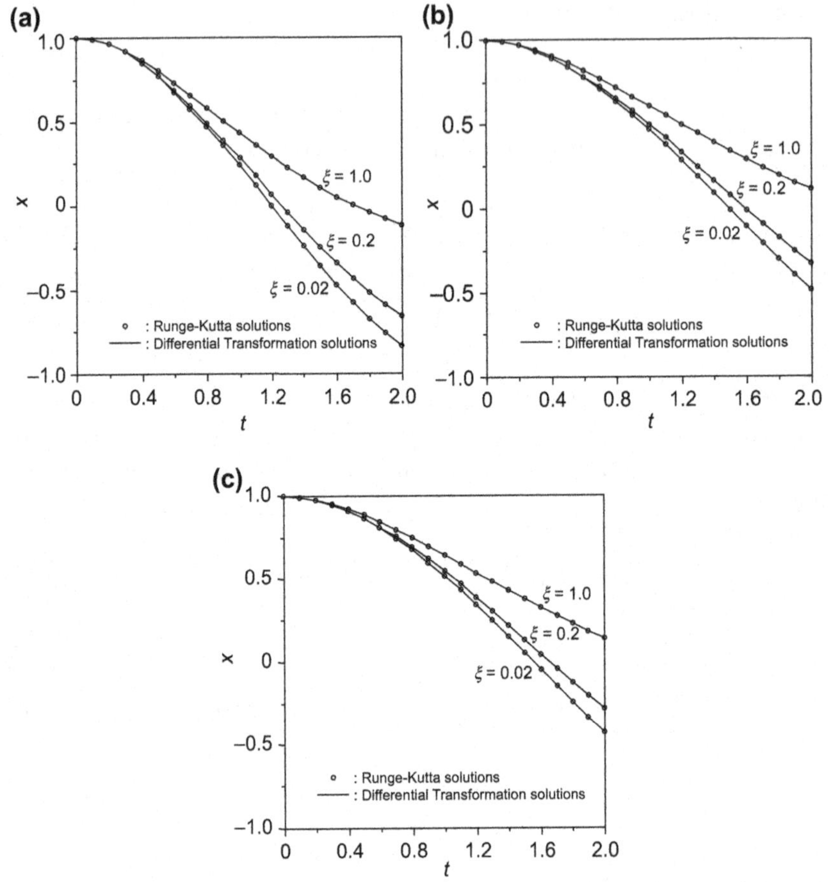

Figure 8.12 Comparison of differential transformation method and Runge–Kutta numerical methods. (a) n = 3. (b) n = 101. (c) n = 1001.

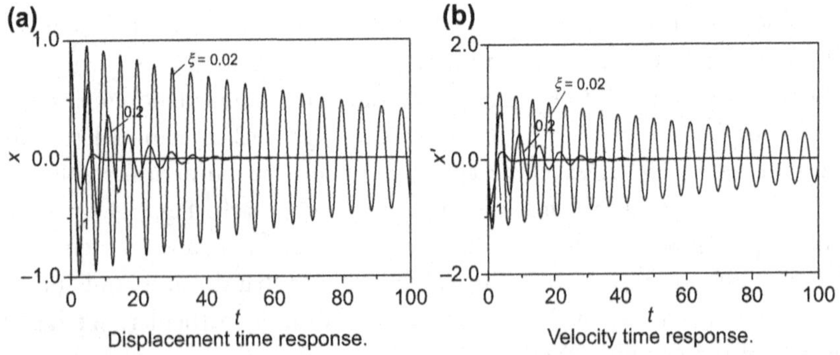

Figure 8.13 Response of damped system for different values of ζ with $n = 3$. (a) Displacement time response and (b) velocity time response ($c = 1.0$).

8.6 FREE VIBRATION OF A CENTRIFUGALLY STIFFENED BEAM

A uniform rotating beam of doubly symmetric cross section is considered and illustrated in Fig. 8.14.

Such a rotating beam vibrates both out-of-plane and in-plane; however, the out-of-plane and in-plane vibrations are uncoupled. The equation of motion for out-of-plane vibration of a centrifugally tensioned uniform Euler–Bernoulli beam is as follows [5];

$$EI_{zz}\frac{\partial^4 y}{\partial x^4} - \rho A\omega^2 y - \frac{\partial}{\partial x}\left(T(x)\frac{\partial y}{\partial x}\right) = 0 \tag{8.76}$$

where $T(x)$ is the centrifugal tension at a distance x from the origin and is given by

$$T(x) = 0.5\rho Ap^2(L^2 + 2R_0L - 2R_0x - x^2) \tag{8.76a}$$

x is the radial coordinate, $y(x)$ is the transverse deflection, R is the radius of hub, L is the length of the beam, A is the cross-sectional area, ρ is the mass density, p is the angular rotational speed, ω is the angular frequency, EI_{zz} is the flexural rigidity for bending in the $x-y$ plane. The dimensionless equation of motion is as follows;

$$\begin{aligned} &D^4\overline{Y}(X) - 0.5\nu(1 + 2\rho_o)D^2\overline{Y}(X) + \nu\rho_o D\left[XD\overline{Y}(X)\right] \\ &+ 0.5\nu D\left(X^2 D\overline{Y}(X)\right) - \mu\overline{Y}(X) = 0 \end{aligned} \tag{8.77}$$

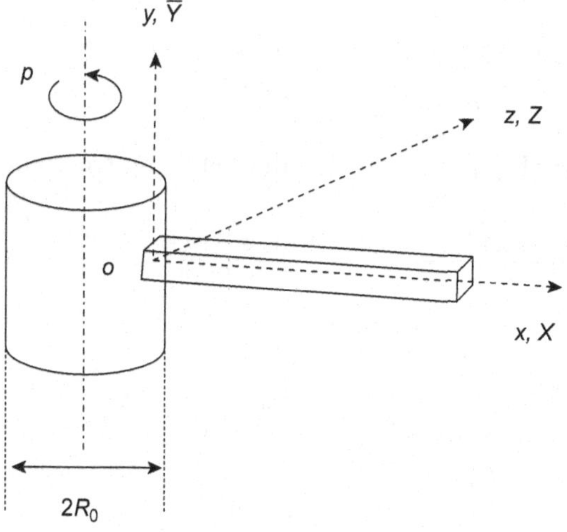

Figure 8.14 A cantilever rotating beam.

where

$$\begin{aligned} &X = x/L, \quad D^n = d^n/dX^n \\ &\overline{Y}(X) = y(x)/L, \quad v = \eta^2 = \rho A p^2 L^4/EI_{zz} \\ &\rho_0 = R_0/L, \quad \mu = \Omega^2 = \rho A \omega^2 L^4/EI_{zz} \end{aligned} \tag{8.78}$$

ρ_0 is the dimensionless offset parameter, η is the dimensionless rotational speed, Ω is the dimensionless frequency, X is the dimensionless radial coordinate, and $\overline{Y}(X)$ is the dimensionless deflection. V and μ are the dimensionless rotational speed and the dimensionless frequency-related parameters. The dimensionless bending moment $M(X)$ and shear force $Q(X)$ are accordingly

$$M(X) = D^2\overline{Y}(X), \quad Q(X) = -DM(X) + \beta(X)D\overline{Y}(X) \tag{8.79}$$

where $\beta(X)$ is the dimensionless tension at coordinate X and is given by

$$\beta(X) = 0.5v(1 + 2\rho_o - 2\rho_o X - X^2) \tag{8.80}$$

The dimensionless boundary conditions of the three most commonly used boundaries are

$$\begin{aligned} &\text{Clamped boundary:} \ \overline{Y}(X) = 0, \ D\overline{Y}(X) = 0; \\ &\text{Free boundary:} \ M(X) = 0, \ Q(X) = 0; \\ &\text{Pinned boundary:} \ \overline{Y}(X) = 0, \ M(X) = 0: \end{aligned} \tag{8.81}$$

From the definition and properties of DTM given in Chapter 1, the DTM of the equation of motion Eq. (8.76) is found as

$$\begin{aligned} &(k+1)(k+2)(k+3)(k+4)Y(k+4) - 0.5v(1+2\rho_o)(k+1) \\ &\times(k+2)Y(k+2) - \mu Y(k) + v\rho_0(k+1)^2 Y(k+1) \\ &+ 0.5vk(k+1)Y(k) = 0 \end{aligned} \tag{8.82}$$

Rearranging Eq. (8.82), one has the following recurrence relation

$$Y(k+4) = \frac{0.5v(1+2\rho_0)(k+1)Y(k+2) + \mu Y(k) - v\rho_0(k+1)^2 Y(k+1) - 0.5vk(k+1)Y(k)}{(k+1)(k+2)(k+3)(k+4)} \tag{8.83}$$

Combining Eq. (8.83) and the appropriate boundary conditions, one obtains solutions to a free vibration problem. As an example, let us consider a rotating cantilever beam. This is because turbine, propeller, and robotic manipulators can all be modeled as cantilevered beams. Performing DTM to the boundary conditions, with origin chosen at the clamped end, one has

$$Y(0) = 0 \tag{8.84}$$

$$Y(1) = 0 \tag{8.85}$$

At the free end, that is, at $X = 1$, one has

$$\sum_{k=0}^{\infty}(k+1)(k+2)Y(k+2) = 0 \tag{8.86}$$

$$\begin{aligned}&\sum_{k=0}^{\infty}(k+1)(k+2)(k+3)Y(k+3)\\&-\sum_{k=0}^{n}0.5\nu(1+2\rho_0)(k+1)Y(k+1)\\&+\sum_{k=0}^{n}\nu\rho_0(k)Y(k)+\sum_{k=2}^{n}0.5\nu(k-1)Y(k-1)=0\end{aligned} \tag{8.87}$$

From Eqs. (8.83)–(8.85), it can be seen that $Y(k)$ is a linear function of $Y(2)$ and $Y(3)$. Eqs. (8.86) and (8.84) can then be written in matrix form as

$$\begin{bmatrix} f_{11} & f_{12} \\ f_{21} & f_{22} \end{bmatrix}\begin{bmatrix} Y(2) \\ Y(3) \end{bmatrix} = 0 \tag{8.88}$$

Setting the determinant of the coefficient matrix of Eq. (8.88) to zero gives the characteristic equation of the structure, from which the natural frequencies are found. The natural frequencies corresponding to various hub offset and rotational speed are computed using Matlab. Though for comparison purpose, the natural frequencies are kept accurate to the fourth decimal place, the precision of the natural frequencies can be as high as the machine precision of the computer used. Both the rotational speed and the offset are seen to affect the natural frequencies greatly. Results for three representative cases are presented in Fig. 8.15.

- Case 1: Dimensionless offset $\rho_0 = 0$ and dimensionless rotational speed $\eta = 0$.
- Case 2: Dimensionless offset $\rho_0 = 3$ and dimensionless rotational speed $\eta = 4$.
- Case 3: Dimensionless offset $\rho_0 = 1$ and dimensionless rotational speed $\eta = 15$.

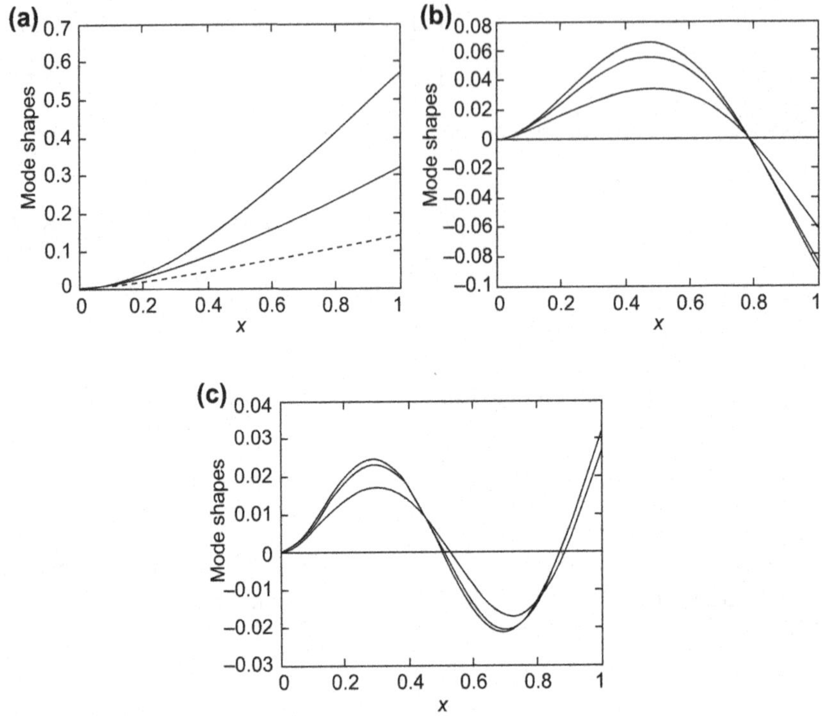

Figure 8.15 Mode shapes of the first three modes corresponding to Case 1 (—), Case 2 (- - -), and Case 3 (. . .). (a) First mode, (b) second mode, and (c) third mode.

8.7 DEFLECTIONS OF ORTHOTROPIC RECTANGULAR PLATE

This example considers large deflections of the orthotropic rectangular plate. The plate has thickness "h" and edge lengths "a" and "b" in the x- and y-directions, respectively. Let O_{xyz} be a Cartesian coordinate system which lies in the midplane of the plate and has its origin at the corner. The lateral loading on the plate is denoted by $p(t)$. Under these conditions, the equations of motion for the plate are given [6].

$$u_{xx} + w_x w_{xx} + c_1\left(u_{yy} + w_x w_{yy}\right) + c_2\left(v_{xy} + w_y w_{xy}\right) = 0 \tag{8.89}$$

$$c_1\left(v_{xx} + w_y w_{xxyy}\right) + c_3\left(v_{yy} + w_y w_{yy}\right) + c_2\left(u_{xy} + w_x w_{xy}\right) = 0 \tag{8.90}$$

$$w_{xxxx} + 2c_4 w_{xxyy} + c_3 w_{yyyy} + \frac{\rho h}{d_1} w_u + \frac{\delta}{d_1} w_t - \frac{p}{d_1} - \frac{12}{h^2}$$

$$\left\{ w_{xx} \left[u_x + v_{21} v_y + \frac{1}{2} w_x^2 + \frac{1}{2} v_{21} w_y^2 \right] + c_3 w_{yy} \left[v_{12} u_{xx} + v_y + \frac{1}{2} w_x^2 \frac{1}{2} w_y^2 \right] \right.$$

$$\left. + 2c_1 \left[u_y + v_x + w_x w_y \right] \right\} = 0,$$

(8.91)

where $c_1 = \frac{\mu g_{12}}{e_1}$, $c_2 = c_1 + v_{21}$, $c_3 = \frac{e_2}{e_1}$, $c_4 = 2c_1 + v_{21}$, $D = \frac{e_1 h^3}{12\mu}$, $\mu = 1 - v_{21} v_{12}$ and $v_{12} e_2 = v_{21} e_1$. Furthermore, $w(x, y, t)$ is the deflection of the neutral plane of the plate, $u(x, y, t)$ and $v(x, y, t)$ are the displacements of the neutral plane of the plate in the x- and y-directions, respectively. Additionally, q is the density of the plate material, h is the plate thickness, and E_1 and E_2 are moduli of elasticity along the principal material axes x and y, respectively, G_{12} is the modulus of rigidity characterizing the change of the angle between the principal material directions, and v_{21} and v_{12} are Poisson's ratios with the first subscript indicating the direction of the tensile force and the second indicating the direction of contraction. d is the damping coefficient of the plate material.

This problem considers two types of boundary conditions:

1. The boundary conditions at all of the edges are clamped and immovable:

$$w = \frac{\partial w}{\partial x} = u = v = 0 \quad \text{at } x = 0, a, \tag{8.92}$$

$$w = \frac{\partial w}{\partial y} = u = v = 0 \quad \text{at } y = 0, a, \tag{8.93}$$

2. The boundary conditions at all of the edges are simply supported and immovable:

$$w = \frac{\partial^2 w}{\partial x^2} = u = v = 0 \quad \text{at } x = 0, a, \tag{8.94}$$

$$w = \frac{\partial^2 w}{\partial y^2} = u = v = 0 \quad \text{at } y = 0, a, \tag{8.95}$$

To reduce the order of the system dynamic equation, this study introduces the following velocity function:

$$g = \frac{\partial w}{\partial t}, \tag{8.96}$$

The system dynamic equation is then given by

$$\frac{\partial w}{\partial t} = g, \tag{8.97}$$

$$w_{xxxx} + 2c_4 w_{xxyy} + c_3 w_{yyyy} + \frac{\rho h}{d_1} w_u + \frac{\delta}{d_1} w_t - \frac{p}{d_1} - \frac{12}{h^2}$$

$$\left\{ w_{xx} \left[u_x + v_{21} v_y + \frac{1}{2} w_x^2 + \frac{1}{2} v_{21} w_y^2 \right] + c_3 w_{yy} \left[v_{12} u_{xx} + v_y + \frac{1}{2} w_x^2 \frac{1}{2} w_y^2 \right] \right.$$

$$\left. + 2c_1 \left[u_y + v_x + w_x w_y \right] \right\} = 0, \tag{8.98}$$

Taking the differential transform of the system equation with respect to time, it can be shown that

$$\frac{k+1}{H} w(x, y, k+1) = g(x, y, k) \tag{8.99}$$

$$w(x, y, k)_{xxxx} + 2c_4 w(x, y, k)_{xxyy} + c_3 w(x, y, k)_{yyyy} +$$

$$\frac{\rho h}{d_1} \frac{k+1}{h} g(x, y, k) + \frac{\delta}{d_1} g(x, y, k) - \frac{p}{d_1} - \frac{12}{h^2}$$

$$\left\{ w(x, y, k)_{xx} \otimes \left[u_x(x, y, k) + v_{21} v(x, y, k)_y + \frac{1}{2} w(x, y, k)_x \otimes w(x, y, k)_x^+ \right. \right.$$

$$\left. \left. + \frac{1}{2} v_{21} w(x, y, k)_y \otimes (x, y, k)_y \right] \right\} -$$

$$\left\{ \otimes c_3 w_{yy} \left[v_{12} u(x, y, k)_{xx} + v(x, y, k)_y + \frac{1}{2} v_{12} w(x, y, k)_y \otimes w(x, y, k)_x^+ \right] \right\}$$

$$+ \left\{ \frac{1}{2} w(x, y, k)_y \otimes w(x, y, k)_y + 2c_1 \left[u(x, y, k)_y + v(x, y, k)_x \right. \right.$$

$$\left. \left. + w(x, y, k)_x \otimes w(x, y, k)_y \right] \right\} = 0, \tag{8.100}$$

where $w(x, y, k)$; $g(x, y, k)$; $u(x, y, k)$; and $v(x, y, k)$ are the differential transforms of $w(x, y, T)$; $g(x, y, T)$; $u(x, y, T)$; and $v(x, y, T)$, respectively.

$$w(x, y, k)_y \otimes w(x, y, k)_X = \sum_{l=0}^{K} w(x, y, k-l)_x w(x, y, l)_x l. \tag{8.101}$$

$$w(x,y,k)_{xx} \otimes w(x,y,k)_X \otimes w(x,y,k)_X$$
$$= \sum_{l=0}^{K} w(x,y,k-l)_{xx} \sum_{m=0}^{l} w(x,y,l-m)_x l, w(x,y,k)_X. \tag{8.102}$$

By taking the differential transform and the finite difference approximation, the associated boundary conditions are transformed to:

1. All edges are clamped

$$w(1,j,k)=0, \quad w(i,1,k)=0, \quad w(m-1,j,k)=0 \quad w(i,n-1,k)=0, \tag{8.103}$$

$$u(1,j,k)=0, \quad u(i,1,k)=0, \quad u(m-1,j,k)=0 \quad u(i,n-1,k)=0, \tag{8.104}$$

$$v(1,j,k)=0, \quad v(i,1,k)=0, \quad v(m-1,j,k)=0 \quad v(i,n-1,k)=0, \tag{8.105}$$

2. All edges are simply supported

$$w(1,j,k)=0, \quad w(i,1,k)=0, \quad w(m-1,j,k)=0 \quad w(i,n-1,k)=0, \tag{8.106}$$

$$u(1,j,k)=0, \quad u(i,1,k)=0, \quad u(m-1,j,k)=0 \quad u(i,n-1,k)=0, \tag{8.107}$$

$$v(1,j,k)=0, \quad v(i,1,k)=0, \quad v(m-1,j,k)=0 \quad v(i,n-1,k)=0, \tag{8.108}$$

$$\frac{w(2,j,k)-2w(1,j,k)+w(0,j,k)}{h_x^2}=0, \quad \frac{w(i,2,k)-2w(i,1,k)+w(i,0,k)}{h_y^2}=0, \tag{8.109}$$

$$\begin{aligned} &\frac{w(m,j,k)-2w(m-1,j,k)+w(m-2,j,k)}{h_x^2}=0, \\ &\frac{w(i,n,k)-2w(i,n-1,k)+w(i,n-2,k)}{h_y^2}=0, \end{aligned} \tag{8.110}$$

In investigating the dynamic motion and large deflections of the orthotropic rectangular plate, this study initially considered the case of a plate with $\frac{E}{E} = 3$, $\frac{G}{E} = 0.5$, $\upsilon_{21} = 0.25$, $\upsilon_{12} = 0.08333$, $\delta = 10$, and $E_1 = 1$, subjected to a variable lateral load of P. The dynamic motion and the large deflections were solved using the proposed hybrid method combining the finite difference method and the differential transform method. Simulations were performed with $H = 0.01$, $k = 5$, $a = 1.0$, and $b = 1.0$. The edges of the plate were divided into 10 equal units such that $m = n = 11$. It was assumed that the edges of the plate were all clamped and that a lateral step force load of $P = 156.34$ were applied.

Fig. 8.16 shows the dynamic motion of the center of the plate at $x = 0.5$ and $y = 0.5$. The corresponding deflections of the center of the orthotropic plate are shown in Fig. 8.17. It can be seen that the deflection varies

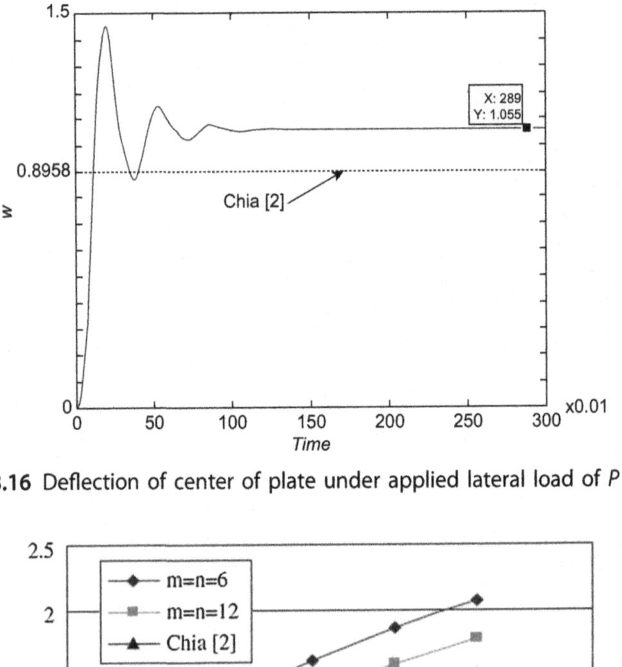

Figure 8.16 Deflection of center of plate under applied lateral load of $P = 100$.

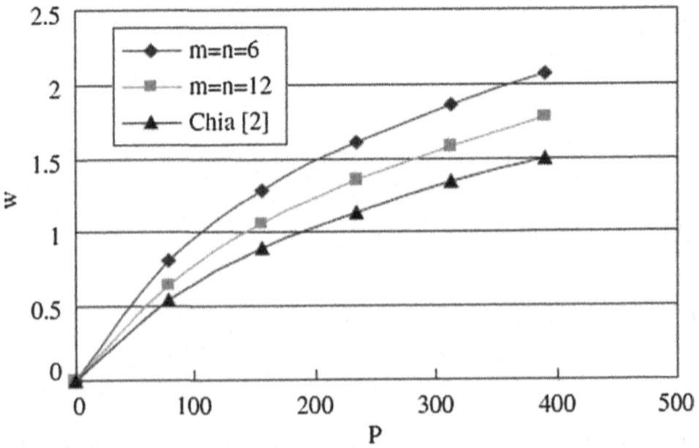

Figure 8.17 Load–deflection curves for clamped rectangular plate under increasing uniform lateral load.

nonlinearly with the lateral force. Furthermore, it is apparent that the current deflection results are in good general agreement with those of Refs. [2] and [6].

8.8 FREE VIBRATION OF CIRCULAR PLATES

The governing differential equation of a thin circular plate undergoing free harmonic vibration in a nondimensional form is as follows [7]:

$$\nabla^4 w = \Omega^2 w \tag{8.111}$$

where ∇^4 is the biharmonic operator, $W = W(r, \theta)$ is dimensionless deflection, r is dimensionless coordinate along the radial axis of the plate, θ is dimensionless coordinate along the tangential axis, and Ω is dimensionless frequency of vibration. From the classical plate vibration theory, deflection of a circular plate in polar coordinates may be expressed as follows:

$$w = f(r)\cos(m\theta), \tag{8.112}$$

where m is the integer number of nodal diameters and $f(r)$ is the radial mode function. Substituting Eq. (8.112) into Eq. (8.111), the governing differential equation of the circular plate becomes:

$$\frac{d^4 f}{dr^4} + \frac{2}{r}\frac{d^3 f}{dr^3} - \frac{B}{r^2}\frac{d^2 f}{dr^2} + \frac{B}{r^3}\frac{df}{dr} + \frac{A}{r^4} f = \Omega^2 f, \tag{8.113}$$

where

$$A = m^4 - 4m^2 \text{ and } B = 2m^2 + 1. \tag{8.114}$$

The boundary conditions at the outer edge ($r = 1$) of the circular plate may be one of the following; simply supported, clamped, and free. These conditions may be written in terms of the radial mode function $f(r)$ as follows:

Simply supported:

$$f(r)|_{r=1} = 0, M_r|_{r=1} = -D\left[\frac{d^2 f}{dr^2} + \nu\left(\frac{1}{r}\frac{df}{dr} + \frac{m^2}{r^2} f\right)\right] = 0. \tag{8.115}$$

Clamped:

$$f(r)|_{r=1} = 0, \left.\frac{df}{dr}\right|_{r=1} = 0 \tag{8.116}$$

Free:

$$M_r\big|_{r=1} = -D\left[\frac{d^2 f}{dr^2} + \nu\left(\frac{1}{r}\frac{df}{dr} + \frac{m^2}{r^2}f\right)\right] = 0,$$

$$V_r\big|_{r=1} = \left[\frac{d^3 f}{dr^3} + \frac{1}{r}\frac{d^2 f}{dr^2} + \left(\frac{m^2\nu - 2m^2 - 1}{r^2}\right)\frac{df}{dr} + \left(\frac{3m^2 - m^2\nu}{r^3}\right)f\right] = 0, \tag{8.117}$$

where M_r is the radial bending moment, V_r is the effective radial shear force, D is the flexural rigidity, and m is the Poisson's ratio. It can easily be noticed that, since Eq. (8.113) is a fourth-order differential equation, four initial conditions are required. One may obtain two of those from the boundary conditions at the outer edge of the circular plate. However, remaining two conditions must be investigated within the regularity conditions at the center of the plate.

Antisymmetric case:

$$f(r)\big|_{r=0} = 0,\ \ M_r\big|_{r=0} = \left.\frac{d^2 f}{dr^2}\right|_{r=0} = 0 \quad \text{for} \quad (m = 1, 3, 5, \ldots) \tag{8.118}$$

Symmetric case:

$$\left.\frac{df}{dr}\right|_{r=0} = 0,\ \ V_r\big|_{r=0} = \left.\frac{d^3 f}{dr^3}\right|_{r=0} = 0 \quad \text{for} \quad (m = 0, 2, 4, 6, \ldots) \tag{8.119}$$

A. Transformation of Free Vibration Equation

Using the transformation operations defined in Chapter 1 and taking the differential transform of Eq. (8.113) at $r_0 = 0$, one may obtain:

$$\begin{aligned}
&AF_k + B\sum_{l=0}^{k}\delta(l-1)(k-l+1)F_{k-l+1} \\
&- B\sum_{l=0}^{k}\delta(l-2)(k-l+1)(k-l+2)F_{k-l+2} + 2\sum_{l=0}^{k}\delta(l-3)(k-l+1) \\
&*(k-l+2)(k-l+3)F_{k-l+3} \\
&+ \sum_{l=0}^{k}\delta(l-4)(k-l+1)(k-l+2)(k-l+3)(k-l+4)F_{k-l+4} \\
&= \Omega^2\sum_{l=0}^{k}\delta(l-4)F_{k-l}.
\end{aligned} \tag{8.120}$$

Simplifying Eq. (8.120) and using the last theorem, the equation of motion in Eq. (8.113) can be transformed into the following recurrence equation:

$$F_{k+4} = \frac{\Omega^2}{A - B(k+4)(k+2) + (k+4)(k+3)^2(k+2)} F_k. \tag{8.121}$$

From Eq. (8.121), the following equations can be obtained for $k = 0$, 1, 2,…,n:

$$F_4 = \frac{\Omega^2 F_0}{A - B(4.2) + (4.3^2.2)},\quad F_5 = \frac{\Omega^2 F_1}{A - B(5.3) + (5.4^2.3)},\quad F_6 = \frac{\Omega^2 F_2}{A - B(6.4) + (6.5^2.4)},$$

$$F_7 = \frac{\Omega^2 F_3}{A - B(7.5) + (7.6^2.5)},\quad F_8 = \frac{\Omega^2 F_4}{A - B(8.6) + (8.7^2.6)},\quad F_9 = \frac{\Omega^2 F_5}{A - B(9.7) + (9.8^2.7)},\ldots$$

and in general they can be formulated as follows:

$$F_{4k} = \frac{\Omega^{2k} F_0}{\prod_{k=1}^{k}\left(A - B(4k)(4k-2) + (4k)(4k-1)^2(4k-2)\right)}, \tag{8.122}$$

$$F_{4k+1} = \frac{\Omega^{2k} F_1}{\prod_{k=1}^{k}\left(A - B(4k+1)(4k-1) + (4k+1)(4k)^2(4k-1)\right)}, \tag{8.123}$$

$$F_{4k+2} = \frac{\Omega^{2k} F_2}{\prod_{k=1}^{k}\left(A - B(4k+2)(4k) + (4k+2)(4k+1)^2(4k)\right)}, \tag{8.124}$$

$$F_{4k+3} = \frac{\Omega^{2k} F_3}{\prod_{k=1}^{k}\left(A - B(4k+3)(4k+1) + (4k+3)(4k+2)^2(4k+1)\right)}, \tag{8.125}$$

In terms of the transforms appearing in Eqs. (8.122)–(8.125), Eq. (8.111) may be written as follows:

$$f(r) = \sum_{k=0}^{\infty} r^{4k} F_{4k} + \sum_{k=0}^{\infty} r^{4k+1} F_{4k+1} + \sum_{k=0}^{\infty} r^{4k+2} F_{4k+2} + \sum_{k=0}^{\infty} r^{4k+3} F_{4k+3}. \tag{8.126}$$

B. Transformation of Boundary/Regularity Conditions

The following equations are obtained by applying the theorems in Chapter 1 to the boundary conditions at the outer edge ($r = 1$);

Simply supported:

$$\sum_{k=0}^{\infty} F_{4k} = 0, \quad \sum_{k=0}^{\infty} (k(k-1) + \nu k - m^2\nu) F_k = 0. \tag{8.127}$$

Clamped:

$$\sum_{k=0}^{\infty} F_k = 0, \quad \sum_{k=0}^{\infty} kF_k = 0. \tag{8.128}$$

Free:

$$\sum_{k=0}^{\infty} (k(k-1) + \nu k - m^2\nu) F_k = 0,$$
$$\sum_{k=0}^{\infty} (k(k-1)(k-2) + k(k-1) + (m^2\nu - 2m^2 - 1)k + (3m^2 - m^2\nu)) F_k = 0. \tag{8.129}$$

At the center of the circular plate ($r = 0$), the boundary conditions which are derived from regularity conditions can be transformed as follows:

Antisymmetric case:

$$F_0 = F_2 = F_4 = \ldots F_{4K} = F_{4k+2} = 0 \quad \text{for } (m = 1, 3, 5, \ldots). \tag{8.130}$$

Symmetric case:

$$F_1 = F_3 = F_5 = \ldots F_{4k+1} = F_{4k+3} = 0 \quad \text{for } (m = 0, 2, 4, 6 \ldots). \tag{8.131}$$

To avoid the unnecessary repeating of derivations, we consider only one particular case, that is, a circular plate simply supported at the outer edge having symmetric modes ($m = 0, 2, 4, \ldots$). The boundary conditions at $r = 1$ introduced in Eq. (8.127) can be written as follows:

$$\sum_{k=0}^{\infty} F_{4k} + \sum_{k=0}^{\infty} F_{4k+1} + \sum_{k=0}^{\infty} F_{4k+2} + \sum_{k=0}^{\infty} F_{4k+3} = 0, \tag{8.132}$$

$$\sum_{k=0}^{\infty}((4k)(4k-1)+v4k-m^2v)F_{4k}+\sum_{k=0}^{\infty}((4k+1)(4k)+v(4k+1)$$
$$-m^2v)F_{4k+1}+\sum_{k=0}^{\infty}((4k+2)(4k+1)+v(4k+2)-m^2v)F_{4k+2}$$
$$+\sum_{k=0}^{\infty}((4k+2)(4k+3)(4k+2)+v(4k+3)-m^2v)F_{4k+3}=0 \tag{8.133}$$

It must be noted that Eqs. (8.132) and (8.133) are valid for both symmetric and antisymmetric cases. For the symmetric modes, Eq. (8.131) must be taken into account. Substituting Eqs. (8.122) and (8.124) into the boundary conditions in Eqs. (8.132) and (8.133), the following expressions are obtained:

$$\sum_{k=0}^{\infty}\frac{\Omega^{2k}F_0}{\prod_{k=1}^{k}\left(A-B(4k)(4k-2)+(4k)(4k-1)^2(4k-2)\right)}$$
$$+\sum_{k=0}^{\infty}\frac{\Omega^{2k}F_2}{\prod_{k=1}^{k}\left(A-B(4k+2)(4k)+(4k+2)(4k+1)^2(4k)\right)}=0, \tag{8.134}$$

$$\sum_{k=0}^{\infty}\frac{((4k)(4k-1)+v4k-m^2v)\Omega^{2k}F_0}{\prod_{k=1}^{k}\left(A-B(4k)(4k-2)+(4k)(4k-1)^2(4k-2)\right)}$$
$$+\sum_{k=0}^{\infty}\frac{((4k+2)(4k+1)+v(4k+2)-m^2v)\Omega^{2k}F_2}{\prod_{k=1}^{k}\left(A-B(4k+2)(4k)+(4k+2)(4k-1)^2(4k)\right)}=0. \tag{8.135}$$

Eqs. (8.134) and (8.135) can be grouped and rewritten as follows:

$$\begin{aligned} X_{11}^{(n)}(\Omega)F_0+X_{12}^{(n)}(\Omega)F_2&=0,\\ X_{21}^{(n)}(\Omega)F_0+X_{22}^{(n)}(\Omega)F_2&=0, \end{aligned} \tag{8.136}$$

where $X_{11}^{(n)}$, $X_{12}^{(n)}$, $X_{21}^{(n)}$, and $X_{22}^{(n)}$ are closed-form polynomials of Ω corresponding to the nth term. It can be clearly seen that $X_{11}^{(n)}$, $X_{12}^{(n)}$ and $X_{21}^{(n)}$, $X_{22}^{(n)}$ terms represent the closed-form series expressions in Eqs. (8.134) and (8.135). Eq. (8.136) can be expressed in the matrix form:

$$\begin{bmatrix} X_{11}^{(n)}(\Omega) & X_{12}^{(n)}(\Omega) \\ X_{21}^{(n)}(\Omega) & X_{22}^{(n)}(\Omega) \end{bmatrix} \begin{Bmatrix} F_0 \\ F_2 \end{Bmatrix} = \begin{Bmatrix} 0 \\ 0 \end{Bmatrix}. \tag{8.137}$$

The frequency equation of the circular plate is obtained by setting the determinant of the coefficient matrix of Eq. (8.137) equal to zero:

$$\begin{vmatrix} X_{11}^{(n)}(\Omega) & X_{12}^{(n)}(\Omega) \\ X_{21}^{(n)}(\Omega) & X_{22}^{(n)}(\Omega) \end{vmatrix} = 0. \tag{8.138}$$

Performing the nontrivial solution in Eq. (8.138), we get $\Omega = \Omega_j^{(n)}$, where $j = 1, 2, 3, \ldots, n$, where $\Omega_j^{(n)}$ is the jth estimated eigenvalue corresponding to n. The value of n can be obtained by the following equation:

$$\left|\Omega_j^{(n)} - \Omega_j^{(n-1)}\right| \le \xi, \tag{8.139}$$

where ξ is the tolerance parameter that is taken as $\xi = 0.00001$ in this example. If Eq. (8.139) is satisfied, then we get jth eigenvalue Ω_j. The corresponding eigen function, $f(r)$, describing the instantaneous deflected shape of the circular plate can be obtained by

$$f(r) = \sum_{k=0}^{\infty} \frac{r^{4k}\Omega^{2k}}{\prod_{k=1}^{k}\left(A - B(4k)(4k-2) + (4k)(4k-1)^2(4k-2)\right)}$$
$$- \left\{ \frac{\sum_{k=0}^{\infty} \frac{\Omega^{2k}}{\prod_{k=1}^{k}\left(A - B(4k)(4k-2) + (4k)(4k-1)^2(4k-2)\right)}}{\sum_{k=0}^{\infty} \frac{\Omega^{2k}}{\prod_{k=1}^{k}\left(A - B(4k+2)(4k) + (4k+2)(4k+1)^2(4k)\right)}} \right\} \tag{8.140}$$
$$* \left\{ \sum_{k=0}^{\infty} \frac{r^{4k+2}\Omega^{2k}}{\prod_{k=1}^{k}\left(A - B(4k+2)(4k) + (4k+2)(4k+1)^2(4k)\right)} \right\}$$

Following the similar procedure, the frequency equations for other types of boundary and regularity conditions are derived as follows:

Simply supported and symmetric case:

$$
\begin{aligned}
&\left\{\sum_{k=0}^{\infty}\frac{\Omega^{2k}}{\prod_{k=1}^{k}\left(A-B(4k)(4k-2)+(4k)(4k-1)^2(4k-2)\right)}\right\}\\
&\cdot\left\{\sum_{k=0}^{\infty}\frac{((4k+2)(4k+1)+v(4k+2)-m^2v)\Omega^{2k}}{\prod_{k=1}^{k}\left(A-B(4k+2)(4k)+(4k+2)(4k+1)^2(4k)\right)}\right\}\\
&-\left\{\sum_{k=0}^{\infty}\frac{\Omega^{2k}}{\prod_{k=1}^{k}\left(A-B(4k+2)(4k)+(4k+2)(4k+1)^2(4k)\right)}\right\}\\
&\cdot\left\{\sum_{k=0}^{\infty}\frac{((4k)(4k-1)+v(4k)-m^2v)\Omega^{2k}}{\prod_{k=1}^{k}\left(A-B(4k)(4k-2)+(4k)(4k-1)^2(4k-2)\right)}\right\}=0.
\end{aligned}
\tag{8.141}
$$

Simply supported and antisymmetric case:

$$
\begin{aligned}
&\left\{\sum_{k=0}^{\infty}\frac{\Omega^{2k}}{\prod_{k=1}^{k}\left(A-B(4k+1)(4k-1)+(4k+1)(4k)^2(4k-1)\right)}\right\}\\
&\cdot\left\{\sum_{k=0}^{\infty}\frac{((4k+3)(4k+2)+v(4k+3)-m^2v)\Omega^{2k}}{\prod_{k=1}^{k}\left(A-B(4k+3)(4k+1)+(4k+3)(4k+2)^2(4k+1)\right)}\right\}\\
&-\left\{\sum_{k=0}^{\infty}\frac{\Omega^{2k}}{\prod_{k=1}^{k}\left(A-B(4k+3)(4k+1)+(4k+3)(4k+2)^2(4k+1)\right)}\right\}\\
&\cdot\left\{\sum_{k=0}^{\infty}\frac{((4k+1)(4k)+v(4k+1)-m^2v)\Omega^{2k}}{\prod_{k=1}^{k}\left(A-B(4k+1)(4k-1)+(4k+1)(4k)^2(4k-1)\right)}\right\}=0.
\end{aligned}
\tag{8.142}
$$

Clamped and symmetric case:

$$\begin{aligned}&\left\{\sum_{k=0}^{\infty}\frac{\Omega^{2k}}{\prod_{k=1}^{k}\left(A-B(4k)(4k-2)+(4k)(4k-1)^2(4k-2)\right)}\right\}\\&\cdot\left\{\sum_{k=0}^{\infty}\frac{(4k+2)\Omega^{2k}}{\prod_{k=1}^{k}\left(A-B(4k+2)(4k)+(4k+2)(4k+1)^2(4k)\right)}\right\}\\&-\left\{\sum_{k=0}^{\infty}\frac{\Omega^{2k}}{\prod_{k=1}^{k}\left(A-B(4k+2)(4k)+(4k+2)(4k+1)^2(4k)\right)}\right\}\\&\cdot\left\{\sum_{k=0}^{\infty}\frac{(4k)\Omega^{2k}}{\prod_{k=1}^{k}\left(A-B(4k)(4k-2)+(4k)(4k-1)^2(4k-2)\right)}\right\}=0.\end{aligned}\tag{8.143}$$

Clamped and antisymmetric case:

$$\begin{aligned}&\left\{\sum_{k=0}^{\infty}\frac{\Omega^{2k}}{\prod_{k=1}^{k}\left(A-B(4k+1)(4k-1)+(4k+1)(4k)^2(4k-1)\right)}\right\}\\&\cdot\left\{\sum_{k=0}^{\infty}\frac{(4k+3)\Omega^{2k}}{\prod_{k=1}^{k}\left(A-B(4k+3)(4k+1)+(4k+3)(4k+2)^2(4k+1)\right)}\right\}\\&-\left\{\sum_{k=0}^{\infty}\frac{\Omega^{2k}}{\prod_{k=1}^{k}\left(A-B(4k+3)(4k+1)+(4k+3)(4k+2)^2(4k+1)\right)}\right\}\\&\cdot\left\{\sum_{k=0}^{\infty}\frac{(4k+1)\Omega^{2k}}{\prod_{k=1}^{k}\left(A-B(4k+1)(4k-1)+(4k+1)(4k)^2(4k-1)\right)}\right\}=0.\end{aligned}\tag{8.144}$$

Free and symmetric case:

$$\left\{\sum_{k=0}^{\infty}\frac{((4k)(4k-1)+v4k-m^2v)\Omega^{2k}}{\prod_{k=1}^{k}(A-B(4k)(4k-2)+(4k)(4k-1)^2(4k-2))}\right\}$$
$$\cdot\left\{\sum_{k=0}^{\infty}\frac{((4k+2)(4k+1)^2+(m^2v-2m^2-1)(4k+2)+(3m^2-m^2v))\Omega^{2k}}{\prod_{k=1}^{k}(A-B(4k+2)(4k)+(4k+2)(4k+1)^2(4k))}\right\}$$
$$-\left\{\sum_{k=0}^{\infty}\frac{((4k+2)(4k+1)+v(4k+2)-m^2v)\Omega^{2k}}{\prod_{k=1}^{k}(A-B(4k+2)(4k)+(4k+2)(4k+1)^2(4k))}\right\}$$
$$\cdot\left\{\sum_{k=0}^{\infty}\frac{((4k)(4k-1)^2+(m^2v-2m^2-1)4k+(3m^2-m^2v))\Omega^{2k}}{\prod_{k=1}^{k}(A-B(4k)(4k-2)+(4k)(4k-1)^2(4k-2))}\right\}=0. \tag{8.145}$$

Free and antisymmetric case:

$$\left\{\sum_{k=0}^{\infty}\frac{((4k+1)(4k)+v(4k+1)-m^2v)\Omega^{2k}}{\prod_{k=1}^{k}(A-B(4k+1)(4k-1)+(4k+1)(4k)^2(4k-1))}\right\}$$
$$\cdot\left\{\sum_{k=0}^{\infty}\frac{((4k+3)(4k+2)^2+(m^2v-2m^2-1)(4k+3)+(3m^2-m^2v))\Omega^{2k}}{\prod_{k=1}^{k}(A-B(4k+3)(4k+1)+(4k+3)(4k+2)^2(4k+1))}\right\}$$
$$-\left\{\sum_{k=0}^{\infty}\frac{((4k+3)(4k+2)+v(4k+3)-m^2v))\Omega^{2k}}{\prod_{k=1}^{k}(A-B(4k+3)(4k+1)+(4k+3)(4k+2)^2(4k+1))}\right\}$$
$$\cdot\left\{\sum_{k=0}^{\infty}\frac{((4k+3)(4k+2)^2+(m^2v-2m^2-1)(4k+3)+(3m^2-m^2v))\Omega^{2k}}{\prod_{k=1}^{k}(A-B(4k+1)(4k-1)+(4k+1)(4k)^2(4k-1))}\right\}=0. \tag{8.146}$$

In Fig. 8.18, the convergence of the first four natural frequencies with respect to the number of terms considered is presented for the free end condition and $m = 1$. It is observed from this figure that evaluating the

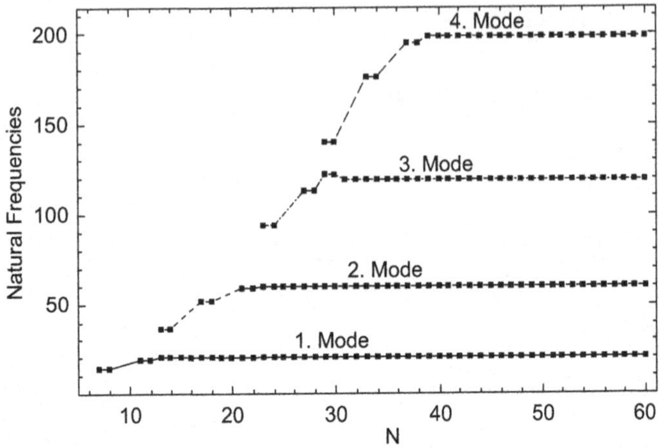

Figure 8.18 Convergence of the natural frequencies for $m = 1$ and free end boundary conditions.

higher natural frequencies needs more number of terms to be considered. Also note that the accuracy of the method increases dramatically with the number of terms taken into consideration.

Fig. 8.19 shows the first three mode shapes for the clamped, simply supported, and free end conditions. Only two cases of the nodal diameter, $m = 0$ and $m = 1$, are considered as examples of symmetric and antisymmetric regularity conditions.

8.9 VIBRATION OF PIPES CONVEYING FLUID

Consider an elastic, straight, fluid-conveying pipe with length l between two ends, as shown in Fig. 8.20, the linear equation of motion is given by [8].

$$EI\frac{\partial^4 w}{\partial x^4} + MU^2\frac{\partial^2 w}{\partial x^2} + 2MU\frac{\partial^2 w}{\partial x \partial t} + (M+m)\frac{\partial^2 w}{\partial t^2} = 0, \tag{8.147}$$

where EI is the flexural rigidity, M and m are the mass per unit length of fluid and the pipe, respectively, U is the fluid flow velocity, $w(x, t)$ is the transverse deflection of the pipe, x is the horizontal coordinate along the centerline of the pipe, and t is the time. Introducing the following nondimensional quantities

$$\xi = \frac{x}{l}, \quad \beta = \frac{M}{M+m}, \quad \tau = \frac{t}{L^2}\left(\frac{EI}{M+m}\right)^{\frac{1}{2}}, \quad u = UL\left(\frac{M}{EI}\right)^{\frac{1}{2}}. \tag{8.148}$$

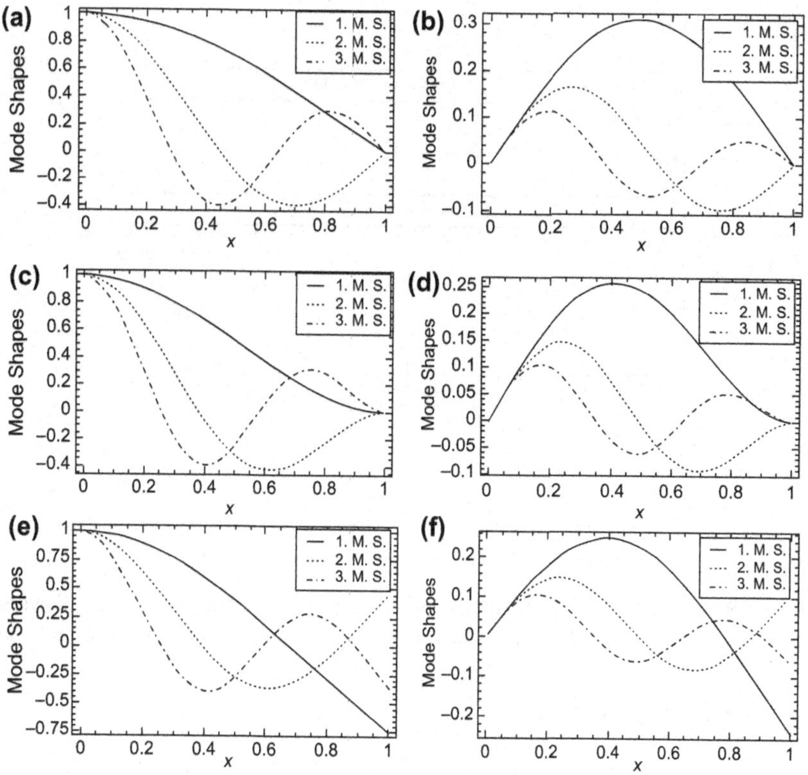

Figure 8.19 Mode shapes for (a) simply supported and $m = 0$; (b) simply supported and $m = 1$; (c) clamped and $m = 0$; (d) clamped and $m = 1$; (e) free and $m = 0$; (f) free and $m = 1$.

Eq. (8.147) can be rewritten as a dimensionless form

$$\frac{\partial^4 \eta}{\partial \xi^4} + u^2 \frac{\partial^2 \eta}{\partial \xi^2} + 2u\sqrt{\beta}\ \frac{\partial^2 \eta}{\partial \xi \partial \tau} + \frac{\partial^2 \eta}{\partial \tau^2} = 0. \tag{8.149}$$

The dimensionless boundary conditions considered in this paper include the following:

1. Cantilevered pipe

$$\begin{aligned} \eta(0, \tau) = \eta'(0, \tau) = 0, \\ \eta''(1, \tau) = \eta'''(1, \tau) = 0, \end{aligned} \tag{8.150}$$

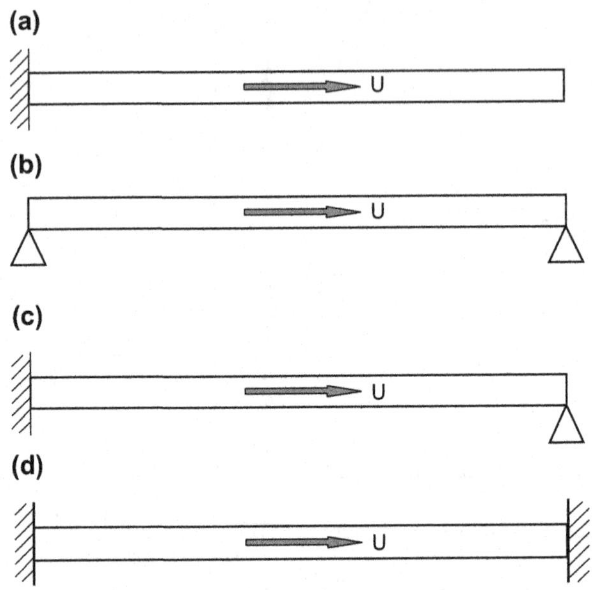

Figure 8.20 Pipe conveying fluid with different boundary conditions. (a) Cantilevered pipe, (b) pinned–pinned pipe, (c) clamped–pinned pipe, (d) clamped–clamped pipe.

2. Pinned–pinned pipe

$$\begin{aligned} \eta(0,\tau) &= \eta''(0,\tau) = 0, \\ \eta(1,\tau) &= \eta''(1,\tau) = 0, \end{aligned} \tag{8.151}$$

3. Clamped–pinned pipe

$$\begin{aligned} \eta(0,\tau) &= \eta'(0,\tau) = 0, \\ \eta(1,\tau) &= \eta''(1,\tau) = 0, \end{aligned} \tag{8.152}$$

4. Clamped–clamped pipe

$$\begin{aligned} \eta(0,\tau) &= \eta'(0,\tau) = 0, \\ \eta(1,\tau) &= \eta'(1,\tau) = 0. \end{aligned} \tag{8.153}$$

The solution of Eq. (8.7) may be expressed as

$$\eta(\xi,\tau) = \bar{w}(x)e^{\omega\tau}. \tag{8.154}$$

Substituting Eq. (8.154) into Eq. (8.149) yields

$$\frac{\partial^4 \bar{w}(x)}{\partial \xi^4} + u^2 \frac{\partial^2 \bar{w}(x)}{\partial \xi^2} + 2\omega u\sqrt{\beta}\,\frac{\partial \bar{w}(x)}{\partial \xi} + \omega^2 \bar{w}(x) = 0. \tag{8.155}$$

From Chapter 1, the differential transformation form of Eq. (8.155) is found as

$$(k+1)(k+2)(k+3)(k+4)W(k+4) + u^2(k+1)(k+2)W(k+2) + 2u\omega\sqrt{\beta}(k+1)W(k+1) + \omega^2 W(k) = 0. \tag{8.156}$$

Rearranging Eq. (8.156), a simple recurrence relation is obtained as follows

$$W(k+4) = \frac{-\left[u^2(k+1)(k+2)W(k+2) + 2u\omega\sqrt{\beta}\,(k+1)W(k+1) + \omega^2 W(k)\right]}{(k+1)(k+2)(k+3)(k+4)}. \tag{8.157}$$

Similarly, the differential transformation form of boundary conditions can be written as:

1. Cantilevered pipe

$$W(0) = W(1) = 0, \tag{8.158}$$

$$\sum_{k=0}^{\infty} k(k-1)W(k) = 0, \tag{8.159}$$

$$\sum_{k=0}^{\infty} k(k-1)(k-2)W(k) = 0, \tag{8.160}$$

2. Pinned–pinned pipe

$$W(0) = W(2) = 0, \tag{8.161}$$

$$\sum_{k=0}^{\infty} W(k) = 0, \tag{8.162}$$

$$\sum_{k=0}^{\infty} k(k-1)W(k) = 0, \tag{8.163}$$

3. Clamped–pinned pipe

$$W(0) = W(1) = 0, \tag{8.164}$$

$$\sum_{k=0}^{\infty} W(k) = 0, \tag{8.165}$$

$$\sum_{k=0}^{\infty} k(k-1)W(k) = 0, \tag{8.166}$$

4. Clamped–clamped

$$W(0) = W(1) = 0, \tag{8.167}$$

$$\sum_{k=0}^{\infty} W(k) = 0, \tag{8.168}$$

$$\sum_{k=0}^{\infty} kW(k) = 0. \tag{8.169}$$

It ought to be pointed out that, in Ref. [8], the number of terms, N, of DTM is chosen to be $N = 55$, showing sufficiently accurate results. In the calculations, convergence of the first four natural frequencies is shown in Fig. 8.21. Here, it can be seen that with the increasing of N, the precision of

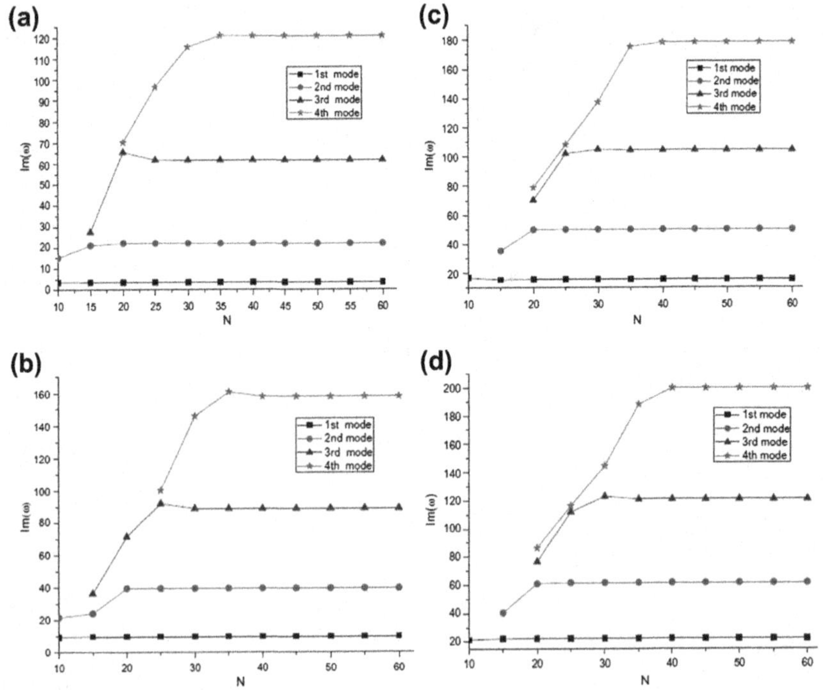

Figure 8.21 Convergence of the first four dimensionless natural frequencies. (a) Cantilevered pipe, (b) pinned–pinned pipe, (c) clamped–pinned pipe, and (d) clamped–clamped pipe.

DTM increases. Compared with the lower-order modes, the higher-order natural frequency requires more terms adopted in the DTM. In the following analysis, therefore, we choose $N = 60$. In the calculations based on DQM, 17 sampling points were used to obtain the convergence solutions shown. It is found that even the sampling points are chosen more than 17, the precision of DQM does not increase. If, however, the number of terms (N) is chosen to be sufficiently large (e.g., $N = 60$), the solution of DTM would infinitely approach to the exact solution. In the case of $u \neq 0$, Fig. 8.22 represents the natural frequencies of pipes conveying fluid for a simply supported pipe conveying fluid. Very good agreement is found between the results obtained by DTM and DQM.

8.10 PIEZOELECTRIC MODAL SENSOR FOR NONUNIFORM EULER—BERNOULLI BEAMS WITH RECTANGULAR CROSS SECTION

Consider the free vibration of an Euler—Bernoulli beam with length L_x, varying thickness $h(x)$ and varying width $b(x)$. A shaped PVDF film of constant thickness is attached onto the top surface and spanned across the entire length of the beam, as shown in Fig. 8.23. Assuming that the PVDF sensor thickness is much smaller than beam thickness, the mass and stiffness of the sensor is then negligible compared to the properties of the beam. The sensor thickness is typically 28—110 μm. As referred to Ref. [9], the output charge $Q(t)$ of the PVDF sensor shown in Fig. 8.23 can be expressed as:

$$Q(t) = \int_0^L \frac{h(x)}{2} F(x) \cdot \left(e_{31} \frac{d^2 w(x,t)}{dx^2} \right) dx \qquad (8.170)$$

where $h(x)$ is the beam thickness function, e_{31} is the PVDF sensor stress/charge coefficient, $w(x, t)$ is the displacement of the beam, and $F(x)$ is the PVDF film shape function. The displacement distribution of the vibrating beam may be represented by a series expansion:

$$w(x,t) = \sum_{n=1}^{N} \eta_n(t)\phi_n(x) \qquad (8.171)$$

where $Z_n(t)$ and $f_n(x)$ are the nth modal coordinates and structural mode shape function. N is the index for the highest order structural mode.

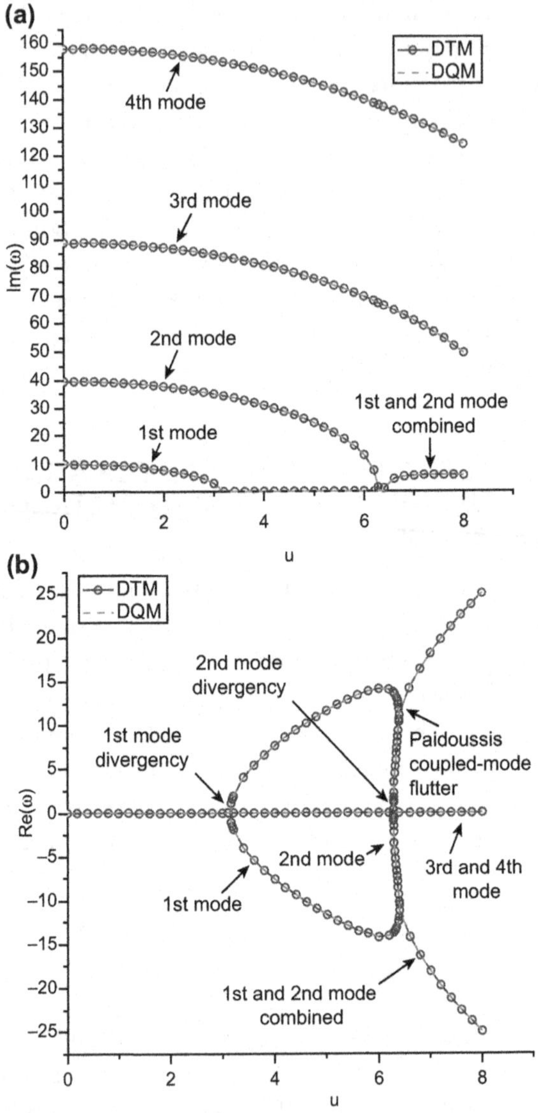

Figure 8.22 (a) The imaginary component and (b) the real component of the dimensionless frequency, as a function of fluid velocity, u, for the lowest four modes of a simply supported pipe conveying fluid, $b = 0.1$.

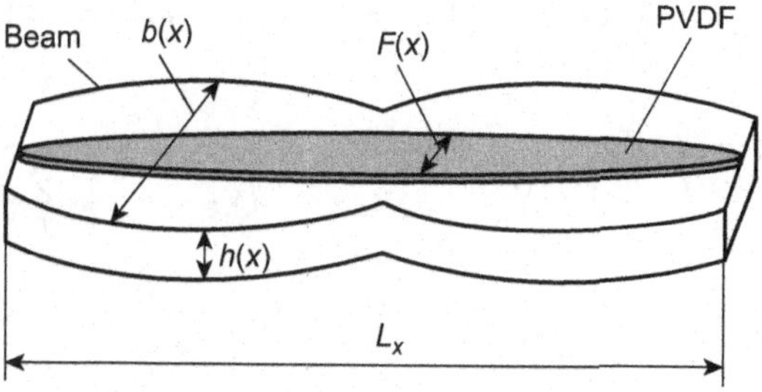

Figure 8.23 A shaped polyvinylidene fluoride (PVDF) film bonded on a nonuniform beam.

Substituting Eq. (8.171) into Eq. (8.170), and expressing in dimensionless form, one obtains:

$$Q(t) = \frac{e_{31}}{2L_x}\sum_{n=1}^{N}\eta_t(t)\int_0^1 h(X)F(X)\cdot\frac{d^2\phi_n(x)}{dX^2}dX = \frac{e_{31}}{2L_x}\sum_{n=1}^{N}\eta_n(t)P(n) \tag{8.172}$$

where $X = (x/L)$, $\phi_n(X) = \phi_n(x)$, $p_n = \int_0^1 h(X)F(X)\cdot(d^2\phi_n(X)/dX^2)dX$ is designated as the modal sensitivity [5,6]. To construct a sensor that accurately measures the target kth structural mode, a shaped PVDF film is used, the output signal of the sensor should be directly proportional to the target kth modal information. According to Eq. (8.172), a sufficient condition for $Q(t) \infty \eta_k$ is given by

$$P_n = \int_0^1 h(X)F(X)\cdot\frac{d^2\phi_n(x)}{dX^2}dX = \begin{cases}\alpha & n = k\\ 0 & n \neq k\end{cases} \quad \text{for } n = 1, \ldots, N \tag{8.173}$$

where a is a nonzero scaling factor, a is given as unity for convenience. To design the appropriate PVDF film shape function $F(X)$, which satisfies the above modal sensing condition equation, we make an approximation by expanding $F(X)$ as a linear function of the second derivative of the mode shape function $\frac{d^2\phi_j(x)}{dX^2}dX = dX$ and beam physical parameters, such as

$$F(X) = \sum_{j=1}^{N} B_j\frac{E(X)I(X)}{h(X)}\cdot\frac{d^2\phi_j(X)}{dX^2} = \sum_{j=1}^{N} B_j\frac{E(X)b(X)h^2(X)}{12}\,\frac{\partial^2\phi_j(X)}{\partial x^2} \tag{8.174}$$

where B_j are the unknown shape coefficients for the PVDF sensor. $E(X)$ is Young's modulus, $I(X) = b(X)h^3(X)/12$ is the cross section second moment of the beam.

Substituting Eq. (8.174) into Eq. (8.173), the modal sensitivity P_n can be rewritten as

$$P_n = \sum_{j=1}^{N} B_j \int_0^1 E(X)I(X)\frac{d^2\phi_j(X)}{dX^2}\cdot\frac{d^2\phi_n(X)}{dX^2}dX = \sum_{j=1}^{N} B_j K(n,j) \tag{8.175}$$

where

$$k(n,j) = \int_0^1 E(X)I(X)\frac{d^2\phi_j(X)}{dX^2}\cdot\frac{d^2\phi_n(X)}{dX^2}dX \tag{8.176}$$

Using Eqs. (8.174)–(8.176), the PVDF sensor shape coefficient B_j can be solved by:

$$B_j = \alpha K^{-1}(k,j) \tag{8.177}$$

For a beam with classical boundary conditions (such as clamped, free, simply supported, or sliding), thanks to the orthogonality of the second derivative of the mode shapes, we get

$$K^{-1}(k,j) = 0 \quad j \neq k \tag{8.178}$$

Thus, the modal sensor shape coefficients B_j and shape function $F(X)$ under the classical boundary conditions for the kth structural mode can be simplified as:

$$B_j = \begin{cases} \dfrac{1}{K_{kk}} & j = k \\ 0 & j \neq k \end{cases} \qquad F(X) = \frac{E(X)b(X)h^2(X)}{12}\frac{\partial^2\phi_j(X)}{\partial x^2} \tag{8.179}$$

In accordance with the above analysis, to design a modal sensor, the second spatial derivative of the mode shapes must be obtained. This will be discussed in the next section by using the DTM technique. Consider the free vibration of a stepped Euler–Bernoulli beam consisting of J uniform sections elastically restrained at both ends, as shown in Fig. 8.24. The stepped beam is divided into J sections with the J mirror systems of reference $x_j (j = 1,2,\ldots,J)$. The dimensionless ordinary differential equation describing the free vibration in each section of the stepped beam is as follows.

$$\frac{d^4\phi_j(X_j)}{dX_j^4} - \Omega_j^4\phi_j(X_j) = 0 \quad X_j \in [0 \quad R_j] \tag{8.180}$$

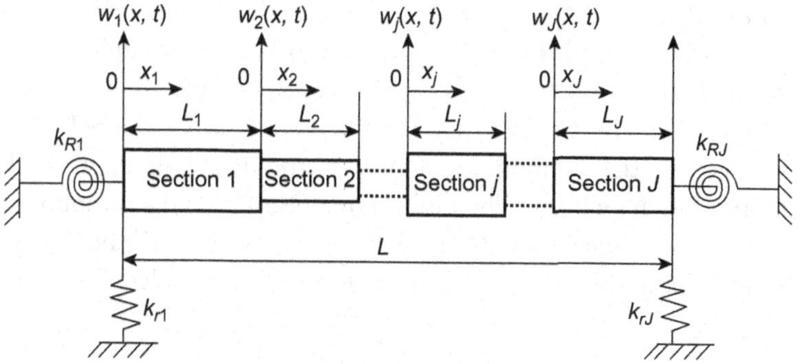

Figure 8.24 The coordinate system for a multiple-stepped beam elastically restrained at both ends.

where subscript j denote the section-j of the stepped beam $\Omega_j^4 = \rho_j A_j \omega^2 L^4 / E_j I_j$, $R_j = (L_j/L)$. I is the cross-section second order area moment of the section-j, $I_j = \left(b_j h_j^3/12\right)$, $A_j = b_j h_j$ is the cross-section area. E_j, ρ_j, L_j, b_j, and h_j are the Young's modulus, density, length, width and thickness of section-j, respectively.

$$\text{clearly} \quad \Omega_j^4 = \frac{\rho_j A_j L^4}{\rho_1 A_1 L^4} \frac{E_1 I_1}{E_j I_j} \Omega_j^4 = \mu_j \Omega_j^4 \tag{8.181}$$

where $\mu_j = (\rho_j A_j/\rho_1 A_1)(E_1 I_1/E_j I_j)$, Ω_1 is the dimensionless natural frequency, and the nth dimensionless natural frequency is denoted as $\Omega_1(n)$. According to Eq. (8.179), $\phi_j(X_j)$ in Eq. (8.180) can be expressed in differential transformation formulation:

$$\phi_j(X_j) = \sum_{m=0}^{M} X_j^m \overline{\phi}_j(m) \tag{8.182}$$

And the second spatial derivative of the mode shapes can be expressed as

$$\frac{d^2\phi_j(X_j)}{dX_j^2} = \sum_{m=0}^{M-2} (m+1)(m+2) X_j^m \overline{\phi}_j(m+2) \tag{8.183}$$

By applying the DTM to Eq. (8.180) and using the transformation operations in Chapter 1 and using Eqs. (8.181) and (8.182), the following recurrence equation can be obtained

$$\overline{\phi}_j(m+4) = \frac{\mu_j \Omega_1^4}{(m+1)(m+2)(m+3)(m+4)} \overline{\phi}_j(m) \tag{8.184}$$

To calculate $d^2\phi_j(X_j)/dX_j^2$ in Eq. (8.17), the differential transformation $\phi_j(m)$ should be solved. From Eq. (18), it can be seen $\phi_j(m)$ $(m \geq 4)$ are a function of $\overline{\phi}_j(0)$, $\overline{\phi}_j(1)$, $\overline{\phi}_j(2)$, $\overline{\phi}_j(3)$ and Ω_1. These $2J+1$ unknown parameters, namely Ω_1 and $\overline{\phi}_j(s)$ $(s = 0, 1, 2, 3$ and $j = 1 \sim J)$ can be determined by the continuity conditions and the boundary condition equations of each section of the beam. For a beam with discontinuous cross sections, stress concentrations at the step locations of the beam are neglected. Then, the continuity conditions in dimensionless form are

$$\phi_{j+1}(0) = \phi_j(R_j), \quad \frac{d\phi_{j+1}(0)}{dX_{j+1}} = \frac{d\phi_j(R_j)}{dX_j} \tag{8.185}$$

$$\frac{d^2\phi_{j+1}(0)}{dX_{j+1}^2} = \frac{E_J I_J}{E_{J+1} I_{J+1}} d^2\phi_j(R_j) \tag{8.186}$$

where $R_j = (L_j/L)$ is denoted as dimensionless length of the jth section and $\sum_{j=1}^{j} R_j = 1$.

Substituting Eq. (8.180) into Eqs. (8.182) and (8.183), the differential transformation for the section-j $(j \geq 2)$ can be written as

$$\overline{\phi}_{j+1}(0) = \sum_{m=0}^{M} R_j^m \overline{\phi}_j(m) \tag{8.187}$$

$$\overline{\phi}_{j+1}(1) = \sum_{m=0}^{M} (m+1) R_j^m \overline{\phi}_j(m+1) \tag{8.188}$$

$$\overline{\phi}_{j+1}(2) = (1/2) \frac{E_j I_j}{E_{j+1} I_{j+1}} \sum_{m=0}^{M} (m+1)(m+2) R_j^m \overline{\phi}_j(m+2) \tag{8.189}$$

$$\overline{\phi}_{j+1}(3) = (1/6) \frac{E_j I_j}{E_{j+1} I_{j+1}} \sum_{m=0}^{M} (m+1)(m+2)(m+3) R_j^m \overline{\phi}_j(m+3) \tag{8.190}$$

Notice that there are only five unknown parameters $\overline{\phi}_1(0)$, $\overline{\phi}_1(1)$, $\overline{\phi}_1(2)$, $\overline{\phi}_1(3)$ and Ω_1 in Eqs. (8.187)–(8.190) through a recursive way. It means that the mode shape of the Jth section of the beam can be expressed as linear functions of $\overline{\phi}_1(0)$, $\overline{\phi}_1(1)$, $\overline{\phi}_1(2)$, $\overline{\phi}_1(3)$, such as

$$\overline{\phi}_j(m) = f_{0m}(\Omega_1)\overline{\phi}_1(0) + f_{1m}(\Omega_1)\overline{\phi}_1(1) + f_{2m}(\Omega_1)\overline{\phi}_1(2) + f_{3m}(\Omega_1)\overline{\phi}_1(3) \quad m = 0, 1, 2, 3 \tag{8.191}$$

The boundary conditions at the ends of the beam shown in Fig. 8.24 can be expressed in dimensionless form

$$\frac{d^2\phi_1(0)}{dX_1^2} - K_{R1}\frac{d\phi_1(0)}{dX_1} = 0, \quad \frac{d^3\phi_1(0)}{dX_1^3} + K_{T1}\phi_1(0) = 0 \tag{8.192a}$$

$$\frac{d^2\phi_j(R_j)}{dX_j^2} - K_{R1}\frac{d\phi_j(R_j)}{dX_j} = 0, \quad \frac{d^3\phi_j(R_j)}{dX_j^3} + K_{T1}\phi_1(R_j) = 0 \tag{8.192b}$$

where $K_{R1} = (K_{R1}L/E_1I_1)$, $K_{T1} = (K_{T1}L^3/E_1I_1)$, $K_{Rj} = (K_{Rj}L/E_1I_1)$, $K_{Tj} = (K_{Tj}L^3/E_1I_1)$, and $R_j = (L_j/L)$; K_{T1} and K_{Tj} are the stiffness of the translational springs; and K_{R1} and K_{Rj} are the stiffness of the rotational springs at $x_1 = 0$ and $x_j = L_j$, respectively. The differential transformations of Eq. (8.192) are obtained with the definition Eq. (8.179) as

$$2\overline{\phi}_1(2) - K_{R1}\overline{\phi}_1(1) = 0 \quad 6\overline{\phi}_1(3) - K_{T1}\overline{\phi}_1(0) = 0 \tag{8.193}$$

$$2\overline{\phi}_J(2) - K_{RJ}\overline{\phi}_J(1) = 0 \quad 6\overline{\phi}_J(3) - K_{TJ}\overline{\phi}_J(0) = 0 \tag{8.194}$$

Substituting Eq. (8.189) into Eq. (8.192) and then rewriting Eqs. (8.193) and (8.194) in matrix form, we obtain:

$$\begin{bmatrix} 0 & -K_{R1} & 2 & 0 \\ K_{T1} & 0 & 0 & 6 \\ D_{31} & D_{32} & D_{33} & D_{34} \\ D_{41} & D_{42} & D_{43} & D_{44} \end{bmatrix} \begin{bmatrix} \overline{\phi_1}(0) \\ \overline{\phi_1}(1) \\ \overline{\phi_1}(2) \\ \overline{\phi_1}(3) \end{bmatrix} = 0 \tag{8.195}$$

where $D_{3m} = k_{Rj}f_{1m}(\Omega_1)$, $D_{4m} = 6f_m(\Omega_1) - k_{Tj}f_{0m}(\Omega_1)$ with $m = 0, 1, 2, 3$. From Eq. (8.195), the dimensionless frequency parameter Ω_1 can be solved by:

$$\det\begin{bmatrix} 0 & -K_{R1} & 2 & 0 \\ K_{T1} & 0 & 0 & 6 \\ D_{31} & D_{32} & D_{33} & D_{34} \\ D_{41} & D_{42} & D_{43} & D_{44} \end{bmatrix} = 0 \tag{8.196}$$

Notice that the matrix in Eq. (8.196) is singular at each frequency parameter Ω_1, and the unknown parameters $\overline{\phi}_1(m)$ ($m = 0, 1, 2, 3$) cannot be directly determined. However, one may choose one quantity of $\overline{\phi}_1(m)$ as the arbitrary nonzero constant, and the remaining three as functions of this arbitrary constant. Without loss of generality, one may choose

$\overline{\phi}_1(0) = 1$. Hence, the remaining three can be solved as functions by using Eq. (8.196):

$$\begin{bmatrix} 0 & 0 & 6 \\ D_{32} & D_{33} & D_{34} \\ D_{42} & D_{43} & D_{44} \end{bmatrix} = \begin{bmatrix} \overline{\phi_1}(1) \\ \overline{\phi_1}(2) \\ \overline{\phi_1}(3) \end{bmatrix} = \begin{bmatrix} K_{TL} \\ D_{31} \\ D_{41} \end{bmatrix} \tag{8.197}$$

By using the solved $\overline{\phi}_1(m)$, the closed-form series solution for the second spatial derivative of the mode shapes $\left(\partial^2\phi_J(X)/\partial X_j^2\right)$ for each section can be obtained. The second spatial derivative of the mode shape function for the entire beam can be written as

$$\frac{d\phi^2(X)}{dX^2} = \left[\frac{d\phi_1^2(X)}{dX_1^2}\ \frac{d\phi_2^2(X)}{dX_2^2} \cdots \frac{d\phi_j^2(X)}{dX_j^2}\right] \tag{8.198}$$

Substituting $(d^2\phi(X)/dX^2)$ into Eqs. (8.176) and (8.177), the PVDF sensor shape coefficients B_j are determined: Then substituting B_j into Eq. (8.174), the shape of the modal sensor can be obtained. It should be noticed that the proposed DTM can be used to obtain the second spatial derivative of the mode shape function $(d^2\phi(X)/dX^2)$ of beams consisting of an arbitrary number of steps through a recursive way. Consequently, the complexity of the problem is reduced to the same order of a beam without any steps. $(d^2\phi(X)/dX^2)$ can be obtained by solving a set of algebraic equations with only five unknowns. It is well known that any type of nonuniform beam can be approximated by a stepped beam with a suitable number of uniform sections. It means that the proposed method can be used to design the piezoelectric modal sensors for any type of nonuniform Euler–Bernoulli beams.

A. Simply Supported Beam With Two Steps

In this example, the shaped piezoelectric modal sensor of the two-step simply supported beam is studied using the DTM. The beam parameters are as shown in Fig. 8.25. For simply supported boundary condition, the stiffness of the translational and rotational springs in Eq. (8.192) can be set to 1×10^9 and 0, respectively. It is important to check how rapidly the dimensionless natural frequencies $\Omega_1(n)$ computed through the DTM converge toward the exact value as the series summation limit M is increased, because the closed-form series solution of the second spatial derivative of the mode shape functions $(d^2\phi_j(X_j)/dX_j^2)$ in Eq. (8.182) will have to be truncated in numerical calculations. Fig. 8.26 shows the first five

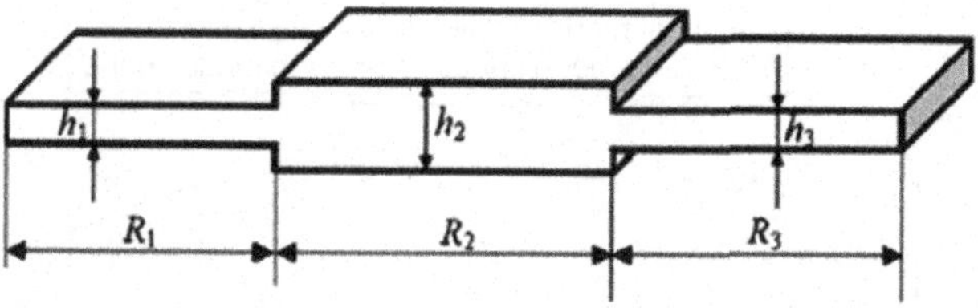

Figure 8.25 A simply supported two-step beam with constant width and step varying thickness when $R_1 = 17/38$, $R_2 = 4/38$, $R_3 = 17/38$, $h_2/h_1 = 2$, $h_3/h_1 = 1$ (dimensions are not scaled).

dimensionless natural frequencies $\Omega_1(n)$ as the function of the series summation limit M. The dimensionless natural frequencies $\Omega_1(1) - \Omega_1(5)$ converge to 3.16, 6.28, 9.68, 12.57, and 16.31 very quickly as the series summation limit M is increased. The excellent numerical stability of the solution can also be observed in Fig. 8.26. For simplicity, the DTM solutions are truncated to $M^{1}/_{4}20$ in Eq. (8.182) in all the subsequent calculations. Fig. 8.27 shows the first four mode shapes and the corresponding second spatial derivative of the mode shapes. Fig. 8.28 shows the shapes of

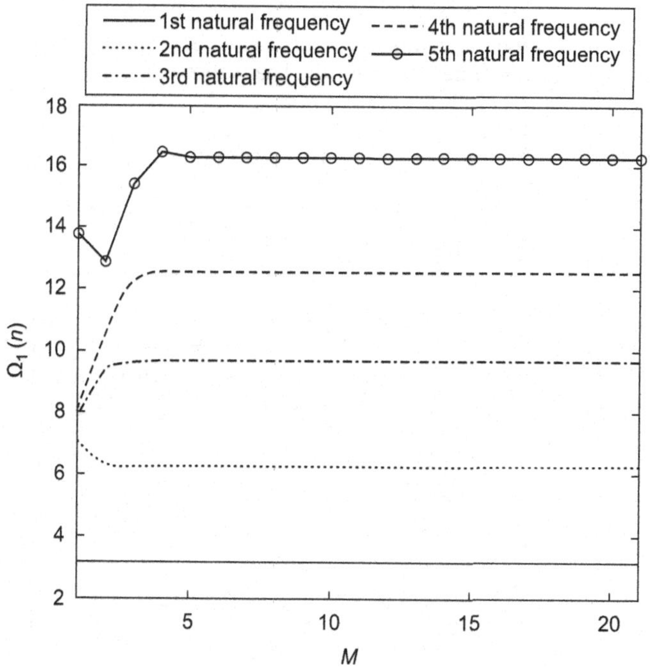

Figure 8.26 The first five dimensionless natural frequencies $\Omega_1(n)$ as the function of the series summation limit M for the beam shown in Fig. 8.25.

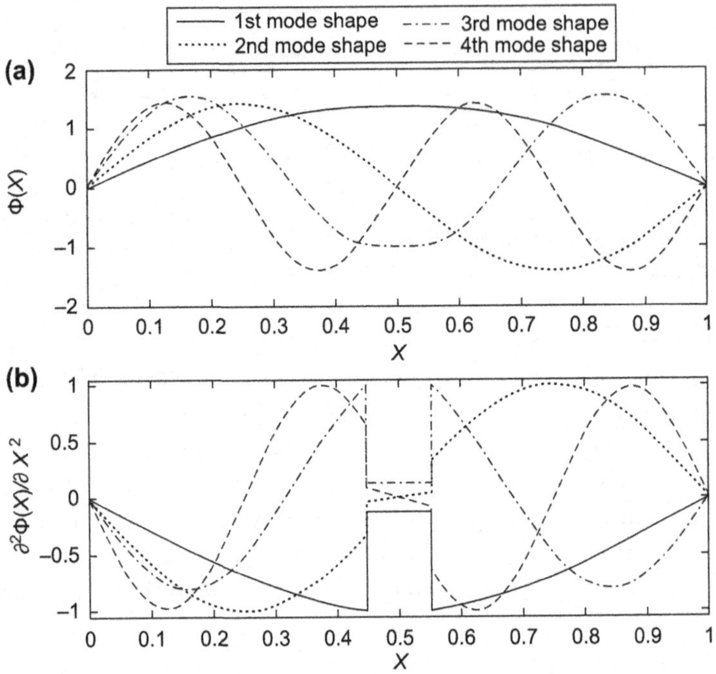

Figure 8.27 The first four (a) normalized mode shapes and (b) corresponding second spatial derivative of the mode shapes for the beam shown in Fig. 8.25.

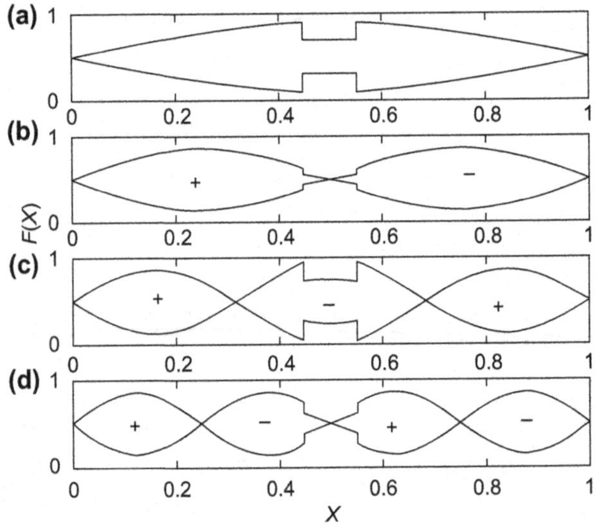

Figure 8.28 The modal sensor shapes of (a) the first; (b) the second; (c) the third; (d) the fourth mode for the beam shown in Fig. 8.25.

the PVDF modal sensors for the first four structural modes. It can be found that the second spatial derivative of the mode shape and modal sensor shape of the third mode shown in Figs. 8.26(b) and 8.27(c) agree well with those shown in Ref. [9].

B. Tapered Beam With Abrupt Changes of Cross Section

In this second, a tapered cantilever beam with abrupt changes of cross section is considered to verify the present method, the beam cross section has constant width but with linearly varying thickness at each segment. The dimension of the beam is shown in Fig. 8.29. Figs. 8.30 and 8.31 show the mode shape, the corresponding second spatial derivative of the mode shapes, and the modal sensor shapes for the first four structural modes.

8.11 FREE VIBRATIONS OF OSCILLATORS

It is known that the free vibrations of an autonomous conservative oscillator with inertia and static type fifth-order nonlinearities are expressed by [10].

$$\frac{d^2u(t)}{dt^2} + \lambda u(t) + \varepsilon_1 u^2(t)\frac{d^2u(t)}{dt^2} + \varepsilon_1 u(t)\left(\frac{du(t)}{dt}\right)^2 + \varepsilon_2 u^4(t)\frac{d^2u(t)}{dt^2}$$
$$+ 2\varepsilon_2 u^3(t)\left(\frac{du(t)}{dt}\right)^2 + \varepsilon_3 u^3(t) + \varepsilon_4 u^5(t) = 0. \tag{8.199}$$

The initial conditions for Eq. (8.199) are given by $u(0) = A$ and $du(0)/dt = 0$, where A represents the amplitude of the oscillation. Motion is assumed to start from the position of maximum displacement with zero

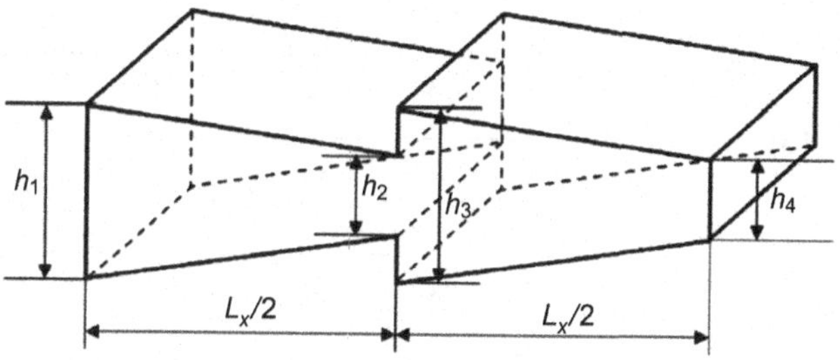

Figure 8.29 A stepped tapered cantilever beam (clamped at left end) with constant width when $h_2/h = 0.8$, $h_3/h_1 = 1$, $h_4/h_1 = 0.8$ (dimensions are not scaled).

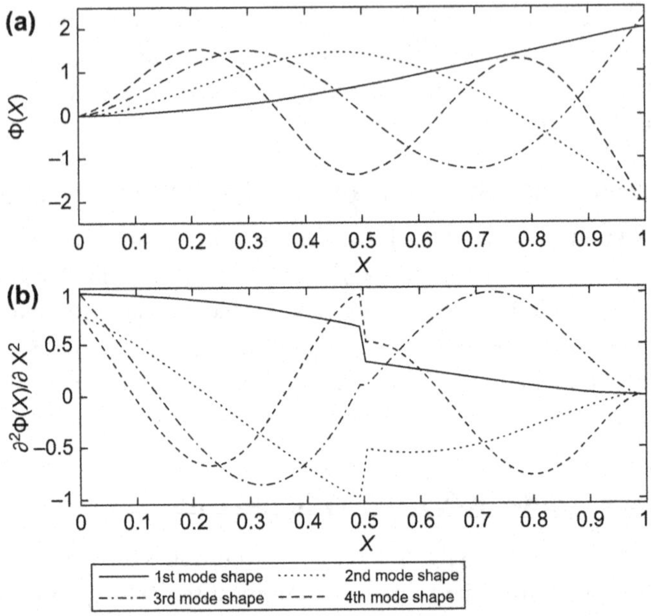

Figure 8.30 The first four (a) normalized mode shapes and (b) corresponding second spatial derivative of the mode shapes for the beam shown in Fig. 8 29.

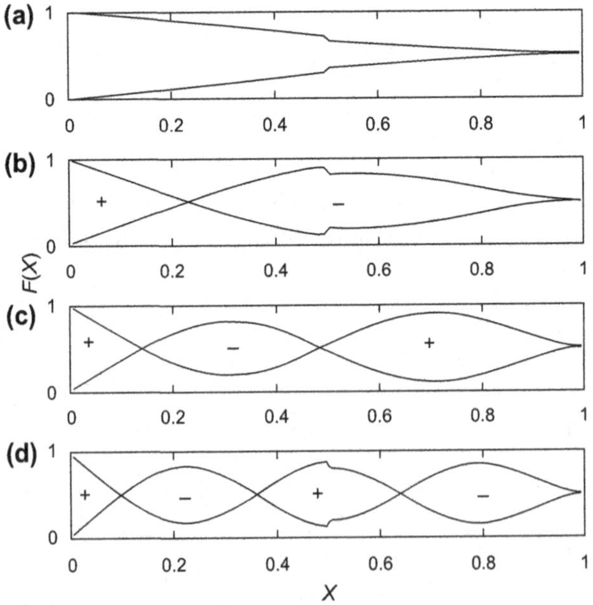

Figure 8.31 The modal sensor shapes of (a) the first; (b) the second; (c) the third; (d) the fourth mode for the beam shown in Fig. 8.29.

initial velocity. Eq. (8.199) can be expressed as two simultaneous first-order differential equations written in terms of $u(t)$ and $v(t)$, i.e.,

$$\frac{du(t)}{dt} = v(t) \tag{8.200}$$

$$\frac{dv(t)}{dt} + \lambda u(t) + \varepsilon_1 u^2(t)\frac{dv(t)}{dt} + \varepsilon_1 u(t)v^2(t) + \varepsilon_2 u^4(t)\frac{dv(t)}{dt} + 2\varepsilon_2 u^3(t)v^2(t) + \varepsilon_3 u^3(t) + \varepsilon_4 u^5(t) = 0 \tag{8.201}$$

Taking the differential transform of Eq. (8.201) with respect to time t gives

$$\frac{k+1}{H}U(k+1) = V(k) \tag{8.202}$$

where $U(k)$ and $V(k)$ are the differential transformations of functions $u(t)$ and $v(t)$, respectively. Taking the differential transform of Eq. (8.201) with respect to time t yields

$$\begin{aligned}
&\frac{k+1}{H}V(k+1) + \lambda U(k) + \varepsilon_1 \sum_{\lambda=0}^{k} \frac{k+1}{H} V(k-\lambda+1) \sum_{m=0}^{\lambda} U(\lambda-m)U(m) \\
&+ \varepsilon_1 \sum_{\lambda=0}^{k} U(k-\lambda) \sum_{m=0}^{\lambda} V(\lambda-m)V(m) + \varepsilon_2 \sum_{\lambda=0}^{k} \frac{k+1}{H} V(k-\lambda) \\
&+ \sum_{m=0}^{\lambda} U(\lambda-m) \sum_{p=0}^{m} U(m-p) \sum_{q=0}^{p} U(p-q)U(q) + 2\varepsilon_2 \sum_{\lambda=0}^{k} U(k-\lambda) \\
&+ \sum_{m=0}^{\lambda} U(\lambda-m) \sum_{p=0}^{m} U(m-p) \sum_{q=0}^{p} V(p-q)V(q) \\
&+ \varepsilon_3 \sum_{\lambda=0}^{k} U(k-\lambda) \sum_{m=0}^{\lambda} U(\lambda-m)U(m) + \varepsilon_4 \sum_{\lambda=0}^{k} U(k-\lambda) \\
&\quad + \sum_{m=0}^{\lambda} U(\lambda-m) \sum_{p=0}^{m} U(m-p) \sum_{q=0}^{p} U(p-q)U(q) = 0
\end{aligned} \tag{8.203}$$

Eqs. (8.202) and (8.203) can be rewritten as follows:

$$U(k+1) = \frac{H}{k+1} V(k) \tag{8.204}$$

$$V(k+1) = \frac{-H}{(k+1)[1+\varepsilon_1 U^2(0) + \varepsilon_2 U^4(0)]}$$

$$\times \left[\begin{array}{l} \lambda U(k) + \varepsilon_1 \sum_{\lambda=1}^{k} \frac{k+1}{H} V(k-\lambda+1) \sum_{m=0}^{\lambda} U(\lambda-m)U(m) \\ + \varepsilon_1 \sum_{\lambda=0}^{k} U(k-\lambda) \sum_{m=0}^{\lambda} V(\lambda-m)V(m) \\ + \varepsilon_2 \sum_{\lambda=1}^{k} \frac{k+1}{H} V(k-\lambda) + \sum_{m=0}^{\lambda} U(\lambda-m) \sum_{p=0}^{m} U(m-p) \sum_{q=0}^{p} U(p-q)U(q) \\ 2\varepsilon_2 \sum_{\lambda=0}^{k} U(k-\lambda) + \sum_{m=0}^{\lambda} U(\lambda-m) \sum_{p=0}^{m} U(m-p) \sum_{q=0}^{p} V(p-q)V(q) \\ + \varepsilon_3 \sum_{\lambda=0}^{k} U(k-\lambda) \sum_{m=0}^{\lambda} U(\lambda-m)U(m) \\ + \varepsilon_4 \sum_{\lambda=0}^{k} U(k-\lambda) + \sum_{m=0}^{\lambda} U(\lambda-m) \sum_{p=0}^{m} U(m-p) \sum_{q=0}^{p} U(p-q)U(q) \end{array} \right] = 0 \tag{8.205}$$

where the initial conditions are given by $U(0) = A$ and $V(0) = 0$. The difference equations presented in Eqs. (8.204) and (8.205) describe the free vibrations of a conservative oscillator with inertia and static fifth-order nonlinearities. From a process of inverse differential transformation, it can be shown that the solutions of each subdomain take $n+1$ terms for the power series of the DTM principle, i.e.,

$$u_i(t) = \sum_{k=0}^{n} \left(\frac{t}{H_i}\right)^k U_i(k), \quad 0 \le t \le H_i \tag{8.206}$$

$$v_i(t) = \sum_{k=0}^{n} \left(\frac{t}{H_i}\right)^k V_i(k), \quad 0 \le t \le H_i \tag{8.207}$$

where $k = 0, 1, 2, \ldots, n$ represents the number of terms of the power series; $i = 0, 1, 2, \ldots$ expresses the ith subdomain; and H_i is the subdomain interval.

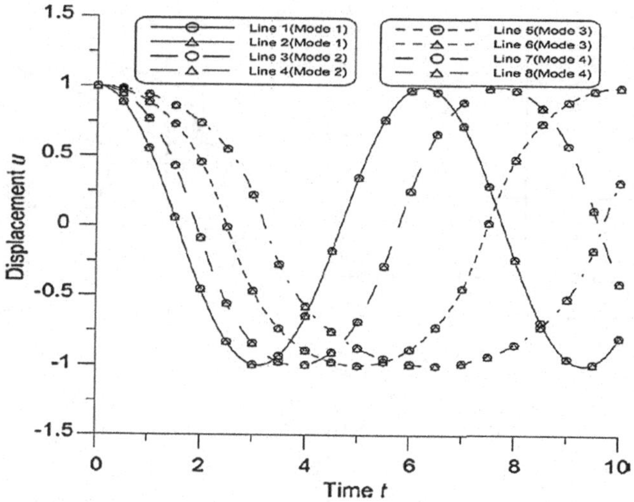

Figure 8.32 Comparison of the differential transformation method (lines 2, 4, 6, 8) and Runge–Kutta method (lines 1, 3, 5, 7) results with $\lambda = 1$ and $A = 1$, for Modes 1–4.

To verify the effectiveness of the proposed DTM, the fourth-order Runge–Kutta numerical method is used to compute the displacement response of the nonlinear oscillator for initial amplitude under four different modes. Fig. 8.32 compares the results for an example of amplitude conditions of the proposed DTM with those of the Runge–Kutta method for each of the four calculation modes. Note that Mode 1 results are given by lines 1 and 2, while the results for Modes 2, 3, and 4 are represented by lines 3 and 4, 5 and 6, and 7 and 8, respectively.

8.12 COMPOSITE SANDWICH BEAMS WITH VISCOELASTIC CORE

The assumptions used to derive the kinematic relations and the governing equations are as follows [11]:

1. The shear angle of the top and bottom face plates is neglected.
2. The core layer is relatively soft and viscoelastic with a complex modulus.
3. The contribution from the core layer is only by transverse shear stresses.
4. Layers are assumed to be incompressible through the thickness.
5. Transverse displacement does not change between the layers.
6. The beam deflection is small.
7. There is no slip between the layers.

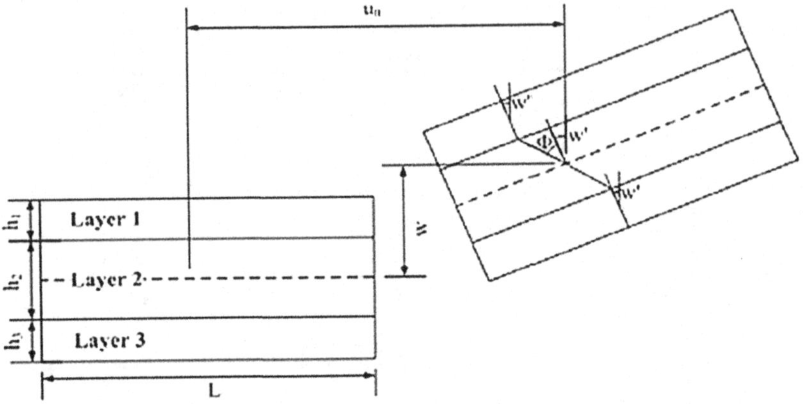

Figure 8.33 Geometry and displacement field.

The configuration of the sandwich beam is presented in Fig. 8.33. Using the geometry in Fig. 8.33, the kinematic relations are derived as follows:

$$u_1 = u_0(x,t) - \frac{h_2}{2}\varphi(x,t) - \left(z + \frac{h_1}{2}\right)\frac{\partial}{\partial x}w(x,t), \quad -\frac{h_1}{2} \le z \le \frac{h_1}{2} \tag{8.208}$$

$$u_2 = u_0(x,t) - z\varphi(x,t), \quad -\frac{h_1}{2} \le z \le \frac{h_1}{2} \tag{8.209}$$

$$u_3 = u_0(x,t) + \frac{h_2}{2}\varphi(x,t) - \left(z - \frac{h_3}{2}\right)\frac{\partial}{\partial x}w(x,t), \quad -\frac{h_3}{2} \le z \le \frac{h_3}{2} \tag{8.210}$$

$$w_i = w(x,t), \quad i = 1,2,3 \tag{8.211}$$

where $\varphi(x,t) = \Phi(x,t)w'(x,t)$ is the total angular displacement, $u_0(x, t)$ is the longitudinal displacement, $w(x, t)$ is the transverse displacement of the centroid of viscoelastic core. Also, u_i and w_i correspond to the longitudinal and transverse displacements of the ith layer. The strain–displacement relations for the sandwich beam can be evaluated from Eqs. (8.208)–(8.211) as follows:

$$\varepsilon_{xx}^{(1)} = \frac{\partial u_0}{\partial x} - \frac{h_2}{2}\frac{\partial \varphi}{\partial x} - \left(z + \frac{h_1}{2}\right)\frac{\partial^2 w}{\partial x^2} \tag{8.212}$$

$$\gamma_{xz}^{(2)} = \frac{\partial w}{\partial x} - \varphi \tag{8.213}$$

$$\varepsilon_{xx}^{(3)} = \frac{\partial u_0}{\partial x} + \frac{h_2}{2}\frac{\partial \varphi}{\partial x} - \left(z - \frac{h_3}{2}\right)\frac{\partial^2 w}{\partial x^2} \tag{8.214}$$

The stress–strain relations can be given as follows:

$$\sigma_{xx}^{(1)} = E_1 \varepsilon_{xx}^{(1)} \tag{8.215}$$

$$\sigma_{xz}^{(2)} = G_2^* \gamma_{xz}^{(2)} \tag{8.216}$$

$$\sigma_{xx}^{(3)} = E_3 \varepsilon_{xx}^{(3)} \tag{8.217}$$

where $G_2^* = G_2(1 + i\eta)$ is the complex shear modulus of the viscoelastic core and E_1, E_3 corresponding to the Young's moduli of the constraining layer and base layer, respectively. Let us consider the face layers as orthotropic. Then, the Young's moduli of these layers can be calculated as follows:

$$E_i = Q_{11}\cos^4(\theta_i) + Q_{22}\sin^4(\theta_i) + 2(Q_{12} + 2Q_{66})\sin^2(\theta_i)\cos^2(\theta_i) \tag{8.218}$$

where θ_i is the angle of lamination of the ith layer and

$$Q_{11} = \frac{E_{11}}{1 - \nu_{12}\nu_{21}}, \quad Q_{12} = \frac{\nu_{12}E_{22}}{1 - \nu_{12}\nu_{21}}, \quad Q_{22} = \frac{E_{22}}{1 - \nu_{12}\nu_{21}},$$
$$Q_{66} = G_{12}, \quad \nu_{21} = \nu_{12}\frac{E_{22}}{E_{11}} \tag{8.219}$$

are the material properties of the composite face layers. For the free vibrations of the beam, Hamilton's principle can be expressed as follows:

$$\int_o^T (\delta U - \delta K)dt = 0 \tag{8.220}$$

where U and K correspond to the elastic strain energy and the kinetic energy, respectively. For the problem considered, Eq. (8.220) becomes:

$$\begin{aligned}
&\int_0^T\int_0^L\int_{-h_1/2}^{h_1/2}\left[\sigma_{xx}^{(1)}\delta\varepsilon_{xx}^{(1)} - \rho_1\left(\frac{\partial w}{\partial t}\delta\frac{\partial w}{\partial t} + \frac{\partial u_1}{\partial t}\delta\frac{\partial u_1}{\partial t}\right)\right]dzdxdt \\
&+ \int_0^T\int_0^L\int_{-h_2/2}^{h_2/2}\left[\sigma_{xz}^{(2)}\delta\gamma_{xz}^{(2)} - \rho_2\left(\frac{\partial w}{\partial t}\delta\frac{\partial w}{\partial t} + \frac{\partial u_2}{\partial t}\delta\frac{\partial u_2}{\partial t}\right)\right]dzdxdt \\
&+ \int_0^T\int_0^L\int_{-h_3/2}^{h_3/2}\left[\sigma_{xx}^{(3)}\delta\varepsilon_{xx}^{(3)} - \rho_3\left(\frac{\partial w}{\partial t}\delta\frac{\partial w}{\partial t} + \frac{\partial u_3}{\partial t}\delta\frac{\partial u_3}{\partial t}\right)\right]dzdxdt = 0
\end{aligned} \tag{8.221}$$

From Eq. (8.221), the governing equations are obtained as follows:

$$(E_1h_1+E_3h_3)\frac{\partial^2 u_0}{\partial x^2}+\frac{h_2}{2}(E_3h_3+E_1h_1)\frac{\partial^2 \varphi}{\partial x^2}+\frac{1}{2}\left(E_3h_3^2+E_1h_1^2\right)\frac{\partial^3 w}{\partial x^3}$$
$$=(h_1\rho_1+h_2\rho_2+h_3\rho_3)\frac{\partial^2 u_0}{\partial t^2}+\frac{h_2}{2}(h_3\rho_3-h_1\rho_1)\frac{\partial^2 \varphi}{\partial t^2}$$
$$\frac{1}{2}\left(h_3^2\rho_3-h_1^2\rho_1\right)\frac{\partial^3 w}{\partial t^2\partial x} \tag{8.222}$$

$$(E_3h_3+E_1h_1)\frac{\partial^2 u_0}{\partial x^2}+\frac{h_2}{2}(E_1h_1+E_3h_3)\frac{\partial^2 \varphi}{\partial x^2}+\frac{1}{2}\left(E_1h_1^2+E_3h_3^2\right)\frac{\partial^3 w}{\partial x^3}$$
$$+2G_2\left(\frac{\partial w}{\partial x^2}-\varphi\right)$$
$$=(h_3\rho_3-h_1\rho_1)\frac{\partial^2 u_0}{\partial t^2}+\frac{h_2}{6}(3h_1\rho_1+h_2\rho_2+3h_3\rho_3)\frac{\partial^2 \varphi}{\partial t^2} \tag{8.223}$$

$$\frac{1}{2}\left(h_1^2\rho_1+h_3^2\rho_3\right)\frac{\partial^3 w}{\partial t^2\partial x}$$
$$+G_2h_2\left(\frac{\partial \varphi}{\partial x}-\frac{\partial w}{\partial x^2}\right)+\frac{1}{2}\left(E_3h_3^2-E_1h_1^2\right)\frac{\partial^3 u_0}{\partial x^3}+\frac{h_2}{4}\left(E_1h_1^2+E_3h_3^2\right)\frac{\partial^3 \varphi}{\partial x^3}$$
$$+\frac{1}{3}\left(E_1h_1^3+E_3h_3^3\right)\frac{\partial^4 w}{\partial x^4}$$
$$-\frac{1}{2}\left(h_3^2\rho_3-h_1^2\rho_1\right)\frac{\partial^3 u_0}{\partial^2 t\partial x}-(h_1\rho_1+h_2\rho_2+h_3\rho_3)\frac{\partial^2 w}{\partial t^2}$$
$$+\frac{h}{4}\left(h_1^2\rho_1+h_3^2\rho_3\right)\frac{\partial^3 \varphi}{\partial^2 t\partial x}+\frac{1}{3}\left(h_1^3\rho_1+h_3^3\rho_3\right)\frac{\partial^4 w}{\partial x^2\partial t^2} \tag{8.224}$$

Notice that, Eq. (8.222) that governs the axial motion of the sandwich beam is uncoupled with the shear angle and the transverse displacement for a symmetrically sectioned beam, where the material and geometric properties of layers 1 and 3 are identical. The boundary conditions are also evaluated as follows:

$$\left(N_x^{(1)}+N_x^{(3)}\right)\delta u_0\Big|_0^L=0 \tag{8.225}$$

$$\frac{h_2}{2}\left(N_x^{(3)} + N_x^{(1)}\right)\delta\varphi\Big|_0^L = 0 \tag{8.226}$$

$$\frac{h_2}{2}\left(Q_x^{(2)}\frac{\partial}{\partial x}M_x^{(1)} + \frac{\partial}{\partial x}M_x^{(3)} + \frac{h_1}{2}\frac{\partial}{\partial x}N_x^{(1)} - \frac{h_3}{2}\frac{\partial}{\partial x}N_x^{(3)}\right)\delta w\Big|_0^L = 0 \tag{8.227}$$

$$\left(\frac{h_3}{2}N_x^{(3)} - \frac{h_1}{2}N_x^{(1)} - M_x^{(1)} - M_x^{(3)}\right)\delta\left(\frac{\partial w}{\partial x}\right)\Big|_0^L = 0 \tag{8.228}$$

where

$$\begin{aligned} N_x^{(i)} &= \int_{-hi/2}^{hi/2} \sigma_{xx}^{(i)} dz, \quad Q_x^{(i)} = \int_{-hi/2}^{hi/2} \sigma_{xx}^{(i)} dz, \\ M_x^{(i)} &= \int_{-hi/2}^{hi/2} z\sigma_{xx}^{(i)} dz, \quad i = 1, 2, 3 \end{aligned} \tag{8.229}$$

For the problem considered, the sectional moment and forces are obtained from Eqs. (8.212)–(8.217) together with Eq. (8.229) as follows:

$$N_x^{(1)} = E_1\left(h_1\frac{\partial u_0}{\partial x} - \frac{1}{2}h_1h_2\frac{\partial\varphi}{\partial x} - \frac{1}{2}h_1^2\frac{\partial^2 w}{\partial x^2}\right) \tag{8.230}$$

$$N_x^{(3)} = E_3\left(h_3\frac{\partial u_0}{\partial x} + \frac{1}{2}h_2h_3\frac{\partial\varphi}{\partial x} - \frac{1}{2}h_3^2\frac{\partial^2 w}{\partial x^2}\right) \tag{8.231}$$

$$Q_x^{(2)} = G_2h_2\left(\frac{\partial w}{\partial x} - \varphi\right) \tag{8.232}$$

$$M_x^{(1)} = -\frac{1}{12}E_1h_1^3\frac{\partial^2 w}{\partial x^2} \tag{8.233}$$

$$M_x^{(3)} = -\frac{1}{12}E_3h_3^3\frac{\partial^2 w}{\partial x^2} \tag{8.234}$$

For the harmonic vibrations of the sandwich beam, the displacement field can be assumed as follows:

$$\begin{aligned} u_0(x,t) &= \overline{u}_0(x,\omega)e^{i\omega t}, \\ w(x,t) &= \overline{w}(x,\omega)e^{i\omega t}, \\ \varphi(x,t) &= \overline{\varphi}(x,\omega)e^{i\omega t}. \end{aligned} \tag{8.235}$$

Then, Eqs. (8.22)–(8.24) become:

$$(E_1h_1 + E_3h_3)\overline{u}''_0 + \frac{h_2}{2}(E_3h_3 - E_1h_1)\overline{\varphi}'' + \frac{1}{2}\left(E_3h_3^2 - E_1h_1^2\right)\overline{w}''$$
$$= -\omega^2\left[(h_1\rho_1 + h_1\rho_2 + h_3\rho_3)\overline{u}_0 + \frac{h_2}{2}(h_3\rho_3 - h_1\rho_1)\overline{\varphi}' + \frac{1}{2}\left(h_3^2\rho_3 - h_1^2\rho_1\right)\overline{w}'\right] \tag{8.236}$$

$$(E_3h_3 - E_1h_1)\overline{u}''_0 + \frac{h_2}{2}(E_1h_1 - E_3h_3)\overline{\varphi}'' + \frac{1}{2}\left(E_1h_1^2 - E_3h_3^2\right)\overline{w}''' + 2G_2(\overline{w}' - \overline{\varphi})$$
$$= -\omega^2\left[(h_3\rho_3 - h_1\rho_1)\overline{u}_0 + \frac{h_2}{6}(3h_1\rho_1 + h_2\rho_2 + 3h_3\rho_3)\overline{\varphi} + \frac{1}{2}\left(h_1^2\rho_1 + h_3^2\rho_3\right)\overline{w}'\right] \tag{8.237}$$

$$G_2h_2(\overline{\varphi}' - \overline{w}'') + \frac{1}{2}\left(E_3h_3^2 - E_1h_1^2\right)\overline{u}'''_0 + \frac{h_2}{4}\left(E_1h_1^2 + E_3h_3^2\right)\overline{\varphi}'''$$
$$+\frac{1}{3}\left(E_1h_1^3 + E_3h_3^3\right)\overline{w}^{iv}$$
$$= -\omega^2\left[\frac{1}{2}\left(h_3^2\rho_3 - h_1^2\rho_1\right)\overline{u}_0 - (h_1\rho_1 + h_2\rho_2 + h_3\rho_3)\overline{w}\right.$$
$$\left. + \frac{h_2}{4}\left(h_1^2\rho_1 + h_3^2\rho_3\right)\overline{\varphi}' + \frac{1}{3}\left(h_1^3\rho_1 + h_3^3\rho_3\right)\overline{w}''\right] \tag{8.238}$$

The resulting differential equation system that is presented in Eqs. (8.236)–(8.238) are transformed by using the basic rules of DTM,

$$U_{k+2} = \frac{1}{4}(k+3)(h_1 - h_3)w_{k+3}$$
$$-\frac{2G_2(E_1h_1 - E_3h_3) + \omega^2h_1h_3(E_1h_3\rho_3 - E_3h_1\rho_1)}{4(k+2)E_1E_3h_1h_3}w_{k+1}$$
$$+\frac{12G_2(E_1h_1 - E_3h_3) + \omega^2h_2[E_3h_3](6h_1\rho_1 + h_2\rho_2) - E_1h_1(h_2\rho_2 + 6h_3\rho_3)}{24(k+1)(k+2)E_1E_3h_1h_3}\Psi_k$$
$$-\frac{\omega^2[E_3h_3(2h_1\rho_1 + h_2\rho_2) + E_1h_1(h_2\rho_2 + 2h_3\rho_3)]}{4(k+1)(k+2)E_1E_3h_1h_3}U_k \tag{8.239}$$

$$\Psi_{k+2} = \frac{12G_2(E_1h_1+E_3h_3) - \omega^2 h_2(6E_3h_1h_3\rho_1 + E_1h_1h_2\rho_2 + E_3h_2h_3\rho_2 + 6E_1h_1h_3\rho_3)}{12(k+1)(k+2)E_1E_3h_1h_2h_3}\Psi_k$$
$$-\frac{(k+3)(h_1-h_3)}{2h_2}w_{k+3} - \frac{2G_2(E_1h_1+E_3h_3)+\omega^2h_1h_3(E_3h_1\rho_1+E_1h_3\rho_3)}{2(k+2)E_1E_3h_1h_3}w_{k+1}$$
$$+\frac{\omega^2[E_3h_3(2h_1\rho_1+h_2\rho_2)-E_1h_1(h_2\rho_2+2h_3\rho_3)]}{2(k+1)(k+2)E_1E_3h_2h_3}U_k \tag{8.240}$$

Boundary conditions for the sandwich beam:

$$w_{k+4} = \frac{1}{2(k+1)(k+2)(k+3)(k+4)\left(E_1h_1^3+E_3h_3^3\right)}$$
$$\times\{6\omega^2h_2(h_3-h_1)\rho_2(k+1)U_{k+1}$$
$$+2\left[6G_2(h_1+2h_2+h_3)-\omega^2\left(h_1^3\rho_1+h_3^3\rho_3\right)\right](k+1)(k+2)w_{k+2}$$
$$+24\omega^2(h_1\rho_1+h_2\rho_2+h_3\rho_3)w_k\left[\omega^2h_2^2(h_1+h_3)\rho_2-12G_2(h_1+2h_2+h_3)\right]$$
$$(k+1)\Psi_{k+1}\} \tag{8.241}$$

The boundary condition and matrix of solution can be found in detail in Ref. [11]. The real and imaginary parts of normalized mode shapes for the transverse displacement are presented in Fig. 8.34 for the first three modes. The convergence of the loss factor for the first four modes with increasing number of terms considered is presented in Fig. 8.35. The figure presents a rapid convergence for the loss factors. Another important point is that, it is necessary to consider more number of terms in the DTM calculations to evaluate higher modes.

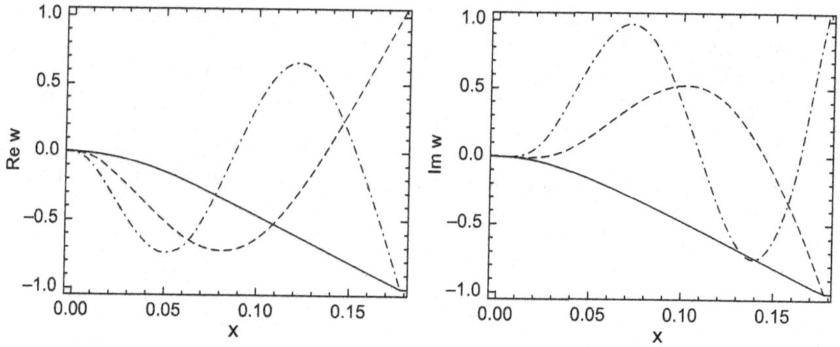

Figure 8.34 The real and imaginary parts of the first three mode shapes (-, first mode; –, second mode; -.- third mode).

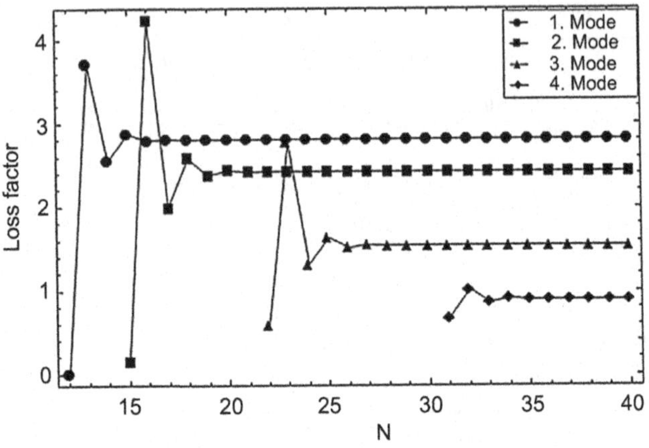

Figure 8.35 Convergence of the modal loss factor with *N*.

REFERENCES

[1] Hatami M, Vahdani S, Ganji DD. Deflection prediction of a cantilever beam subjected to static co-planar loading by analytical methods. HBRC Journal 2014;10(2):191–7.

[2] Suddoung K, Charoensuk J, Wattanasakulpong N. Vibration response of stepped FGM beams with elastically end constraints using differential transformation method. Applied Acoustics 2014;77:20–8.

[3] Mao Q. Design of shaped piezoelectric modal sensors for cantilever beams with intermediate support by using differential transformation method. Applied Acoustics 2012;73(2):144–9.

[4] Kuo B-L, Lo C-Y. Application of the differential transformation method to the solution of a damped system with high nonlinearity. Nonlinear Analysis: Theory, Methods & Applications 2009;70(4):1732–7.

[5] Mei C. Application of differential transformation technique to free vibration analysis of a centrifugally stiffened beam. Computers & Structures 2008;86(11):1280–4.

[6] Yeh Y-L, Chi Wang C, Jang M-J. Using finite difference and differential transformation method to analyze of large deflections of orthotropic rectangular plate problem. Applied Mathematics and Computation 2007;190(2):1146–56.

[7] Yalcin HS, Arikoglu A, Ozkol I. Free vibration analysis of circular plates by differential transformation method. Applied Mathematics and Computation 2009;212(2):377–86.

[8] Ni Q, Zhang ZL, Wang L. Application of the differential transformation method to vibration analysis of pipes conveying fluid. Applied Mathematics and Computation 2011;217(16):7028–38.

[9] Mao Q. Design of piezoelectric modal sensor for non-uniform Euler–Bernoulli beams with rectangular cross-section by using differential transformation method. Mechanical Systems and Signal Processing 2012;33:142–54.

[10] Chen S-S. Application of the differential transformation method to the free vibrations of strongly non-linear oscillators. Nonlinear Analysis: Real World Applications 2009;10(2):881–8.

[11] Arikoglu A, Ozkol I. Vibration analysis of composite sandwich beams with viscoelastic core by using differential transform method. Composite Structures 2010;92(12):3031–9.

INDEX

'*Note*: Page numbers followed by "f" indicate figures, "t" indicate tables.'

E

F

R

S

T

V

Printed in the United States
By Bookmasters